DAIRY CATTLE FEEDING AND NUTRITION

ANIMAL FEEDING AND NUTRITION
A Series of Monographs and Treatises

Tony J. Cunha, Editor

Distinguished Service Professor Emeritus
University of Florida
Gainesville, Florida

and

Dean, School of Agriculture
Professor of Animal Science
California State Polytechnic University
Pomona, California

DAIRY CATTLE FEEDING AND NUTRITION

W. J. Miller
Animal and Dairy Science Department
The University of Georgia College of Agriculture
Athens, Georgia

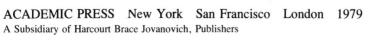

ACADEMIC PRESS   New York   San Francisco   London   1979
A Subsidiary of Harcourt Brace Jovanovich, Publishers

ACADEMIC PRESS, INC.
111 Fifth Avenue, New York, New York 10003

United Kingdom Edition published by
ACADEMIC PRESS, INC. (LONDON) LTD.
24/28 Oval Road, London NW1 7DX

Library of Congress Cataloging in Publication Data

Miller, William Jack, Date
    Dairy cattle feeding and nutrition.
    (Animal feeding and nutrition)
    Includes bibliographies.
    1.   Dairy cattle--feeding and feeds.   I.   Title.
SF203.M56   636.2'1'4   78–51234
ISBN 0–12–497650–6

PRINTED IN THE UNITED STATES OF AMERICA
79 80 81 82   9 8 7 6 5 4 3 2 1

To the late Professor Olin T. Fosgate
a great teacher of Dairy Science

and to my wife Marianna and family

this book is dedicated in appreciation

Contents

4   The Use of Nonprotein Nitrogen (NPN) for Dairy Cattle

5   Mineral and Trace Element Nutrition of Dairy Cattle

6 Vitamin Requirements of Dairy Cattle

7 Fat (Lipids) and Water Requirements and Utilization by Dairy Cattle

8 Fiber Utilization and Requirements of Dairy Cattle

9   Forages and Roughages for Dairy Cattle

10   Concentrates, By-Products, and Other Supplements for Dairy Cattle

11   Nonnutritive Additives and Constituents

12   Evaluation of Feeds for Dairy Cattle

13 Feeding the Milking Herd

14 Feeding and Raising the Young Dairy Calf

15 Feeding and Management of Heifers, Dry Cows, and Bulls

16 Feeding and Nutrition of Veal Calves

17   Raising and Feeding Dairy Beef for Meat Production

18   Nutritional and Metabolic Disorders of Dairy Cattle

19   Integrating the Feeding and Nutrition of Dairy Cattle into Practical Feeding Programs for Individual Farms

Foreword

This is the second in a series of books on animal feeding and nutrition. The first was "Swine Feeding and Nutrition" which appeared in 1977. New developments in feeding and nutrition have been quite numerous in recent years. This has resulted in new feed-processing methods, changes in diets, new uses of by-product feeds, and more supplementation with vitamins, minerals, and amino acids. New developments in the use of feed additives have made them even more essential under certain situations as intensification of animal production occurs. Moreover, farmers are developing animals which produce and reproduce at a higher rate. This places more pressure on the adequacy of the diet. Proper feeding and nutrition are very important since feed constitutes a major share of the cost of animal production. The volume of scientific literature is rapidly increasing and is so large and its interpretation so complex that there will be a continuing need for summarizing and interpreting these new developments in up-to-date books. This necessities that top authorities in the field collate all available information in one volume for each species of farm animal.

There is now an increasing awareness of the seriousness of the world's food supply problem. At least 500 million people are seriously deficient in protein and calorie intake. Two billion people, half the world's population, suffer from some form of malnutrition. Every two and one-half years, the world's population increases by a little over 200 million people. This is the equivalent of another United States to feed. Animals provide a very important share of the world's food intake. In the United States, for example, 44% of the food and about 56% of the nutrients come from animal products. All of the developing countries have 60% of the world's livestock and poultry but produce only 22% of the world's meat, milk, and eggs. Better feeding, breeding, and management of their animal production enterprises would increase their feed supply greatly. It is hoped that this series of books on animal feeding and nutrition will be of some assistance in the United States and in world food production.

This second book in the series, Dairy Cattle Feeding and Nutrition, is very well written by Dr. W. J. Miller, a distinguished scientist who is recognized worldwide for his outstanding work in animal nutrition. The book will be of

considerable value to all who are concerned with dairy cattle production throughout the world. It is my pleasure to advise on this project and to encourage the
other authors working on books. Special thanks go to the staff of Academic
Press, who have been superb to work with on this series.

Tony J. Cunha

Preface

This book was designed to provide information needed by those interested in the feeding and nutrition of dairy cattle. It contains basic information for students in courses on feeds and feeding, dairy cattle production, and animal nutrition. The simple presentation of numerous topics of importance in dairy cattle feeding and nutrition should be valuable to all those needing a quick summary, while the many key references will be helpful to those wishing to pursue a particular subject further. The book will be especially valuable to all individuals who plan dairy cattle feeding programs. Feed manufacturers, dealers, sales personnel, and others concerned with producing the various ingredients used in dairy cattle should find it most helpful. This book will also be useful to dairy farm owners, managers, and herdsmen both as a book to read in total and as a reference text. It should be of value to county agents, farm advisors and consultants, veterinarians, and teachers of agricultural education as a ready reference in dairy cattle feeding and management.

In the first chapter, the importance of the dairy industry is discussed, and a summary is presented of the basic mechanisms by which dairy cattle digest feed, obtain nutrients, and utilize feedstuffs. Chapters 2 through 8 contain concise, up-to-date summaries of the requirements, utilization, and problems related to energy, protein, nonprotein nitrogen, minerals, vitamins, fats, water, and fiber. The new concepts being developed in protein and nonprotein nitrogen utilization are reviewed, and their relationship to older concepts is explained. The coverage of minerals is especially thorough, including an explanation of several new concepts and applications. Fiber as an essential nutrient and the different types of fiber are reviewed. Forages, grains, other concentrates and supplements as sources of nutrients are discussed in Chapters 9 and 10. The many specialized ingredients and substances sometimes used or found in dairy cattle feeds are the subject of Chapter 11. The evaluation of feeds is reviewed in Chapter 12, including a candid description of the applications and limitations of the methods now in use and of those being developed.

Chapters 13 through 15 deal with feeding and managing the milking herd, young calves, dry cows, bulls, and growing heifers. These chapters and the two which follow make use of the basic information covered in the first twelve chapters. Veal production and dairy beef are discussed in Chapters 16 and 17. The nutritional and metabolic problems of dairy cattle are reviewed in Chapter 18. The last chapter concerns the coordination of all the basic and applied information in practical dairy cattle operations.

It is hoped that this book will be of worldwide use. The requirements, metabolism, and utilization of nutrients by dairy cattle are the same everywhere. Feedstuffs used as sources of nutrients vary throughout the world. However, much of the basic information in these areas also will be applicable in most countries. The concepts related to metabolic disorders and the necessity to incorporate the information into practical operations are universal.

During the preparation of this book, I have obtained considerable assistance and numerous helpful suggestions from many eminent scientists. I wish to express my sincere appreciation to them and to those who provided photographs and other materials. I am especially grateful to the following: O. T. Fosgate, W. P. Flatt, M. W. Neathery, D. R. Mertens, J. W. Lassiter, A. E. Cullison, R. P. Gentry, P. R. Fowler, E. R. Beaty, R. H. Whitlock and D. M. Blackmon (Georgia); R. B. Becker, C. B. Ammerman, J. K. Loosli and B. Harris (Florida); N. J. Benevenga, L. D. Satter and E. C. Meyer (Wisconsin); E. E. Bartley and J. L. Morrill (Kansas); L. S. Bull and R. W. Hemken (Kentucky); E. W. Swanson, J. K. Miller and M. C. Bell (Tennessee); P. W. Moe, H. K. Goering and H. F. Tyrrell (USDA); R. L. Kincaid and I. A. Dyer (Washington State); W. Chalupa and R. S. Adams (Pennsylvania); H. R. Conrad and G. H. Schmidt (Ohio); D. L. Bath and T. J. Cunha (California); J. Hartmans and J. Kroneman (Netherlands); M. Kirchgessner (West Germany); G. D. O'Dell (South Carolina); J. Kubota, S. E. Smith, P. J. Van Soest, R. G. Warner, D. G. Braund, J. T. Reid and D. V. Frost (New York); J. L. Evans (New Jersey); A. C. Field (Scotland); J. E. Oldfield and O. H. Muth (Oregon); J. L. Shupe and A. E. Olson (Utah); P. E. Stake and L. R. Brown (Connecticut); D. A. Morrow (Michigan); C. H. Noller (Indiana); C. R. Walker (Maine); A. M. Smith (Vermont); G. M. Jones and W. R. Murley (Virginia); and J. B. Holder (New Hampshire).

I am especially grateful to Ruth Harris, Sheila Heinmiller, and Linda Parten for conscientiously typing the manuscript in several drafts. Likewise, I am appreciative to E. R. Quillan for checking many of the references and to Ida Ho, S. L. Burgess, Evelyn Barnes, Stephanie Knopp, D. V. Cleveland, and W. D. Stowe for technical assistance on the illustrative material. Also, I am indebted to The University of Georgia for providing the opportunity and support for this project, especially the cooperation of W. P. Flatt and L. J. Boyd.

W. J. Miller

1

Introduction and Utilization of Nutrients by Dairy Cattle

1.0. INTRODUCTION

1.0.a. Importance of Dairy Cattle and the Dairy Industry

Throughout recorded history, the dairy cow has made great contributions to the health and feeding of mankind. The value of dairy cattle and the dairy industry can be viewed from several vantage points which aid in understanding the great importance of these animals and this industry.

Economically, the dairy industry is the largest single component of livestock agriculture in much of the world. Income to United States farmers from the sale of milk and milk products was $11.8 billion in 1977 (USDA, Crop Reporting Board, 1978). Likewise, a substantial portion of the beef produced in the United States comes from dairy cattle (Schmidt and Van Vleck, 1974). In many countries, especially in some Western European nations, a much higher percentage of beef is from dairy animals. Over a period of many years, approximately one-fifth of the total cash farm receipts in the United States has come from dairy products and from the sale of dairy cattle as meat animals (Foley *et al.*, 1972; Schmidt and Van Vleck, 1974).

The economic contribution of dairy cattle can also be considered in terms of the unique advantages afforded both individual farmers and society by the utilization of agricultural resources. In numerous countries, a high percentage of the land on many farms is best suited for the production of forages which have little direct utility for humans (Figure 1.1). Forages, as well as many by-products from food processing and other industries, can be utilized effectively only by ruminants. In the United States, about 68% of the feed and 63% of nutrients eaten by dairy cattle is forage, with by-products supplying another large component (Cunha, 1975; Hodgson, 1977).

The lactating cow is far more efficient in converting feed nutrients into human food nutrients than any other ruminant or other type of farm animal commonly grown for meat (Table 1.1). If the need arises, almost all of the feeds consumed

1

Fig. 1.1 (A)

Fig. 1.1 (B)

by dairy cattle can come from ingredients unsuitable for human consumption, thereby substantially increasing, rather than reducing, the total supply of food nutrients for humans (Ely and McCullough, 1975; Harshbarger, 1975). Thus, the dairy cow can be a key factor in enlarging the total supply of human food produced on a given land area.

For the individual farmer with land primarily suited for the growing of forages, a dairy operation often provides considerably more in terms of gross

Fig. 1.1(C)

Fig. 1.1(D)

Fig. 1.1 Large acreages in many countries are well adapted to growing forages but often are unsuitable for crops requiring intensive cultivation. (A) Eastern United States, (B) Queensland, Australia, (C) reclaimed land in Netherlands, and (D) Bavaria, West Germany. [(A), (B), and (D) courtesy of W. J. Miller, University of Georgia.]

TABLE 1.1
**Efficiency of Livestock in Converting Feed Nutrients to
Edible Products**

Animal	Conversion efficiency (%)[a][b]	
	Protein	Energy
Dairy cattle	25	17
Beef cattle	4	3
Lambs	4	—
Swine	14	14
Broilers	23	11
Turkeys	22	9

[a] Based on lifetime production and feed consumption.
[b] Adapted from Harshbarger (1975).

income than most alternative farming or livestock enterprises. To large numbers of farmers and their families this is a tremendous advantage.

On land that is best suited for forage and pasture production, a dairy is often of great value in soil conservation. Where erosion is a problem, the pollution of streams is much lower when forages, rather than cultivated crops, are grown. Similarly, ruminants are very beneficial for the preservation of a beautiful countryside without major special expenses. There is no better way, without substantial cost, to maintain land as beautifully as with a well-kept pasture.

Undoubtedly, the greatest contribution of the dairy cow lies in the value of dairy products to human health and well-being. The high nutritional value of milk for humans has been known and appreciated for centuries. In spite of recent criticisms of its specific constituents, there is still a relatively close association between the consumption of appreciable amounts of dairy products and the degree of human health (Dairy Council Digest, 1976a,b). For example, in the United States, dairy products supply the human diet with 75% of its calcium, a critical element (Hodgson, 1977). It is possible that, past infancy, optimum nutrition of most humans could be achieved without the use of milk and milk products. However, near optimum nutrition is seldom attained without the use of dairy products. In the United States milk and milk products are certainly invaluable to the health of millions of people. Numerous other nations would benefit from increased milk consumption. In the developing countries, poor health due to inadequate nutrition is believed to be a major obstacle to economic and social progress. Increased consumption of dairy products would measurably improve the nutrition and health of these populations.

1.0.b. The Cost of Feeding Dairy Cattle in Relation to Other Costs

When calculating the total cost of feed in the production of milk, the feed eaten by calves, heifers, dry cows, and bulls must be considered in addition to what is given to lactating cows. Likewise, production costs, in terms of fair market values, of feed ingredients grown on the farm must be added. In computing total production expenses, such items as taxes, equipment, labor, land use, and interest are included. As well, it seems appropriate that the expense associated with the physical feeding of the animals be added to the total cost of meeting the nutritional needs of dairy cattle.

Although quite variable under different conditions, typically, the total cost of providing feed for dairy cattle substantially exceeds all the other expenses combined (Smith, 1976).

1.0.c. Agriculture: A Rapidly Changing Industry

That agricultural practices are constantly evolving is well known. The fact that virtually every major development (and many seemingly minor ones) has the potential for significant practical effects on the feeding and nutrition of dairy cattle is not as well appreciated. For instance, when a plant breeder develops a new superior crop variety or strain, the composition and nutritional character of the crop is often measurably altered (see Section 5.6.b).

Some types of altered agricultural practices that can influence practical dairy cattle feeding are: different cultural and fertilization programs in growing feed crops; new crop pests and diseases; altered methods of manufacturing fertilizer ingredients; new methods of food processing whereby new or different by-product feed ingredients are produced; and new ways of storing, transporting, and handling feed ingredients. Often—although not always—new research discoveries suggest better feeding practices; however, added government regulations frequently impose new restrictions limiting the uses of the newest and best technologies.

Perhaps, over the short run, variations in price relationships will have the greatest effect on the practical feeding and nutrition of dairy cattle. The varying relative prices involve not only the value of the products sold and the feed ingredients purchased, but also all other expenses, ranging from energy, taxes and interest, to labor, equipment and fertilizer. Some of the implications are rather far reaching. For instance, the higher energy prices, after 1973–1974, greatly increased the price of nitrogen fertilizer. It thus became much more profitable to grow legume forages relative to grass forages. The supplemental feed needed when legumes are fed often is greatly different from those indicated with nonlegume forages.

1.0.d. Necessity of Understanding Principles of Dairy Cattle Nutrition in Order to Adjust to Changing Agricultural Practices and Economic Conditions

Because of the rapidity with which economic conditions and agricultural practices change, practical dairy cattle feeding programs require constant modification. If these modifications are to result in optimum efficiency of the dairy operation, the individuals making the decisions and/or supplying the technical advice must possess a thorough knowledge of the principles of dairy cattle feeding and nutrition.

It is often the case that widely recommended changes in feeding practices are sound and useful for some farms but not advantageous for others. Likewise, new methods and products that are not useful or economical are frequently suggested. Only those who understand the fundamentals of dairy cattle feeding can critically evaluate many of the new developments.

1.0.e. Role of Feeding Costs and Level of Milk Production in a Profitable Dairy Operation

As discussed earlier (see Section 1.0.b), the total cost of feeding dairy cattle exceeds all other costs combined. Frequently, there are huge variations in the prices of identical amounts of required nutrients from different ingredients and feed sources. Thus, the profit or loss on a dairy farm is influenced considerably by the particular feeding program used. Similarly, the feeding of expensive, nonessential nutrients increases costs without elevating income.

In developing practical feeding programs, cattle performance and efficiency must be considered in addition to the costs. The importance of this concept as it relates to various classes of dairy cattle under different conditions will be discussed in greater detail in later chapters. At this point, it seems important to emphasize that when the lifetime efficiency of converting feed nutrients to milk nutrients is considered, the dairy cow is most efficient when she is fed for a level of milk production that approaches her maximum genetic potential. For cows with the potential for high production, this usually involves the feeding of locally available forages, a method often requiring substantial supplementation with concentrates. When nutrients from concentrates and forages have similar relative prices, the practical approach is to feed for a high level of production. This has long been customary in many areas of the United States, largely due to the abundance of relatively low-priced grains and by-product feeds. Concordantly, numerous studies in the United States have shown that the income above feed costs as well as the total profits on dairy farms, are closely related to the level of milk production (see Section 13.3). In many countries throughout the world, dairy cattle produce milk under many and varying conditions. Particularly in developing countries, nutrients from concentrates are often enormously more expensive than those from forages, or are unavailable, due to other needs or for reasons of national policy. With these conditions prevailing, a lower level of milk production and reduced overall efficiency in converting feed nutrients to milk usually must be accepted. Of the total nutrients consumed over the lifetime of the cow, a high proportion goes toward maintenance (see Section 1.6.a). Thus, if a dairy cow is underfed, and produces at a low level, the proportion of the total feed nutrients used for milk production is inefficiently directed, and comprises a much smaller percentage of the total.

1.0.f. Ways to Feed Dairy Cattle—The Importance of Using Those That Are Both Nutritionally and Economically Sound

As will be explained more fully later, dairy cattle are very adaptable and versatile animals. Even though their nutrient requirements are fairly specific, they can obtain these nutrients from an enormous number of different feed combinations. At different times and in various places, numerous combinations can be nutritionally adequate at a minimal cost. Often, only a few feeding programs on a given dairy farm will meet a herd's nutritional needs at a particular time, at or near the lowest cost. Making the best choice(s) has a great effect on the amount of profit or loss accruing from the dairy operation.

1.1. DIGESTION IN THE RUMINANT STOMACH

The remarkable ability of ruminants, including dairy cattle, to digest and utilize forages such as pasture, hays, straws, and silages to meet their nutritional needs is mentioned above (see Section 1.0.a). This capability is possible because of the anatomy and function of the ruminant stomach.

1.1.a. General Anatomy and Role of the Ruminant Stomach

Whereas man, swine, rats, dogs, and numerous other animals have a simple or true stomach with only one compartment, cattle and other ruminants have a complex four-compartment stomach. The four compartments are the rumen or paunch, reticulum, omasum, and abomasum or true stomach (Figures 1.2, 1.3, 1.4, and 1.5). The reticulum is often called the honeycomb because the lining has a honeycomb-like appearance. Similarly, the omasum is termed the manyplies because its numerous leaves resemble the pages of a book. Often the rumen, reticulum, and omasum are referred to as the forestomach.

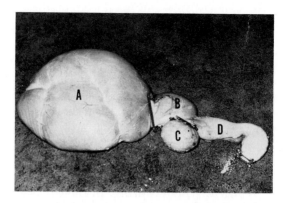

Fig. 1.2　The four compartments of the stomach of a mature cow after dissection and removal from the body. This shows the direction of feed passage but does not represent the location arrangements in the live animal. (A) rumen, (B) reticulum, (C) omasum, and (D) abomasum. (Courtesy of R. B. Becker, Univ. of Fla. Agric. Exp. Sta.)

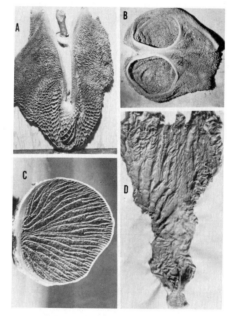

Fig. 1.3　This shows interior views of the compartments of the cow's stomach. (A) the rumen wall, divided at the entrance of the esophagus and esophageal groove which has been stretched open along the reticulum wall to the opening to the omasum. Note the papillae which line the rumen wall (upper portion) and the honeycomb-like structure of the reticulum (lower portion); (B) posterior of the rumen showing muscular pillars; (C) interior of the omasum showing the leaves or manyplies; (D) interior of the abomasum from the fundic or site of digesta entrance (top) to the pyloric valve (bottom) where digesta goes to the small intestine. (Courtesy of R. B. Becker, Univ. of Fla. Agric. Exp. Sta.)

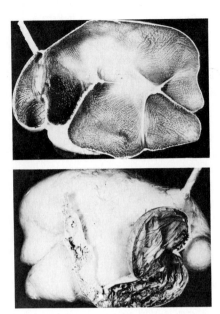

Fig. 1.4 Top. Interior (sagittal section) view of the rumen and reticulum showing the esophageal groove. Bottom. View of the ruminant stomach from the right side showing much of the exterior but also a cross section of the omasum, abomasum and duodenum. (See Benevenga *et al.*, 1969.) (Courtesy of N. J. Benevenga, Univ. of Wisconsin.)

Fig. 1.5 Interior view of the reticulum, forepart of the rumen, and esophogeal groove. (Courtesy of N. J. Benevenga, Univ. of Wisconsin.)

Initially, feed enters the rumen and reticulum, which, since they are separated only by an incomplete partition, essentially function in tandem. Indeed, these compartments are often referred to as the *reticulorumen*. The major difference in the ability of ruminants and most nonruminants, or simple stomached animals, to utilize forages is made possible by the microbial digestion and other vital processes taking place in the reticulorumen. Thus, to understand the practical feeding and nutrition of dairy cattle, it is necessary to appreciate the effects of the extraordinary events occurring in these compartments.

The unabsorbed portion of the partially digested "feed," or digesta, passes from the rumen and reticulum to the omasum and from there to the abomasum or true stomach. Digestion in the abomasum and beyond is similar to that of simple-stomached animals. There are, however, differences, including a more continuous digestion process in these parts of the tract in the ruminant because of the constant dribbling of partially digested food into the abomasum. In contrast, digestion in the abomasum of monogastric animals is more cyclic and more dependent on the frequency of eating.

The very large capacity of the ruminant stomach, especially of the rumen compartment, is a key feature in the ability of dairy cattle to utilize forages (Fig. 1.2). When filled, the reticulorumen comprises about 13% of the total weight of a mature cow (Warner and Flatt, 1965).

1.1.b. Development of the Ruminant Stomach from Birth to Maturity

Because the rumen and reticulum are not functional at birth, the nutritional requirements of the newborn calf are comparable to those of nonruminants. Other nutritional differences between the newborn calf and older cattle, such as the presence of the esophageal groove, are described in Chapter 14. Within a few days after birth, the rumen and reticulum begin growing very rapidly.

In mature cattle, the rumen and reticulum contain about 86% of the total stomach contents, with about 5% in the reticulum, as compared to some 7–11% for the omasum and 3–7% in the abomasum (Warner and Flatt, 1965). At birth, the abomasum, or true stomach possess about 60% of the total capacity of the stomach compartments. In the young calf, the relative capacity of the rumen and reticulum increases very rapidly, and constitutes about 60% of the total by 6 weeks of age. By 3–4 months of age, the relative capacities of the compartments approximate those in the mature animal (Warner and Flatt, 1965). Thus, in this short time span, the relative capacity of the rumen and reticulum have increased greatly.

The exceedingly rapid growth of the reticulorumen is accompanied by comparable development in functional capacity. This development is stimulated by the presence of dry feed containing at least some forage, roughage, or their fermentation products—in particular, volatile fatty acids (see Section 1.1.c; Fig-

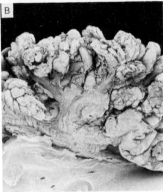

Fig. 1.6 Rumen papillae as seen with scanning electron microscopy in 6-week-old calves. (A) well developed papillae in calf fed concentrates plus long alfalfa hay *ad libitum;* (B) abnormal heavily keratinized papillae having multiple branching in calf fed the same low fiber pelleted concentrate. Inadequate forage or all finely ground forage can lead to unhealthy rumen papillae. (Magnifications: A × 5, B × 6.5) (see McGavin and Morrill, 1976) (Courtesy of M. D. McGavin and J. L. Morrill, Kansas Agric. Exp. Sta.)

ure 1.6) (Warner and Flatt, 1965). The expanded rumen function greatly increases the ability of the developing calf to rely on forages, or other fibrous feeds, and microbial synthesis for much of its nutritional needs.

1.1.c. Digestion of Feeds in the Rumen (Reticulorumen) and the Synthesis of Nutrients for Utilization by Cattle

The ruminant stomach probably evolved during prehistoric times as an effective adaptation for survival. Development of the forestomach (rumen, reticulum, and omasum) enabled the animal to consume a large amount of feed very rapidly with a minimum of chewing before swallowing and then move to a resting place to rechew the feed. Thus, the ruminant could spend most of its time hiding from predators. Several features suggest an evolutionary development of this type. Because of their specialized stomach, cattle can not only utilize a wide variety of feeds—in particular, either green or dried forages—but also seeds or grains, and a host of other materials. Further, the microorganisms of the reticulorumen synthesize many of the essential nutrients required by the body, making it much easier for the animal to obtain a balanced diet.

Cattle eat rapidly, and little chewing is done before the feed is mixed with saliva and swallowed to the rumen and reticulum. Ingested feed is thoroughly mixed in the reticulorumen by a rhythmic sequence of muscular contractions. Mixing aids in inoculating the fresh ingesta with microorganisms (microbes), spreads saliva throughout the reticulorumen, and enhances absorption of the products of digestion. After eating, cattle spend considerable time rechewing the feed. This rechewing process is termed rumination or "chewing the cud." Although the time spent ruminating depends, among other factors, on the type of feed, typically it involves about one-third of the cow's 24-hour day.

Rumination is a well-synchronized process or chain of events. In essence, a small bolus of feed and liquid is regurgitated, passing from the rumen through the esophagus to the mouth, where it is rechewed for about a minute, mixed with saliva, and reswallowed into the rumen. Following a brief pause, the process is repeated with another bolus. The rechewing decreases the size of the feed particles, thereby greatly increasing the surface area for microbial action and digestion. Chewing the cud is a more relaxed function than the original eating and often is associated with contented cows.

Cattle produce huge quantities of saliva (about 100 quarts/day in a cow), which lubricates the feed and aids swallowing. Substantial amounts of buffering compounds, especially bicarbonates and phosphates, present in saliva are key factors in keeping the pH of the rumen within the suitable range, between 5.5 and 7.2, for effective functioning of the microorganisms. Saliva also provides several nutrients needed by the microbes.

One of the most remarkable and important symbiotic relationships in all of nature is that which exists between the ruminant animal and the microbes of its reticulorumen. The rumen can be viewed as a huge, exceedingly complex, and efficient liquid or semiliquid fermentation vat containing billions of microbes per milliliter. One ounce is equivalent to about 28 ml. These microorganisms include a great variety of bacteria and protozoa that conduct a multitude of decomposition, transformation, and synthetic reactions. While bacteria are present in the greatest numbers, protozoa are much larger in size. Relative numbers of various microbes change substantially when different types of feeds are eaten. Because this adaptation in microbial populations is time consuming, it is more efficient to avoid drastic and sudden changes in feeds. Major changes should be made gradually.

In the practical feeding and nutrition of dairy cattle, the most important functions of the rumen microbes are (1) digestion of fibrous feeds, (2) synthesis of essential nutrients such as B vitamins and essential amino acids, and (3) the utilization of compounds such as nonprotein nitrogen, which are otherwise almost unusable.

The most abundant carbohydrates produced by plants are starches and celluloses. To an appreciable extent, the carbohydrate in grains is starch, while stems and leaves, which are the structural parts of plants, contain large amounts of cellulose and related complex carbohydrates. Animals such as man, who, with their simple stomachs, are able to digest the starch and simple sugars, cannot, however, digest the cellulose. In this respect, without the rumen microbes, cattle would be little different from simple-stomached animals. Rumen microbes very efficiently convert cellulose, and related compounds such as hemicellulose, to products that can be used for energy by the animal. The products formed are largely acetic, propionic, and butyric acids, which are given the general term of volatile fatty acids (VFA or VFA's).

Since much of the carbohydrate in forages consists of cellulose and related compounds, cattle can utilize these feeds as a major source of energy. Generally, about 60–80% of the energy available to cattle comes from the VFA's. While not always in the best interest of the animal, the rumen microbes also convert simple sugars and starces to VFA's. In certain situations, large amounts of lactic acid may be produced to the detriment of the ruminant animal involved (see Section 18.8).

Synthesis of the essential amino acids, B vitamins, and vitamin K is the second great contribution of rumen microorganisms. Because of the rumen microbes, cattle, except for young calves, do not require a dietary source of these essential nutrients, as do simple-stomached animals.

The third great practical contribution of the rumen microorganisms is the use of nonprotein nitrogen compounds along with carbohydrate energy sources to synthesize protein or amino acids. This enables cattle to eat synthetic or waste nitrogen products, thereby creating high-quality protein, of great value in human nutrition.

The rumen microbes also produce large amounts of carbon dioxide and methane gas, which normally are removed from the rumen by belching, along with some absorption into the blood. If the gases are not removed, the severe rumen distention known as bloat ensues (see Section 18.3).

1.2. DIGESTION IN THE ABOMASUM (TRUE STOMACH) AND INTESTINE OF DAIRY CATTLE

Digesta, or partially digested feed, passes from the reticulorumen into the omasum, where much of the water is removed, and then into the abomasum. In ruminants, digestion in the abomasum closely resembles that in the true stomach of simple-stomached animals (see Section 1.1.a). The abomasum secretes substantial amounts of gastric juice containing hydrochloric acid, giving the abomasal contents a low pH around 2.5. This low pH is a necessary part of a good environment for the enzymes, pepsin and rennin, that digest proteins to simpler compounds. Rennin is important in curdling and digesting milk in the young calf but is of little significance in mature cattle.

From the abomasum, the digesta goes to the small intestine that in cattle functions similarly to that of nonruminants. The small intestine is especially well equipped for both digesting materials and absorbing nutrients. In mature cattle, the highly coiled small intestine is about two inches in diameter and some twenty times the length of the animal (around 130 feet) (Sisson and Grossman, 1953). Internally, the small intestine has a huge amount of surface area in the form of tiny, irregular fingerlike projections called villi. The surface of the small intestine, including the villi, is lined with a single layer of muscosal cells, which are extremely active, versatile, selective, and adaptable in their contributions to digestion and absorption (Ingelfinger, 1967).

Digest or digestion are the terms used to describe the changes to the feed in the digestive tract prior to absorption, including the breaking down of complex compounds into simpler ones for absorption from the digestive tract into the body of the animal. Undigested residue is eliminated in the feces.

The digesta passing from the abomasum into the small intestine is sometimes called the chyme. Due to the low pH of the abomasal medium, the digesta is very acidic as it enters the upper small intestine. Alkaline secretions from the pancreas and liver largely neutralize the contents in the forepart of the small intestine.

Many secretions, including pancreatic juice from the pancreas, bile from the liver by way of bile duct and gall bladder, and numerous substances from the intestinal mucosa greatly aid digestion in the small intestine. The three major enzymes in the pancreatic juice, namely, trypsin, chymotrypsin, and carboxypeptidase, are involved in further breaking down the partially digested proteins into amino acids for absorption and utilization by the animal.

Lipase, an enzyme from the pancreas, breaks fats into fatty acids and glycerol for absorption. One of the primary roles of the bile secretion is emulsifying fat globules so they can be digested by lipase. Bile also aids in absorption of lipids or fatty materials.

A third major type of feed substance digested in the small intestine is starches, which are broken down (hydrolyzed) into glucose, with the aid of the enzymes amylase and maltase, prior to absorption. Some sugars, such as lactose (milk sugar), must be digested before they can be absorbed. The very young calf does not have enzymes for the digestion of starch or several important sugars, including sucrose (ordinary cane sugar), and thus cannot effectively utilize these carbohydrates.

1.3. ABSORPTION AND TRANSPORT OF NUTRIENTS IN DAIRY CATTLE

The rumen is an important site for the absorption of various organic compounds, including volatile fatty acids (VFA's) and ammonia. In fact, a major portion of the VFA's, acetic, propionic, and butyric, is absorbed through the rumen wall, with most of the remainder absorbed from the omasum and abomasum (Church, 1975). For most of the other nutrients utilized by cattle, the small intestine apparently is the most important site, and often the only absorption site.

With many nutrients, the absorption sites are not clearly established because of conflicting results obtained with different types of techniques. Considerable amounts of many nutrients, especially water and minerals such as sodium, potassium, and chlorine, are secreted back into the digestive tract at various points. These secretions play a key role in the overall digestive and absorption processes. The huge amounts of saliva are one example (see Section 1.1.c). Likewise, the digestive enzymes, which are essential to much of the digestion in the abomasum and small intestine, are themselves proteins. These enzymes may have one or more mineral element as essential cofactor(s) with the protein part of the molecule.

Absorption of nutrients occurs by processes varying from simple diffusion to those requiring a carrier system. Many nutrients and substances are almost totally absorbed whereas with others, the percentage is very low. As will be discussed later, the ability to excrete some ingested substances without absorbing them into the tissues is an important defense mechanism cattle have against many toxic substances as well as for some essential nutrients. Often essential nutrients would be toxic if a high percentage of those eaten were absorbed.

Except for fats taken up by the lymphatic system, most of the nutrients absorbed from the rumen and small intestine enter the blood stream and are transported to the liver and to all tissues of the body for utilization by the cells. The blood also provides a transportation system for elimination of metabolic waste products.

Some digestion and absorption occur in the cecum and large intestine primarily via the action of microbes. In cattle this has much less importance than digestion in the rumen.

1.4. NUTRIENT REQUIREMENTS OF DAIRY CATTLE

From the above discussions it is apparent that the nutrients which dairy cattle have available in feed and those required by the tissues and cells of the body are very different. The cells and tissues need usable forms of energy; 9 or 10 essential amino acids; vitamins A, D, E, K; several B vitamins; about 21 essential mineral elements; and water.

Since they are synthesized by rumen microbes, dairy cattle do not require dietary sources of essential amino acids, the B vitamins, and vitamin K. Likewise, the microbes convert forms of energy that are not usable by the animal into those which can be utilized.

1.4.a. Variability in Nutrient Requirements

The required amounts of the various essential nutrients for cattle are greatly influenced by several key factors with age being one of the most important. Since the newborn calf does not have a functional rumen, it requires about the same dietary nutrients as a nonruminant. The calf rapidly develops a functional rumen and with it much less exacting dietary needs.

In addition to the influence of age on rumen development, the amounts of many nutrients required per 100 lb of body weight decrease appreciably with advancing age due to a changing growth pattern. The rate of growth relative to the size of the animal declines with age. Likewise, the composition of the growth changes with the young calf depositing a higher percentage of protein, mineral, and water in new tissue. As animals mature, fat deposition increases relative to other substances, especially water.

Both the total quantity and relative proportions of the various nutrients required are greatly affected by the growth rate, lactation, the level of milk production, and other physiological differences. Reproduction and excessive exercise also increase the nutrients needed.

As is true of all life, some individuals require higher amounts of nutrients than other, apparently comparable, animals. Thus, when planning a practical feeding program, variability among similar animals needs some consideration. This is one of the reasons a small safety margin is included in the tables of the nutrient requirements for dairy cattle (see Appendix, Tables 1, 2, and 3) (National Research Council, 1978).

1.4.b. Biological Availability of Feed Nutrients

Often it is much easier to determine the nutrient composition of feed ingredients than to measure the availability of the nutrients to a dairy animal. Accordingly, there is much more information on the amount of nutrients in feeds than on their availability. Dairy cattle need nutrients that can be absorbed and utilized.

Since the biological availability of many nutrients varies greatly among feed ingredients, such information is a key item in expressing and using the nutrient requirement information in the practical feeding of dairy cattle. The availability of certain nutrients is much more variable than that of others. For instance, evaluating the usable or available energy in feeds is crucial, whereas for most feeds, the availability of protein is much less variable. As another contrast, most of the dietary sodium is absorbed, but with magnesium the percentage is quite variable. The role of availability will be discussed in subsequent chapters.

1.5. DETERMINATION OF NUTRIENT REQUIREMENTS OF DAIRY CATTLE

When using the nutrient requirement data in the practical feeding of dairy cattle, often it is helpful to know how the values are obtained. The two basic but independent approaches used in determining the amount of each nutrient required are (1) the factorial method; and (2) the feeding experiment method. Insofar as possible, results from both approaches are compared. Ideally, the results should agree, but because of numerous complexities, often they do not. In such situations it is important to understand the reasons for the disagreement and arrive at decisions based on the best information known.

Determining nutrient needs by the factorial method involves two major steps with some modifications for different nutrients (see Sections 3.2 and 5.8). First, the amount of the available nutrient used by the animal for each function is established. Second, the availability of that nutrient is measured. The efficiency with which the available nutrient is utilized for different functions also must be considered.

The amount of the available nutrient needed is determined for (1) maintenance, (2) growth and/or fattening, (3) milk synthesis and secretion, (4) reproduction, and (5) work or exercise above and beyond the normal. All dairy cattle have a maintenance requirement, but whether other functions are involved depends on the specific situation of the animal. Amounts needed for each function are totaled to arrive at the quantity of the available nutrient required. This amount must be combined with information on the percentage available in feeds to determine the nutrient needed in the ration.

Where pertinent, variations in the efficiency with which a nutrient is utilized for different functions is considered in determining requirements. For instance, digestible energy is used more efficiently for milk production than for fattening.

Generally, the availability of a nutrient in a feed source is measured in metabolism studies. While the nature of the study needed varies with the different nutrients, usually the proportion in the feed which can be absorbed and/or utilized by the tissue is measured or estimated. This aspect will be considered further in later sections.

For many nutrients, especially certain minerals and vitamins, the experimental data needed to calculate requirements by the factorial method have not been obtained. Likewise, determining the maintenance requirement for some nutrients, especially certain minerals, presents difficult technical problems.

The feeding experiment method for determining the requirement of a nutrient consists of feeding various known quantities of the nutrient and studying the response and performance of the animals. The smallest amount which gives optimum health and performance is the minimum requirement. A satisfactory estimate of the amount needed depends on a suitable criteria of optimum performance and health. Unfortunately, for many of the nutrients, sensitive and accurate criteria are unknown or have not been developed. Lack of satisfactory criteria of adequacy, especially for several of the minerals, is one of the major weaknesses of the feeding experiment method. In developing the nutrient requirements (see Appendix Tables 1, 2, and 3), information from both the factorial and the feeding experiment methods was utilized.

1.6. UTILIZATION OF NUTRIENTS IN THE TISSUES OF DAIRY CATTLE

A dairy animal is an exceedingly complex creature involving innumerable biochemical processes each requring many different nutrients. Thus, it is neither feasible nor possible to determine the amount of nutrient for each process or reaction. However, it is convenient and helpful to divide the ways dairy cattle utilize nutrients into broad functionally related categories. The major ones are maintenance, growth and/or fattening, milk secretion, reproduction, and work.

1.6.a. Nutrients Required for Maintenance; Overhead Costs

A substantial part of the total nutrients used by dairy animals is for maintenance of a normal state of health. Even an animal that is not growing, reproducing, or lactating, requires an appreciable amount of feed nutrients to carry on the innumerable processes that are a part of life. All of life's processes require nutrients. Most of the cells of the body are alive with numerous energy-using reactions going on constantly. Even though the average composition of cells may not change rapidly, there is a constant turnover or exchange of the chemical components. Just as there is a cost for changing parts on an automobile, these reactions require energy.

Likewise, it is an expense to transport nutrients, the new parts, largely in the blood from the digestive tract to the cells, and to return discarded ones, the waste products, for excretion. The dairy animal has a fabulous transport or circulatory system. For example, typically the heart of a 1000-lb dairy cow pumps about 60 tons of blood daily (Smith, 1959).

Keeping the body temperature constant is one part of the maintenance requirement for nutrients. Although tremendously helpful to dairy cattle, the rumen microbes use energy and other nutrients in their life processes, thus adding to the maintenance cost. Likewise, the digestion and absorption of nutrients by the animal has an energy cost.

Even though the greatest emphasis in studying maintenance needs often has been on the energy used, all processes involve complex biochemical reactions and require other nutrients. Thus, there is a maintenance requirement for all the essential nutrients.

It is appropriate to think of the nutritional needs for maintenance as a part of the overhead cost in a dairy operation. One of the most effective ways to reduce cost in most manufacturing enterprises is to reduce the overhead. A higher level of milk production, a longer productive life for the cow, efficient reproduction, and early calving reduce the amount of feed nutrients (per 100 lb of milk produced) going for overhead.

1.6.b. Nutrients for Growth and Fattening

The body of a dairy animal contains large amounts of proteins, fats, and minerals, indicating a substantial need for many nutrients for growth. As with all biological processes, deposition of nutrients during growth is not 100% efficient. Over and above the maintenance needs, a growth requires more nutrients than those contained in the tissues produced.

The relative amounts of nutrients required varies with age and the type of tissue being deposited. For example, an older animal which is fattening does not need as high a percentage of protein and minerals as a young calf that is rapidly building muscle and bone.

1.6.c. Nutrients for Reproduction

A major function of all animals is reproduction. For most efficient milk production, cows should calve about once per year because annual milk production declines with longer calving intervals.

For the first 6–7 months of pregnancy, the added nutrients required is quite small. However, during the last 2–3 months, the fetus grows very fast and has a rapidly increasing need for nutrients. Likewise, substantial amounts of placental and related tissues, and fluids are involved in reproduction. Even so, in dairy cattle the overall nutrient requirements for reproduction are much lower than those for maintenance, growth, and lactation.

1.6.d. Importance of Nutrients for Milk Secretion

Because the primary purpose of keeping dairy cattle is for the milk, in practical feeding and nutrition great emphasis is placed on supplying the nutrients needed for optimum milk production. With the high producing dairy cow at the peak of lactation, a tremendous quantity of high quality nutrients for human consumption are secreted into the milk each day. Accordingly, with high production, much larger amounts of most nutrients are needed for milk synthesis than that for all other functions combined. In effect, high production reduces the relative amounts needed for maintenance and thus is a key reason for the high efficiency of the lactating dairy cow compared to meat animals. Another factor in the high efficiency of the cow is the lower loss in the conversion of many dietary nutrients to milk nutrients.

The overall efficiency of converting specific feed nutrients to milk nutrients varies appreciably. Typically milk will contain about 17% of the feed energy and 25% of the protein eaten by the cow during the course of its lifetime (Table 1.1). If only the lactation period is considered, substantially higher percentages of the feed nutrients are secreted into the milk.

1.6.e. Work (Exercise) and Nutrient Requirements

A moderate amount of exercise is a necessary part of the maintenance cost. The exercise associated with eating, going to the milking parlor and related light activity is unavoidable and probably necessary for good health. In some situations dairy animals may walk considerable distances in grazing and going to and from pastures. When unusually large amounts of exercise are involved, additional nutrients must be provided (see Section 2.2.f).

1.7. EXCESS NUTRIENTS: ECONOMICS, EFFECTS ON THE ANIMAL, TOXICITY, AND TOLERANCE

In studying the nutritional needs of dairy cattle, most of the emphasis is on the minimum amounts necessary to obtain optimum performance. Even so, it is important to understand the effects of excessive consumption of the various nutrients. Although this aspect will be considered for each nutrient in succeeding chapters, perhaps mention of a few general principles is appropriate here.

The disadvantages of feeding more nutrients than needed can be divided into the economic cost and adverse effects on the health or performance of the animals. Since feeding an excess of any nutrient generally is of no benefit to the animal, any additional financial cost is wasted and reduces profits. With nutrients required in large amounts, such as energy and protein, the economic loss can be considerable. In contrast, the economic loss due to excessive intake is minor for some of the vitamins and minerals that are used in minute amounts.

Dairy cattle can eat far more than the minimum requirements of most nutrients without adverse effects on health or performance. Only with energy does a moderately excessive intake have a measurable effect on the animal, namely, extra fat deposition. In terms of excessive intakes, energy also is relatively unique, in that the amount of the excess consumption tends to be self-limiting as the animal usually will not eat a tremendous excess. Thus a real toxicity from excess energy does not occur in the sense that it can with most other essential nutrients.

REFERENCES

Benevenga, N. J., S. P. Schmidt, and R. C. Laben (1969). *J. Dairy Sci.* **52**, 1294–1295.

Church, D. C. (1975). "Digestive Physiology and Nutrition of Ruminants," 2nd ed., Vol. 1. Oregon State Univ. Bookstore.

Cunha, T. J. (1975). *Feedstuffs* **47**(53), 8 and 37.

Dairy Council Digest (1976a). **47**, No. 1.

Dairy Council Digest (1976b). **47**, No. 5.

Ely, L. O., and M. E. McCullough (1975). *Ga. Agric. Res.* **17**, 9–13.

Foley, R. C., D. L. Bath, F. N. Dickinson, and H. A. Tucker (1972). "Dairy Cattle: Principles, Practices, Problems, Profits." Lea & Febiger, Philadelphia, Pennsylvania.

Harshbarger, K. E. (1975). *Nutr. News* **38**, 9 and 12.

Hodgson, H. J. (1977). *Hoard's Dairyman* **122**, 591, 599, and 600.

Ingelfinger, F. J. (1967). *Nutr. Today,* pp. 2–10.

McGavin, M. D., and J. L. Morrill (1976). *Am. J. Vet. Res.* **37**, 497–508.

National Research Council (NRC) (1978). "Nutrient Requirements of Dairy Cattle," 5th rev. ed. Natl. Acad. Sci., Washington, D.C.

Schmidt, G. H., and L. D. Van Vleck (1974). "Principles of Dairy Science." Freeman, San Francisco, California.

Sisson, S., and Grossman, J. D. (1953). "The Anatomy of the Domestic Animals." Saunders, Philadelphia, Pennsylvania.

Smith, N. E. (1976). *J. Dairy Sci.* **59**, 1193–1199.

Smith, V. R. (1959). "Physiology of Lactation," 5th ed. pp. 103–104. Iowa State Univ. Press, Ames.

USDA, Crop Reporting Board (1978). "Milk—Production, Deposition, Income, 1975–77," SRS Da 1-2(78). USDA, Washington, D.C.

Warner, R. G., and W. P. Flatt (1965). *In* "Physiology of Digestion in the Ruminant" (R. W. Dougherty *et al.,* eds.), pp. 24–38. Butterworth, London.

2

Energy Requirements of Dairy Cattle

2.0. THE PRACTICAL IMPORTANCE OF MEETING THE ENERGY NEEDS; ENERGY: A CENTRAL ROLE IN FEEDING DAIRY CATTLE

Much more feed is required to supply the energy needs of a dairy cow than for all other nutrients combined. Thus, a major part of the total cost of feeding dairy cattle is for energy. Other nutrients usually cost more per pound. Every action, function, process, and biochemical reaction of a dairy animal requires energy. Just as with an automobile, the cow must have energy to function; but, unlike the machine, she needs energy constantly to live whether she is producing milk or is dry.

The practical feeding of dairy cattle, generally, is planned around supplying the energy needs. After the basic energy plan is established, the ration is balanced for the other required nutrients. Although very large, the total quantity of feed a cow can eat is limited (see Appendix, Table 5). When the average usable energy content of the feed is low, modern dairy cows producing large amounts of milk are unable to eat sufficient feed to meet their energy needs. Thus, to produce near their optimum potential, lactating cows must be fed a diet with a fairly high concentration of usable energy.

Unlike most of the other required nutrients, energy is not a specific substance. Rather, numerous organic compounds can be used for energy. However, the amounts of usable energy in a pound of different feeds vary markedly. Thus, measuring the energy value of feeds is a key aspect of dairy cattle nutrition. Often, it has been suggested that the value of feeds depends primarily on the usable energy content (Nehring and Haenlein, 1973).

2.1. MEASURES OF USABLE ENERGY FOR DAIRY CATTLE

Several measures and systems of evaluating the energy content of feeds have been developed and used (National Research Council, 1966). The energy value

22

in feeds, energy requirements of dairy cattle, and practical feeding recommendations are described in terms of these measures. Understanding the major systems of energy evaluation is important to practical dairy cattle nutrition.

In the United States the two most widely used energy measurement systems for dairy cattle are total digestible nurtients (TDN) and net energy (NE). However, digestible energy (DE) is gaining in importance. Methods and procedures for evaluating feeds are discussed further in Chapter 12. Some of the measures of usable energy value, especially TDN, involve determining certain chemical fractions in the feed plus a biological evaluation of the amount of each component utilized.

2.1.a. Digestible Energy (DE)

One of the most easily determined measures of usable energy in feeds is DE. The DE in a feed generally is expressed as megacalories which is abbreviated as Mcal and is equal to 1,000,000 small calories (NRC, 1966). The megacalorie is sometimes called the therm (T).

Digestibility is determined by analyzing the amount of the nutrient in the feed and the feces with the difference being the amount digested (Fig. 2.1 and see Section 12.8). Often it is referred to as apparent digestibility as some of the nutrients in the feces usually have been absorbed and reexcreted into the feces. Thus, the DE content of a feed is the apparent digestible energy (Fig. 2.1) and is

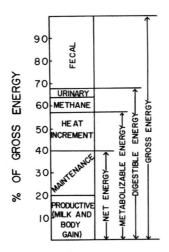

Fig. 2.1 Typical proportions of gross energy lost in feces, gaseous products of digestion, urinary energy, heat increment (work of digestion), and that utilized for maintenance and for productive purpose in a lactating dairy cow. Note the various measures of usable energy, digestible energy (DE), metabolizable energy (ME), and net energy (NE). (See also Flatt, 1966; NRC, 1966.)

somewhat lower than the amount of energy actually absorbed which is the true digestible energy (see Section 12.3). Digestible energy values have most of the same advantages, weaknesses, and applications as TDN (see Section 2.1.d).

2.1.b. Metabolizable Energy (ME)

Since feeds having similar DE contents will provide somewhat different amounts of usable energy, often more sophisticated measures are used. One of these is metabolizable energy (ME), which is determined from DE by subtracting the energy in the urine and in the gaseous products formed during digestion (Fig. 2.1). The gaseous products of digestion are largely methane, which represents over 99% of the energy value, and carbon dioxide produced by the rumen microbes. They supply little or no usable energy to the animals. The amounts of these gaseous and urinary products vary with different feeds. Generally, gaseous energy losses are greater from digestion of forages than of concentrates. In the United States, ME values have not been widely used in the practical feeding of dairy cattle. However, a special modification is used in the United Kingdom (Blaxter, 1967; Agricultural Research Council, 1965) (see Section 2.1.e).

2.1.c. Net Energy (NE)

Of the usable energy measures in general use, net energy (NE) is the most difficult to determine experimentally (Figs. 12.1, 12.2). When different feeds are digested and metabolized, variable amounts of feed energy are converted to heat which generally, except in very cold weather, is of little value to the animal. In hot weather this extra heat may even create stress with additional energy being used in dissipating the heat. This heat energy, known as the heat increment or "work of digestion," is subtracted from ME to obtain NE (Fig. 2.1). In a dairy cattle ration a substantial portion of the ME is lost as the heat increment. The percentage increases with more fibrous rations.

Net energy is that which remains after the energy in feces, gasses, urine, and the heat created during digestion and assimilation of nutrients are subtracted from the total or gross feed energy (GE) (Fig. 2.1). The formula is

$$\text{NE} = \text{GE} - \left(\begin{array}{c} \text{fecal} \\ \text{energy} \\ \, \end{array} + \begin{array}{c} \text{gaseous} \\ \text{product} \\ \text{energy} \end{array} + \begin{array}{c} \text{urinary} \\ \text{energy} \\ \, \end{array} + \begin{array}{c} \text{heat} \\ \text{increment} \\ \text{energy} \end{array} \right)$$

Accordingly, NE is that which can be used by the animal for maintenance, growth, reproduction, milk production, work, and/or fattening (Fig. 2.1).

Net energy is the most accurate way to express the usable energy content of feeds and the energy requirements of dairy cattle. However, there are problems and limitations with this system. Perhaps the most serious is the tremendous amount of effort and expense involved in determining the NE content of feeds. Accordingly, the amount of research data on NE content of feeds is limited. Much of this disadvantage is overcome by using special formulas to calculate NE values from simpler measures, especially TDN (Moore et al., 1953) or DE (Nehring and Haenlein, 1973).

Calculating the NE in feeds from DE or TDN content yields very useful values as 80% of the variation in NE content of various feedstuffs is associated with differences in digestibility (Moe, 1976). It is also possible to estimate NE content of feeds from feeding experiments (see Section 12.10). This was the basis of many of the estimated net energy values calculated by F. B. Morrison for *Feeds and Feeding* (1957).

Dairy cattle do not use energy with the same efficiency for all functions (see Section 1.6 and 2.3). Thus, the NE value of a feed depends somewhat on the function for which the energy is used. Net energy for maintenance ($NE_{maintenance}$ or NE_m) of lactating cows and for lactation ($NE_{lactation}$ or NE_l) are utilized with about equal efficiency; however, NE for growth or gain (NE_{gain} or NE_g) is used less efficiently (National Research Council, 1978). Accordingly, feed ingredients have higher NE values for lactation and maintenance than for growth or fattening. This is a distinct disadvantage of the NE system as different tables are required for different functions. To overcome this disadvantage, the NE requirements for weight gains have been adjusted for the difference in efficiency, so that only one set of feed tables are required (see Appendix, Table 4) (NRC, 1978). Many of the NE values of feeds shown in Table 4 of the Appendix were calculated from TDN values by the formula NE_l (Mcal/lb DM) $= -0.054 + 0.0111$ TDN (% of dry matter) (NRC, 1978). This calculation was made using maintenance TDN values and assumed on 8% decrease in TDN at average production levels. The 8% is based on an assumed 4% decrease in TDN content for each additional multiple of intake above maintenance, or feed intake of three times maintenance (Fig. 2.2).

When a dairy animal is fed a diet that is deficient in one or more nutrients, other than energy, often the energy is utilized less efficiently. Thus, NE values determined on unbalanced diets generally are abnormally low.

2.1.d. Total Digestible Nutrients (TDN)

The TDN system of determining the energy values of feeds and requirements of dairy cattle has been widely used, especially in the United States, for a long time. The TDN content of a feed is the sum of the digestible crude protein, digestible crude fiber, digestible nitrogen free extract, plus 2.25 times the digestible ether extract. In evaluating a feed, TDN is expressed as a percentage, either as fed or on a dry matter basis. However, TDN requirements are given in pounds.

Determining the TDN content of a feed requires a digestion trial (see Section

12.8) and chemical analyses of the feed and feces for crude protein, crude fiber, ether extract, and nitrogen free extract. Generally, TDN values correlate fairly closely with digestible energy values (see Section 2.2). However, TDN partially considers the urinary energy losses. Although protein contains more DE per pound than carbohydrates, most of the additional energy is lost in the urine. Thus TDN more accurately reflects the usable energy value of protein than does DE.

When compared with the NE system, TDN has several important weaknesses. Generally, except for some urinary protein energy, the only losses taken into consideration are those in the feces. Because of this, the energy value of fibrous feeds, such as forages, which result in higher gaseous and heat increment losses, is overvalued by TDN relative to concentrates. Whereas, ether extract from true fats contributes about 2.25 times as much energy as carbohydrates, much of the ether extract in forages is material other than true fat and has a lower energy value (see Sections 12.2, 12.2.b, and 12.2.c). With a higher level of feeding, the

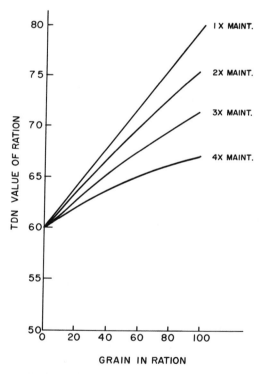

Fig. 2.2 With rations containing concentrates, the digestibility and total digestible nutrients (TDN) content decline as level of feed intake is increased. The assumption is that rate of depression increases 0.052% for each 1% increase in grain. (Adapted from Tyrrell and Moe, 1975.)

TDN content of the same ration declines due to a lower digestibility (see Fig. 2.2) (Tyrrell and Moe, 1975).

Although the TDN system gradually is being replaced by the NE system in the practical feeding of dairy cattle, especially in the United States, TDN is still very useful for several reasons. The relative ease of determining TDN values combined with long time usage has resulted in a great wealth of experimental data. Likewise, these data have been and are the basis for calculating many of the NE values now in use. As mentioned previously, of the variation in NE content among different feedstuffs, 80% is due to change in digestibility (Moe, 1976). Unlike NE values, usually, the digestibility and therefore the TDN content of a feed is not greatly affected by many nutritional deficiencies. The weaknesses in the TDN system are comparatively well known and systematic. Thus, with the appropriate modifications for specific conditions, the system can be used quite effectively in feeding dairy cattle. Other than the wealth of experimental data on TDN, DE has many of the advantages of TDN and is somewhat easier to determine experimentally.

2.1.e. Other Measures of Energy

Several other systems of evaluating the usable energy content of feeds and determining the energy requirements have been widely used in some parts of the world. Kellner's Starch Equivalents, developed in Germany many years ago from respiration experiments, is a type of NE system (Nehring and Haenlein, 1973). Likewise, the Scandinavian Feed Unit system is another way to express NE with information obtained from feeding experiments rather than from calorimetry (see Section 12.9).

The widely recommended modified metabolizable energy system used in the United Kingdom adjusts the values in such a way as to achieve most of the advantages of the NE system (Annison, 1971; ARC, 1965). Factors are used to account for differences in the nature of the ration, the level of feeding, and type of animal function involved.

2.2. ENERGY USES AND REQUIREMENTS

Every dairy animal uses energy continuously for the vast number of biochemical reactions essential to life (see Section 1.6). Thus it is both possible and interesting to consider energy metabolism from the biochemical vantage point. However, a division of energy uses and needs on a functional basis has greater application in the practical feeding of dairy cattle.

Since several measures of energy (see Section 2.1) are widely used, the energy requirements are expressed as NE for maintenance, body gain, and for lactation; as ME; as DE; and as TDN (see Appendix, Tables 1, 2, and 3) (NRC, 1978). The DE requirements were calculated from TDN values on the basis that 1.0 lb of TDN has 2.00 Mcal of DE (NRC, 1978), because many more data are available for TDN. Likewise, 1 kg of TDN has 4.409 Mcal of DE.

The ME requirement values for lactating cows generally are based on experimental data, but the ME requirements of growing dairy cattle were calculated from the NE requirements for maintenance and gain (see Appendix, Tables 1, 2, and 3) (NRC, 1978). Due to inadequate experimental data, the ME values of feeds were calculated from the DE content of feeds (NRC, 1978). The calculations were made using the equation

$$ME(Mcal/lb\ DM) = -0.20 + 1.01\ DE(Mcal/lb\ DM)$$

or by the metric system as

$$ME(Mcal/kg\ DM) = -0.45 + 1.01\ DE(Mcal/kg\ DM)$$

This formula assumes that ME as a percentage of DE varies linearly from 80% at 50% DE to 88% at 80% DE (Moe and Tyrrell, 1977; NRC, 1978). Since many of the DE data are from TDN measurements, the ME values are quite dependent on the TDN information. Although the relationships are not constant with all conditions, ME can be approximated from TDN assuming that 1.0 lb of TDN = 1.640 Mcal of ME, or that 1.0 kg of TDN = 3.615 Mcal of ME.

2.2.a. Energy Requirements for Maintenance

The energy required to maintain dairy cattle is one of the major costs of the dairy industry. Accordingly, considerable research has been conducted to measure and describe this key nutritional requirement. The maintenance requirement for energy is that quantity needed to keep the animal at the same size and same tissue composition, without producing any milk or reproducing. In practice, young dairy cattle are growing and possibly pregnant, while cows are lactating and/or pregnant. Thus, generally, determining the energy needed for maintenance is an academic division of the total energy needs rather than a directly established measure (Moe and Tyrrell, 1975).

The amount of energy required for maintenance increases with the size of the animal, but the increase is not directly proportional to weight. Rather, the increase is proportional to the 0.75 power of the weight (Blaxter, 1967; NRC, 1978). In Tables 1 and 2 of the Appendix the maintenance requirement of growing dairy animals for energy expressed as Mcal of net energy was calculated in metric units with the results difficult to convert to a per pound basis. The formula used was $0.077\ Mcal \times kg^{0.75}$ of body weight (NRC, 1978). For lactating cows the maintenance requirement used in the tables (see Appendix, Table 2) was assumed to be $0.08\ Mcal \times kg^{0.75}$ of body weight. This value includes a 10%

activity factor. The maintenance needs of lactating cows also have been calculated in terms of DE, ME, and TDN (see Appendix, Table 2) (NRC, 1978). The values used were as follows: 0.155 Mcal DE, 0.133 Mcal ME, or 0.0352 kg TDN per $kg^{0.75}$ of body weight.

Generally, the energy required for maintenance varies approximately with the surface area of the body, which is approximated by using weight to the 0.75 power, in mammals ranging in size from a few grams, such as mice or shrew, to elephants weighing several tons (Blaxter, 1967). However, there are some major deviations from this "surface area law" generalization for certain species including dairy cattle. The maintenance requirement of energy for dairy cattle is somewhat larger than would be calculated from the curve for all species.

The energy needed for maintenance of dairy cattle is influenced by some factors in addition to body weight, particularly the amount of activity. With all conditions apparently equal, including activity, the maintenance requirement between comparable individuals will vary by as much as 8–10% (NRC, 1978). Likewise the maintenance needs of lactating cows are about 10–15% higher than for dry, nonpregnant cows (NRC, 1978). Further, there is evidence suggesting a higher maintenance requirement for dairy cattle in the last few weeks of pregnancy (ARC, 1965).

The energy values for maintenance (see Appendix, Table 2) are sufficient for the usual activity of lactating cows fed in dry lot systems or in individual stalls. However, the requirements are increased by 5% for each added mile walked. Thus, the needs are approximately 10% higher for cows grazing lush pasture and as much as 20% more for those grazing sparse pasture (NRC, 1978). Under severe winter conditions in cold climates the maintenance requirements are elevated with the total feed needs increasing by as much as 8%.

2.2.b. Energy Requirements for Growth

Dairy cattle require energy to form tissues. Likewise, the physiological and biochemical reactions use energy necessary to metabolize, and transport the components, and synthesize the new tissue. Thus, growth requires more energy, in addition to maintenance, than the amount in the tissue.

The energy needed for a pound of growth in weight varies with the composition of the gain (Fig. 2.3). In the small calf the increased weight gain contains more protein, water, and minerals and relatively less fat (Blaxter, 1967). In older animals the gain has a higher percentage of fat with less water, protein, and minerals. The energy needed for a pound of fat is far higher than for the other constituents. Accordingly, as the animal becomes larger, energy requirements per pound of weight gain increases (Fig. 2.3). For example, when the higher energy requirement for maintenance, due to the larger size, also is considered, the total amount of feed energy needed for 1.6 lb of gain per day is more than

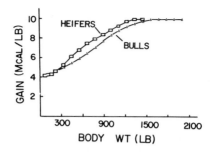

Fig. 2.3   Energy content of body weight gain of growing, nonfattening Holstein dairy heifers and bulls. Note the increase in energy as the animals become more mature. A comparable pattern would apply to other breeds if the scale is adjusted for differences in mature weight. (Adapted from NRC, 1978, and information courtesy of Dr. E. W. Swanson.)

three times as great with a 1000-lb Holstein heifer as for a 200-lb calf gaining at the same rate (see Appendix, Table 1).

Especially in young animals, the amount of fat deposited, relative to protein, mineral, and water, varies somewhat with the rate of gain. When gains are very rapid, the young animal will fatten faster and require more energy per pound of gain, exclusive of maintenance needs.

2.2.c. Energy for Fattening

Although fattening does not have the same role in dairy cattle as with beef cattle, dairy animals are sometimes fattened, either intentionally or as a secondary effect of other objectives. Generally, high producing cows use some of their body fat to supply energy needs during early lactation and replace it in late lactation or the dry period. In the nutrient requirement tables (see Appendix, Tables 2 and 3), energy over and above that needed for maintenance is recommended for the dry and pregnant cow during the last two months of gestation. The added energy is for reproduction. Likewise, if the cow is not in good condition, it may be desirable to provide additional energy to replace body reserves of energy in anticipation of the next lactation. Although overall beneficial, this deposition and subsequent loss of body tissue energy is less efficient than the direct use of feed energy for milk production (see Section 2.2.b). Excessive fattening of dairy cattle kept for reproduction and/or milk production can be detrimental and should be avoided (see Section 2.5, and Chapters 15 and 18).

As dairy cattle grow larger, an increasing percentage of the total tissue deposited is fat. Thus, fattening and growth are overlapping phenomena and not sharply distinguished (Blaxter, 1967). In the production of veal and dairy beef, fattening is a key practical aspect (see Chapters 16 and 17). Likewise, ''cull'' dairy cows are sometimes fattened before being sold for beef.

The fattening process does not involve the deposition of fat to the exclusion of other substances. But the ratio of fat to other constituents increases as the animal becomes fatter. The energy value of fat is substantially higher than protein; and fatty tissue contains less water than muscle. Thus, the energy deposited in a pound of fat tissue is much greater than in a pound of lean. Likewise, as the total weight increases, the maintenance requirement goes up. Contrary to some popular opinions, maintenance of body fat requires energy (Blaxter, 1967).

2.2.d. Energy Needs for Milk Production

Milk synthesis requires energy to form milk constituents and for the biochemical and physiological processes involved (see Section 2.4). The energy requirement for production of a pound of milk is dependent on the energy content of the milk. Generally, there is a relatively close correlation between the amount of fat in milk and the energy content. Frequently milk production is calculated to an approximately equal energy basis using the expression 4% fat-corrected milk or as fat-corrected milk which is abbreviated FCM. The formula used is

$$FCM = 0.4 \text{ milk (lb)} + 15 \text{ fat (lb)}$$

The energy needs to synthesize milk are presented in terms of the fat percentage (see Appendix, Table 2). The energy requirement per pound of 4% fat milk is about 0.34 Mcal of net energy, or 0.33 lb of TDN (NRC, 1978) (see Section 2.3). These values do not include the energy required for maintenance. Thus, for each pound of milk, the total amount of energy for maintenance and milk synthesis decreases as the level of milk increases.

2.2.e. Energy for Reproduction

During the first six months of pregnancy, the added energy requirements are quite small. However, the fetus, fetal membranes, fluids, and the size of the uterus grow at an exponential rate during pregnancy (Fig. 5.6) (NRC, 1978). In Jersey cows, the total weight of these reproductive products was 121 lb at term (Becker *et al.*, 1950; NRC, 1978). Of this total, 64% developed during the last two months and 38% in the last month. The total weight of the reproductive products is about 60–100% greater in the larger breeds, such as Holsteins and Brown Swiss, than in the Jersey.

Amounts of additional energy required during the last several weeks of pregnancy have not been well defined (NRC, 1978). Depending on the size of the cow, 3.0–6.0 Mcal of additional net energy or 1.27–2.53 lb of TDN per day are suggested for the last 4–8 weeks of gestation as adequate to meet the increased energy needs for development of the cow and calf (Flatt *et al.*, 1969; NRC, 1978). These amounts do not allow for fattening. If the cow is in "thin" condition, additional energy should be provided during the last few weeks of the gestation period.

2.2.f. Requirements for Work Energy

Since dairy cattle are rarely used as draft animals in developed countries, the practical importance of muscular work primarily relates to the additional expenditure of energy associated with grazing and other types of everyday activity. Muscular energy used by dairy cattle in routine living such as eating, chewing, and going to the milking parlor is a part of the maintenance requirement (see Sections 1.6.a and 2.2.a). The added energy required to graze and in group feeding and management situations can be viewed as an extra maintenance cost.

2.3. EFFICIENCY OF ENERGY UTILIZATION BY DAIRY CATTLE FOR VARIOUS FUNCTIONS

Compared with most other animals, dairy cows are efficient converters of feed energy to milk energy (see Section 1.0.a). Beneath this overall result are several complex aspects, including variable efficiencies in energy utilization with different functions and types of feeds (see Section 2.1.c). Feed energy is utilized more efficiently for lactation and maintenance than for growth or fattening (NRC, 1978). This is reflected in feeds having a higher net energy value for lactation and maintenance than for depositing nutrients in the form of tissues. The reasons for the lower efficiency of energy utilization for weight gains relative to lactation and maintenance, can be understood if one compares these functions to the manufacture of different products, with some being more complex than others. In the synthesis of milk some components of the blood are converted to entirely different compounds in the milk, whereas others are only transferred. A similar situation exists in the synthesis of new body tissues for weight gain. In both milk synthesis and body weight gains, numerous physiological processes and biochemical reactions are involved, with each requiring energy. With milk synthesis, either fewer reactions or those requiring less energy are necessary compared with weight gains. Thus, it is comparable to the manufacture of different products with those requiring more involved processes resulting in a greater amount of energy and cost above that for the raw material used.

Although there are differences in the utilization efficiency of feed energy for different functions, all are presented in terms of the same units (see Tables 1, 2, and 3 in Appendix). This simplification avoids the need to have different values for different functions. The net energy of the feeds is expressed in terms of the value for lactation (NE_l) which has the same efficiency as maintenance (see Appendix, Table 4). Requirements for weight gain and pregnancy were adjusted to compensate for the differences (see Appendix, Tables 1, 2, and 3).

The level of feeding affects the efficiency of certain aspects of feed digestion and metabolism. Some years ago, it was shown that with a higher level of concentrate feeding, digestibility of the ration declines (Fig. 2.2) (Moe *et al.*, 1965; Tyrrell and Moe, 1975). To compensate for this, the recommended amounts of feed per pound of milk were increased for high producing cows. However, subsequent research indicated that about half of the lower digestibility was compensated for by reduced energy losses in methane and urine (Moe *et al.*, 1966). Thus, the amount of usable feed energy required per pound of milk produced above maintenance, etc., is about the same with low and high producing cows. However, the lower digestibility of the same feed at the higher feed intake necessary for higher production means that more feed is needed per pound of milk. The values used in the tables represent typical production levels and are sufficiently accurate for practical purposes. Feeding the same cow at higher levels often will result in a higher proportion of the feed nutrients going for body fat synthesis (Flatt, 1966).

The relative efficiency of different type feeds varies with the functions involved. A substantial part of this effect is related to changes occurring in the rumen (NRC, 1978). Although an oversimplification, generally, rations that are higher in concentrates, especially starches, result in a lower ratio of acetate to propionate production by the rumen microbes (see Section 1.1.c). Such rations favor body fat deposition relative to milk synthesis and generally are more efficient for fattening.

The efficiency of converting digestible energy, above maintenance, to milk energy is about 55% and ranges from about 53 to 57% as the digestibility of the diet increases (Moe and Tyrrell, 1975). Similarly, metabolizable energy is converted to milk energy with an average efficiency of about 62% and a range of 61 to 64%.

2.4. CAUSES AND EFFECTS OF AN ENERGY DEFICIENCY

The suggested energy requirement recommendations (see Appendix, Tables 1, 2, and 3) (NRC, 1978) are believed to be sufficient for optimum performance under typical conditions in the United States. These energy levels will not necessarily result in either the maximum rates of growth or of milk production which might be achieved with *ad libitum* feeding of a high energy ration, especially over short time periods (NRC, 1978). As is true with all nutrient requirements, there is no sharp demarcation line between adequate energy and insufficient energy intake.

In healthy dairy cattle, a deficiency of usable energy generally is caused either by too little feed or a ration with too low a concentration of usable energy. Although there are exceptions due to low palatability, normally dairy cattle will consume feed to their maximum physical capacity, or to at least fully meet their energy requirements (see Section 2.6 and Fig. 2.5). With good management practices, dairy cattle are usually fed forages free choice. When this is done, a deficiency of energy is caused by too low a content of usable energy in the total diet. The amount of concentrates needed with forages to provide a sufficient average energy content depends on the quality of the forage. Usually the most important factors determining forage quality are content of usable energy and the amount which dairy cattle will eat when fed free choice (see Chapter 12). On dairy farms, probably the most frequent cause of insufficient energy is a combination of poor quality and/or inadequate amounts of forages, and lack of enough supplemental concentrates.

In young dairy cattle, inadequate energy intake decreases growth rate and delays the onset of puberty. The reduced growth rate is accompanied by changes in body composition, including a greater reduction in fat content of tissues than of mineral or protein. Thus, the animal underfed on energy will be leaner, with less effect on skeletal size. Unless the underfeeding of energy is extreme and long extended, there will be little or no permanent damage to health. With only a moderate underfeeding period followed by adequate energy intake, young dairy cattle will, at least partially, make up for the slower growth period. This catching up is known as compensatory growth.

Extreme underfeeding of energy for extended periods delays the time when the heifer is large enough to breed, thus making her older at first calving. Reproductive ability also is delayed in the bull. With insufficient energy intake, the maintenance requirements must be met for a longer time before the animal is productive. If the heifer is bred to calve at a materially smaller size than she

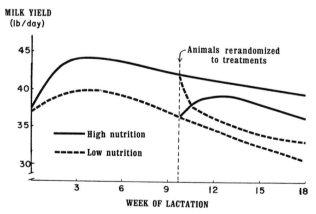

Fig. 2.4 Cows underfed during the first 10 weeks of the lactation and subsequently well fed produce less milk in late lactation than those fed adequately throughout. (From Moe and Tyrrell, 1975, as adapted from Broster *et al.,* 1969.)

would if given the recommended energy intake levels, generally milk production will be reduced, at least in the first lactation. The frequencies of calving difficulties will be greatly increased when heifers are extremely small at calving. Most of the adverse effects of inadequate energy feeding prior to first calving can be overcome by subsequently feeding adequate energy (Swanson and Hinton, 1964).

In the lactating cow, the first effect of inadequate energy intake is lowered milk production, accompanied by loss of body condition. When cows are underfed during early lactation, milk production is lower throughout that lactation (Fig. 2.4). Milk production will be greatly decreased before there is very much effect on reproduction or impairment of the cow's health. Although the thinner body condition decreases energy used for maintenance somewhat, the percentage reduction is small compared with the decreased milk yield. Accordingly, the overall efficiency of converting feed energy to milk is rapidly reduced with an energy deficiency. Likewise, with inadequate energy intake, the solids-not-fat and protein content of milk usually will be reduced.

2.5. CAUSES AND EFFECTS OF EXCESS ENERGY

Since the recommended energy levels (see Appendix, Tables 1, 2, and 3) are those believed to result in optimum rather than maximum growth and milk production, feeding above these levels often will increase growth rate or milk production by a small amount. However, the increased performance usually would not be sufficient to pay for the extra feed and may cause serious health problems (see Section 18.12).

As excessive levels of energy are fed, much of the increase is deposited as body fat. Only a small amount goes for higher milk production or growth of nonfat tissue. Cattle are not able to eat more feed energy than they can convert to fat (Blaxter, 1967). However, as they become excessively fat, the amount of feed voluntarily consumed decreases. Since the maintenance requirement is increased by the larger weight, the excessively fat dairy animal uses a higher percentage of its total energy intake for maintenance.

Excessive fattening of dairy heifers may cause permanent damage to the mammary development and impair reproduction, thus decreasing their lifetime milk producing ability (NRC, 1978; Swanson, 1960). Because excessive fattening of heifers not only adds cost but decreases their value for dairy purpose, it should be avoided (see Section 15.2). The fat deposited in the mammary gland of overfed young heifers may peranently reduce the amount of secretory tissue and impair the subsequent capacity to produce milk (Swanson, 1960).

In dairy cows, excessive fat often adversely affects reproduction and causes other health problems (Trimberger et al., 1972) (see Section 18.12 on the fat cow syndrome). High producing cows often will benefit from "fairly high condition" at the time of calving, as usually they cannot consume enough feed in early lactation to meet their energy needs.

2.6. BUILT-IN MECHANISMS FOR CONTROLLING ENERGY BALANCE IN DAIRY CATTLE (PRINCIPLE OF HOMEOSTASIS)

If the concentration of the functional forms of any essential nutrient in the body tissues of animals reaches either sufficiently high or low levels, death occurs. In contrast, extreme shortages or excesses of most essential nutrients in feeds are common occurrences in nature. Thus, every animal must have mechanisms permitting them to adapt to the highly variable amounts of nutrients encountered. To understand these adaptations it is helpful to use the physiological term homeostasis (or homeostatic control) which is defined as "the condition of relative uniformity which results from the adjustments of living things to changes in their environment" (Cannon, 1929; Miller, 1975). Homeostatic control for every nutrient is essential for survival. The remarkable adaptability of ruminants to a great variety of feeds is possible because of good homeostatic control mechanisms for most nutrients.

The homeostatic mechanisms for energy in dairy cattle are totally different from those of any other essential nutrient. Unlike all others, the efficiency with which dietary energy is absorbed, metabolized, utilized for body functions, or excreted varies remarkably little when the level of intake changes drastically.

When more energy is consumed than needed for optimum performance, dairy animals fatten (see Section 2.5). Likewise, if energy intake is too low, either milk production declines or the animals will have less body fat or both. Anytime usable energy intake changes, there is either a change in the body content of energy or in the total energy of the milk produced. Thus, dairy cattle do not have mechanisms permitting them to perform at the same rate without altering body energy stores when consumption varies. In contrast, the amounts of other essential nutrients eaten can vary substantially without material changes in performance or major variations in the content of functional forms in body tissues.

Accordingly, it becomes obvious that the major mechanism dairy cattle use for homeostatic control of energy is change in feed consumption. It is equally apparent that change in feed intake cannot be the primary homeostatic control mechanism for any other nutrient. It is the old principle that no one can serve two masters. Normally, energy is the master controlling feed intake.

Changes in voluntary feed intake by dairy cattle are controlled in two main ways (Fig. 2.5). With fibrous, poorly digested feeds, dairy cattle generally voluntarily consume feed to the limits of their physical capacity. When physical

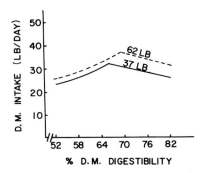

Fig. 2.5 With rations of low digestibility, physical capacity limits dry matter (DM) intake; however, with highly digestible feeds, physiological factors determine maximum consumption. Dry matter intake increases with digestibility to about 65–68% and subsequently declines. The point at which DM intake is highest increases with the energy needs of the cow. In these examples, 1000-lb cows are capable of producing 37 and 62 lb of milk daily. At lower milk production (capacity), physiological factors limit intake at a lower digestibility. (Data adapted from Conrad *et al.*, 1964, with modifications including, especially, an assumed higher DM intake for cows with greater energy needs associated with higher milk production.)

capacity restricts intake, and the dairy animal does not consume as much usable energy as needed, performance is reduced. In contrast, with a highly digestible ration, chemical and/or physiological factors limit the amount of dry matter eaten. With high producing dairy cows, generally physical capacity determines the intake of rations having a dry matter digestibility below about 65–68%, whereas above this digestibility level, physiological and/or chemical factors restrict the amount of usable energy consumed (Conrad *et al.*, 1964) (see Fig. 2.5).

REFERENCES

Agricultural Research Council (ARC) (1965). "The Nutrient Requirements of Farm Livestock," No. 2. ARC, London.

Annison, E. F. (1971). *Feedstuffs* **43** (10), 34–35.

Becker, R. B., P. T. Dix-Arnold, and S. P. Marshall (1950). *J. Dairy Sci.* **33**, 911–917.

Blaxter, K. L. (1967). "The Energy Metabolism of Ruminants." Thomas, Springfield, Illinois.

Broster, W. H., V. J. Broster, and T. Smith (1969). *J. Agric. Sci.* **72**, 229–245.

Cannon, W. B. (1929). *Physiol. Rev.* **9**, 399.

Conrad, H. R., A. D. Pratt, and J. W. Hibbs (1964). *J. Dairy Sci.* **47**, 54–62.

Flatt, W. P. (1966). *J. Dairy Sci.* **49**, 230–237.

Flatt, W. P., P. W. Moe, and L. A. Moore (1969). *Eur. Assoc. Anim. Prod., Publ.* **12**, 123–136.

Miller, W. J. (1975). *J. Dairy Sci.* **58**, 1549–1560.

Moe, P. W. (1976). *Feed Manage.* **27** (6), 28 and 33.

Moe, P. W., and H. F. Tyrrell (1975). *J. Dairy Sci.* **58**, 602–610.

Moe, P. W., and H. F. Tyrrell (1977). "Proc. 1st Int. Symp. on Feed Composition, Animal Nutrient Requirements, and Computerization of Diets," 1976. pp. 232–236. Int. Feedstuffs Inst., Logan, Utah.

Moe, P. W., J. T. Reid, and H. F. Tyrrell (1965). *J. Dairy Sci.* **48,** 1053–1061.

Moe, P. W., W. P. Flatt, and L. A. Moore (1966). *J. Dairy Sci.* **49,** 714 (abstr.).

Moore, L. A., H. M. Irvin, and J. C. Shaw (1953). *J. Dairy Sci.* **36,** 93–97.

Morrison, F. B. (1957). "Feeds and Feeding," 21st ed. Morrison Publ. Co., Ithaca, New York.

National Research Council (NRC) (1966). "Biological Energy Interrelationships and Glossary of Energy Terms." Natl. Acad. Sci., Washington, D.C.

National Research Council (NRC) (1978). "Nutrient Requirements of Dairy Cattle," 5th rev. ed. Natl. Acad. Sci., Washington, D.C.

Nehring, K., and G. F. W. Haenlein (1973). *J. Anim. Sci.* **36,** 949–964.

Swanson, E. W. (1960). *J. Dairy Sci.* **43,** 377–387.

Swanson, E. W., and S. A. Hinton (1964). *J. Dairy Sci.* **47,** 267–272.

Trimberger, G. W., H. F. Tyrrell, D. A. Morrow, J. T. Reid, M. J. Wright, W. F. Shipe, W. G. Henderson (1963). *Proc. Cornell Nutr. Conf. Feed Manuf.* pp. 3–43.

Trimberger, G. W., H. F. Tyrrell, D. A. Morrow, J. T. Reid, M. J. Wright, W. F. Shipe, W. G. Merrill, J. K. Loosli, C. E. Coppock, L. A. Moore, and C. H. Gordon (1972). *N.Y. Food Life Sci. Bull.* **8.**

Tyrrell, H. F., and P. W. Moe (1975). *J. Dairy Sci.* **58,** 1151–1163.

3

Protein Requirements of Dairy Cattle

3.0. Introduction

In popular nutrition, protein content of feed and foods often has been viewed as synonymous with quality and total value. For instance, over a period of many years, the overall quality of forages was evaluated on the basis of the protein percentage. Although this was a good index in many situations, often the primary reason was the high correlation of protein with other important factors rather than just the contributions of protein to the nutritional needs of cattle. This will be discussed more fully in the section on forage evaluation (see Chapter 12).

A major part of all the vital organs and tissues of dairy cattle is protein. Because much of the body is protein, a substantial quantity is required in the diet. Except for energy, fiber, and water, cattle require more protein in the diet than all other nutrients combined.

Protein is a collective term encompassing with some common characteristics, numerous similar but distinctly different compounds. Most protein molecules are quite large, but there is a huge range in their sizes. All are made up of small building units known as amino acids in variable proportions arranged in different configurations. Many contain other substances especially certain lipids, carbohydrates, or minerals such as iron and phosphorus which usually give them special properties and functions. Lipoproteins, phosphoproteins, and metalloproteins are general names given to some of these complex proteins. Other proteins are specific substances. For instance, hemoglobins, which transport oxygen to the tissues and carbon dioxide back to the lungs, are special iron-containing proteins. Enzymes, the biological catalysts regulating most biochemical reactions of the body, usually function in association with one or more essential mineral elements.

There are about 25 different amino acids that are classified as either essential or nonessential (see Section 3.0.b for list). The essential ones must be absorbed into the body of animals for use by the tissues, but the nonessential amino acids can be synthesized in the body tissues. About 60% of the amino acids of the animal body consist of essential ones compared to 40% that are nonessential (Maynard and Loosli, 1969). Whereas essential amino acids must be present in the diet of nonruminants, they are synthesized by the rumen microbes (see Section 1.1.c). Chemically, all of the amino acids contain one or more amino groups (NH_2), one or more carboxylic acid groups (COOH), and a carbon chain. Methionine and cystine contain sulfur.

In addition to true proteins, numerous closely related nitrogenous compounds are of great importance, including nucleic acids such as ribonucleic acid (RNA) and deoxyribonucleic acid (DNA). Synthesis of these nitrogenous compounds requires dietary protein. Likewise, many of these nitrogenous substances in the feed may be utilized by dairy cattle in meeting their protein needs. For example, 10–20% of the protein-nitrogen of rumen bacteria is in various nucleic acids (Chalupa, 1975). Part of these nucleic acids may be degraded in the rumen and utilized, but the nucleic acids going beyond the rumen do not contribute much to the protein needs of the animal (Burroughs *et al.,* 1975b; Chalupa, 1975).

Direct chemical determination of protein in feeds is impractical. Accordingly, nutritionists take advantage of the relatively constant proportion of nitrogen in all proteins. Since the average nitrogen content of protein is 16%, multiplying total nitrogen (N) in feed by the factor 6.25 (100/0.16 = 6.25) yields a reasonable estimate of total protein. Some proteins contain a different N percentage; for instance, milk protein is 15.7% N.

Many feeds, including silages, contain substantial amounts of nonprotein nitrogen (NPN) that is included in the total protein by this common procedure. Therefore, protein determined from nitrogen times 6.25 is often called crude protein. Formerly, the distinction between true protein and crude protein was given considerable emphasis. Since NPN can be utilized by dairy cows, the sharp delineation now receives less attention in dairy cattle nutrition and the terms protein and crude protein often are used interchangeably.

3.0.a. Synthesis of Essential Amino Acids in the Rumen

Although some protein goes through undigested, much of the dietary "protein" is degraded in the reticulorumen by the microbes to amino acids and simple nitrogenous compounds, especially ammonia (NH_3) (see Section 4.1). If the rate of release is moderate and the other needed substances are available, most of the ammonia liberated by the microbes will be utilized by them or other microbes in synthesizing amino acids and proteins for their own tissues. Although other compounds are used, ammonia is a major soluble nitrogen source for synthesis of protein by rumen microbes (Blackburn, 1965; Chalupa, 1975; National Research Council, 1976). If ammonia is released more rapidly in the rumen than it can be utilized, much of the excess will be absorbed into the blood stream and subsequently converted into urea by the liver, for excretion or recycling via saliva or blood for reuse in the rumen. This recycling of urea back to the rumen provides a nitrogen conservation system during periods of low protein intake.

Even though there is a considerable range, the dry weight of rumen microbes apparently averages about 65% protein (Hungate, 1966; Purser, 1970). This protein contains all the essential amino acids. These microbes pass on to the abomasum and small intestine where they are digested and their component nutrients, including the amino acids, are absorbed and utilized. Thus, after digesting and using much of the feed eaten by the dairy cow, the rumen microbes are digested and utilized by the cow. As much as 30% of the bacterial protein produced in the rumen can be digested and utilized by other microbes (Chalupa, 1975). Even so, the cow gains important benefits including the essential amino acids (see Section 1.1.c).

When dietary proteins are converted to ammonia by microbial action in the rumen, a considerable amount of nonnitrogenous carbonaceous or ''carbohydrate like'' material is released. These nonnitrogenous residues or carbon chains of the proteins and amino acids can be utilized for energy, or as parts to be used by the microbes, along with ammonia, in synthesis of new amino acids. Thus, only a portion of the material is discarded rather than being used.

3.0.b. Essential Amino Acids for the Baby Calf

Since very young calves do not have sufficient rumen microbes to synthesize the essential amino acids, these must be supplied in the diet. The amino acids shown to be essential for the tissues of ruminants are (1) histidine, (2) isoleucine, (3) leucine, (4) lysine, (5) methionine, (6) phenylalanine, (7) threonine, (8) tyrosine, and (9) valine (Black *et al.,* 1957; Downes, 1961; Purser, 1970). These are dietary essentials for the baby calf and must be available for absorption from the small intestine of older cattle.

The rumen and reticulum of the calf develop and become functional over a period of several weeks after birth (see Section 1.1.b). Thus, there is a gradual but comparatively rapid decrease in the dependency of the calf on dietary protein as the source of essential amino acids.

3.1. PROTEIN DIGESTIBILITY AND UTILIZATION

By definition, the true digestible protein in a feed is the portion absorbed by the animal. As a practical matter, digestibility is determined in a digestion experiment by measuring the protein (N × 6.25; see Section 3.0) in the feed and the feces. The difference is apparent digestible protein, usually referred to simply as digestible protein. In addition to undigested protein, feces contain protein of metabolic origin. Because of its influence on protein digestibility values and requirements, some understanding of metabolic fecal protein is useful in the practical feeding of dairy cattle.

Metabolic fecal protein by definition comes from within the body and is of endogenous origin. It is derived from unabsorbed digestive juices, sloughing of cells from the digestive tract and microbial protein from the cecum. Since there is no reliable way to separate the metabolic fecal protein from the undigested protein of the feed, indirect approaches have been used (Agricultural Research Council, 1965; Swanson, 1977). One method is to determine the protein (N ×6.25) in the feces, when a protein-free, or almost free, diet is fed. Another indirect way to estimate metabolic fecal protein is from regression equations with diets of widely varying protein contents (ARC, 1965, Swanson, 1977).

Usually true digestibility of protein is not affected materially by the protein content of the diet. However, the apparent digestibility of protein from feeds is greatly influenced by the metabolic fecal protein. Since the metabolic fecal protein remains essentially constant, the percentages of the apparently digested protein increases directly with each additional increment of crude protein in the diet (Glover *et al.*, 1957; Holter and Reid, 1959; Knight and Harris, 1966; National Research Council, 1978; Preston, 1972). Results of numerous digestion studies have shown that this relationship is linear and that the correlation is very high between the total crude protein content of the diet and apparent digestibility of protein (NRC, 1978). Thus, generally, protein requirements of dairy cattle can be expressed in terms of total crude protein. Except in special situations, as discussed below, it is not necessary to determine the protein digestibility of feeds.

Excessive heating, charring, browning, or carmelizing or other comparable processes often substantially reduce the digestibility of feed proteins (Fig. 3.1

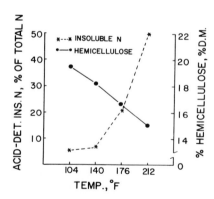

Fig. 3.1   Effect of temperature on browning of forage as indicated by increase in acid-detergent insoluble nitrogen and decrease in hemicellulose. Forages were heated in 10-ml flask for 24 hours at 53% moisture. (Adapted from Goering *et al.*, 1973.)

TABLE 3.1
Effect of Excessive Heating on Crude Protein Digestibility of Various Forages[a]

Forage	No. of samples	Crude protein digestibility (%)		
		Observed	Expected	Reduced by heating
Silages and hays	44	51	71	20
Alfalfa brome silages	3	67	74	7
Alfalfa, dehydrated	1	32	66	34
Hay, brown	5	31	62	31

[a] Adapted from Goering and Waldo (1974).

and Table 3.1) (Goering *et al.*, 1972, 1973; Goering and Waldo, 1974; Gordon *et al.*, 1961; Mertens, 1977; NRC, 1978; Sutton and Vetter, 1971). Low moisture silage (haylage) and hay stored with excessive moisture are damaged frequently as indicated by a caramel-like odor and brown color (NRC, 1978). It has been suggested that only 80% of the actual crude protein content in such moderately heat-damaged feeds should be used in calculating the protein needs of dairy cattle (NRC, 1978). When the forage is deep, dark-brown or black, more than 50% of the crude protein may be unusable (see Sections 3.11 and 12.5).

3.2. DETERMINATION OF PROTEIN REQUIREMENTS OF DAIRY CATTLE

Great progress has been made in understanding the protein nutrition and metabolism of dairy cows within the past few years. This has resulted in several new concepts that should eventually lead to more accurate ways of calculating the protein requirements of dairy cattle than can be done now (see Sections 3.10 and 3.11). As discussed earlier (see Section 1.5), nutrient requirements can be determined by the factorial method or from results of feeding experiments. Generally, it is desirable to determine requirements by both approaches and compare results.

The protein requirements of dairy cattle have been calculated by the factorial method (see Appendix, Tables 1, 2 and 3) (NRC, 1978). Knowledge of the way these results are obtained can be helpful in understanding the feeding and nutrition of dairy cattle. With current information, the most accurate way of calculating the protein needs of dairy cattle from feed composition tables is as total crude protein (see Section 3.1) (NRC, 1978). In earlier revisions of the bulletin on the Nurtient Requirements of Dairy Cattle, the quantities of digestible protein required were presented also (National Research Council, 1971). The digestibility of protein in feeds is directly related to the total protein percentage in the diet (see Section 3.1). Accordingly, it is usually more accurate to calculate digestible protein in a diet from a prediction equation (Section 3.1) based on the total crude protein percentage than to use average digestibility values of components feeds. Correction for unavailable protein because of heat damage is important in some situations (see Section 3.11).

The protein requirements of dairy cattle can be divided into that needed for (1) maintenance, (2) gain in body weight, (3) reproduction, and (4) lactation or milk synthesis. These values represent net protein needs and must be increased for the unabsorbed protein and the fraction of the absorbed protein not utilized for essential functions to arrive at the total crude protein needs of the animal.

The maintenance requirement for protein is determined in three separate components which are added together. These three are (1) the metabolic fecal protein loss, (2) the endogenous urinary protein, and (3) the protein loss in skin secretion, hair and scurf (ARC, 1965; NRC, 1978). The metabolic fecal protein loss was discussed earlier (see Section 3.1). The endogenous protein in the urine is the portion excreted irrespective of whether the animal is fed protein in the diet. It is thought to come from degradation and replacement of protein structures and other nitrogenous components of tissue by irreversible reactions such as the conversion of creatine to creatinine (ARC, 1965).

The detailed procedure used in calculating the protein requirements (NRC, 1978) is rather complex. For each class of dairy cattle the basic formula used (NRC, 1978) in the factorial methods is as follows:

Total crude protein needed in diet =

$$\left(\frac{\text{Maintenance requirement} + \text{Body gain protein} + \text{Reproduction protein}}{\text{Net utilization value}} \right.$$

$$\left. + \frac{\text{Lactation protein}}{\text{Net utilization value (lactation)}} \right)$$

Expressed in grams per day, the following calculations are involved. The maintenance requirement is the sum of three parts (NRC, 1978; Swanson, 1977) as follows: (1) metabolic fecal protein is 6.8% of the fecal dry matter; (2) endogenous urinary protein is calculated as $2.75 W^{0.50}$ (W is body weight in kg) (1 lb = 0.4536 kg); (3) protein lost in skin secretions, scurf, and hair is calculated by $0.2 W^{0.60}$ (W is body weight in kg).

The protein in body weight gain declines from 19% of weight gain for newborn calves to 16% as animals approach maturity without fattening (Fig. 3.2 (NRC, 1978).

The protein for reproduction which is the protein in the products of conception (fetus, placenta, fetal fluids, and the uterus) was calculated with a formula developed by Jakobsen (1957; NRC, 1978). This gave an average of 88 grams of protein daily for the last 60 days of gestation for a 1100-lb cow and varied with

TABLE 3.2
Net Utilization Values of Protein for Various Classes of Dairy Cattle[a]

Type animal	Diet protein absorbed as amino acids (%)	Conversion efficiency (%)	Net utilization value[c] (%)
Baby calf fed only whole milk	91	77	70
110-lb calf fed milk and concentrates	87	75	65
165-lb calf fed mainly concentrates	80	70	56
220-lb calf fed forage and concentrates	75	67	50
330-lb cattle fed forage and concentrates	75	63	47
Cattle over 440 lb fed for maintenance and growth	75	60	45
Lactation[b]	75	70	52

[a] Adapted from NRC (1978).

[b] The lactation factors apply only to the portion of the feed used for lactation and not to that utilized for maintenance of the lactating cow.

[c] Net utilization values used to calculate protein requirements. These values are the product of those in the first two columns.

weight to the 0.70 power. When calculated in grams, protein is equal to 1.136 $W^{0.70}$ (W is weight in kg).

The lactation protein requirement includes milk protein plus the metabolic fecal protein* associated with the extra feed needed for lactation. Milk protein was calculated on the basis of the fat content of milk by the formula: percent protein in milk is equal to $1.9 + 0.4$ times percent fat in milk (NRC, 1978). This formula does not give reasonable values when the milk fat is depressed by special diets such as those containing inadequate fiber or unground roughage (see Chapter 8).

At this point, the net protein required for deposition in tissues and products of conception, for secretion of milk, and the unavoidable losses associated with maintenance will have been determined. To obtain total crude protein requirements in feed from these net protein requuirement values, the unabsorbed protein

---

*In growing, reproducing, or lactating cattle, the maintenance protein requirement is higher than for animals fed at a maintenance level. Thus, the net protein requirements for the growing, reproducing, and lactating functions are higher than the protein in the products alone. This is the additional metabolic fecal protein associated with the extra feed needed above that required for maintenance. Since only one value is presented (see Appendix, Tables 1 and 2) for the functions other than milk, this added protein is included in the calculations shown above.

and that absorbed but not utilized for essential functions must be added. These can be conceived as losses in utilizing crude protein and are called the net utilization value.

The net utilization value is the product of (1) the percentage of the dietary protein absorbed as amino acids and (2) the efficiency of converting absorbed amino acids to protein which can be used for maintenance, body gain, reproduction or milk (Table 3.2). The values used for the percentage of dietary crude protein absorbed as amino acids in calculating the requirements varied from 91% for calves fed milk, to 80% for calves given high concentrate diets, to 75% for all other classes of dairy cattle (Table 3.2) (NRC, 1978). The value used for the efficiency of converting absorbed amino acids to protein was 70% for lactating cows (NRC, 1978). For growing and nonlactating dairy cattle, the value used decreased gradually from 80% for young calves to 60% for those fed only forages (NRC, 1978). The net utilization values as a percentage of dietary protein are presented in Table 3.2.

3.3. USING PROTEIN REQUIREMENT VALUES—SPECIAL CONSIDERATIONS

The formula described in the previous section permits the protein requirements to be calculated by the factorial method. The exact amount, calculated in pounds per day, for the same animal varies somewhat depending on the dry matter digestibility of the diet. Metabolic fecal protein, which is a part of the maintenance loss, increases with fecal dry matter. Thus, the protein required per day, with the same level of feed intake, would be higher when the feed dry matter is 60% digestible than when 75%. Since less total feed is needed to provide sufficient energy, the required percentage of protein in the total feed increases with higher dry matter digestibility.

In preparing the tables of the protein requirements (see Appendix, Tables 1, 2, and 3), the calculations were based on the most likely type of diet to be fed to each class of dairy cattle when liberal forage feeding is the objective (NRC, 1978). These protein requirement values should be viewed as good estimates and not as highly exact answers. Moderate deviations from the assumptions used in preparing the tables would not seriously affect the usefulness of the results. With practical feeding, minor deviations from the estimated protein requirements should not materially affect animal performance or feed costs.

In selecting and preparing the formulas for calculating protein requirements by the factorial method, results of critical feeding experiments with dairy cattle were considered (NRC, 1978). The results (see Appendix, Tables 1, 2 and 3) are consistent with many studies on milk production (Broster, 1972; Chandler, *et al.,* 1976; Gardner and Park, 1973; Huber, 1975; Paquay *et al.,* 1973; Thomas, 1971; Van Horn and Jacobson, 1971) and with growing animals (Brisson *et al.,* 1957;

Broster *et al.*, 1969; Forbes, 1924; Gardner, 1968; Jacobson, 1969; Jahn and Chandler, 1976; Lofgreen *et al.*, 1951; Stobo and Roy, 1973; Stobo *et al.*, 1967).

The protein requirements calculated by the factorial method are those necessary to meet the needs of the animal without regard to certain other effects such as efficiency of feed utilization. For example, the protein needs of the rumen microorganisms also must be considered. If cattle are fed less than 11–12% crude protein in the feed dry matter, rumen fermentation and digestion may decrease resulting in lower voluntary feed consumption and depressed utilization of roughage nutrients (Burroughs *et al.*, 1949; Hungate, 1966; NRC, 1978). The protein percentage needed for optimum rumen fermentation varies somewhat depending on the amount of degradable or fermentable protein compared to undegradable or bypass protein in the diet (see Section 3.10). Because of the effects on rumen fermentation, often it is enconomical to feed at least 11–12% crude protein in the feed dry matter for mixed forage concentrate rations even though this may be more than required by the animal (NRC, 1978).

Protein is required in relatively large amounts and generally is fairly expensive per pound. Accordingly, in many if not most situations, feeding substantially more than needed for optimum performance of the animals is wasteful. However, providing adequate protein to meet all the needs is important. The additional protein needed by high-producing dairy cows in early lactation, which are using body fat as a major part of their energy needs, is discussed more fully later (see Section 3.5). Especially for high-producing cows, it is important to provide sufficient protein in forms which are not degraded by the rumen microbes (see Sections 3.9 and 3.10).

3.4. THE PROTEIN REQUIREMENTS OF BABY CALVES

Because they are so different, the protein needs of baby calves and older dairy cattle are presented as separate topics. The three major differences are that the baby calf must have (1) a dietary source of essential amino acids (see Section 3.0.b), (2) a high percentage of total protein, and (3) a dietary source of protein that is digestible in the abomasum and small intestine. Whereas a great variety of feed proteins and nonprotein nitrogen compounds can be effectively used by older dairy cattle, the number of suitable sources of protein for the baby calf is relatively small (see Chapter 14).

In comparison with its size, the baby calf is growing rapidly. Thus, the percentage of dietary nutrients being deposited in tissue is quite high relative to the maintenance requirement. Likewise, in the baby calf a higher proportion of the total tissue growth is protein (Fig. 2.3 and 3.2). Accordingly, the percentage

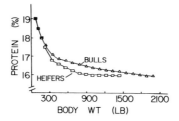

Fig. 3.2   Protein content of body weight gain of growing, nonfattening dairy heifers and bulls. Note the decrease in protein percentage as the animals become more mature. Compare with Fig. 2.3 showing the change in energy content and note the smaller change in protein relative to energy. (These data are for Holsteins. A similar pattern exists for other breeds if body weights are adjusted for differences in mature size.) (Adapted from NRC, 1978; and information courtesy of Dr. E. W. Swanson.)

of protein needed in the diet of the baby calf is substantially higher than for older dairy cattle (NRC, 1978). A minimum of 22% protein is recommended in the dry matter of a milk replacer for baby calves (see Appendix, Table 3). As the calf grows, the percentages of protein needed in the total diet declines quite rapidly, with only 16% needed in starters for calves up to 3–4 months old (NRC, 1978).

3.5. PROTEIN REQUIREMENTS OF OLDER DAIRY CATTLE

Expressed either as a percentage of the dry matter or as a ratio to the usable energy content of the diet, the protein requirements vary with the function for which it is used. The lowest percentage protein is for maintenance. Thus, mature animals just being maintained have lower requirements than other cattle. In most situations only mature bulls are not performing functions requiring substantial amounts of protein. Very little protein is needed by the bulls to produce sperm. Although 8.5% crude protein (see Appendix, Table 3) in the feed dry matter is sufficient for mature bulls, they will digest and utilize their feed more efficiently if given more protein (see Section 3.3).

Since milk is a high-protein food, the ratio of protein to energy required for the cow to synthesize milk is much higher than for maintenance. Accordingly, the larger the milk yield the higher the dietary protein percentage needed (see Appendix, Table 3; Fig. 3.3) (NRC, 1978). Likewise, a small cow producing 40 lb of milk will require a somewhat higher minimum percentage of protein than a large cow for the same amount of milk.

On a practical basis, 16% protein in the ration dry matter is adequate for most lactating cows that are consuming enough feed energy to fully meet their energy needs (see Appendix, Tables 1, 2, and 3; Fig. 3.3). During early lactation many

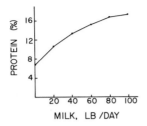

Fig. 3.3 This illustrates the percent crude protein which would be needed in the feed dry matter to meet requirements of 1300-lb cows which were neither gaining nor losing weight. Calculations based on 3.5% fat milk with the TDN content of feeds going from 60% for dry cows to 75% for highest producers for practical feeding situations as presented in Table 2 of the Appendix. Note the large increase in protein percentages with higher milk production.

high-producing cows will not consume sufficient feed to supply the needed energy. The extra energy is obtained from body fat. Along with body fat a small amount of body protein is used, but the ratio of protein obtained from the body tissue to that needed for milk synthesis is low relative to the energy obtained (see Section 1.6.b and 1.6.d). Accordingly, high-producing cows in early lactation need a higher percentage of protein than indicated in Tables 1, 2, and 3 of the Appendix (Coppock, 1969; Crowley, 1975; Hemken, 1975). As much as 20% protein may be required in the total dry matter (see Section 3.8 and Chapter 13).

In contrast to the unusually high dietary protein content needed by high-producing cows in early lactation, during late lactation, when they are depositing fat, a lower protein content will suffice (Hemken, 1975). The amount of protein deposited in the adipose tissue with fat is comparatively low. Higher percentages of protein are recommended for dry, pregnant dairy cows because they are reproducing.

The percentage of protein needed in the diet for growing dairy cattle gradually decreases with age. A part of this decrease is due to the smaller percentage of the total feed used for tissue growth and another part is associated with a higher percentage of the new tissue being deposited as fat (Fig. 2.3 and 3.2). Because feed utilization is higher (see Section 3.3) when at least 11–12% crude protein in the dry matter is fed, often it is desirable to use this level as a minimum. This amount is sufficient for heifers and bulls beyond 8 to 12 months of age.

3.6. EFFECTS OF PROTEIN DEFICIENCY

Because protein is such a key part of so many functions and biochemical processes, insufficient protein has a multitude of effects. However, the severity of symptoms caused by a protein deficiency depends on the extent of the short age. In lactating cows, too little protein reduces milk production (Fig. 3.4) and lowers the protein and solids-not-fat content of the milk. When severely deficient, cows lose an unusually large amount of weight in early lactation and do not regain it normally in late lactation (NRC, 1978).

Inadequate protein reduces the growth rate of the fetus resulting in smaller and weaker calves at birth. Likewise, with a severe deficiency, the unborn fetus may die or be lost prior to birth. The growth rate of calves and other growing cattle is reduced with low-protein diets.

In dairy cattle of all ages, protein deficiency reduces the protein content of blood, organs, and tissues, including smaller amounts of immune and transport proteins as well as lower hormone secretions. Accordingly, the animals have less resistance to many types of diseases, metabolic disorders, and infections. With insufficient protein, wounds heal more slowly; hair and probably hooves and horns grow slower (Martin, *et al.,* 1969).

More protein may be required for maximum fermentation and digestion by the rumen microorganisms than necessary for the other functions in some classes of dairy cattle (see Section 3.3). If protein is deficient, the rate of microbial fermentation is reduced decreasing the amount of feed which can be digested per day. When feed intake is limited by physical factors (see Section 2.6), animals also may be energy deficient. Likewise, if protein is deficient, net energy content of a given ration may be further reduced due to less efficient utilization of digested energy.

3.7. EFFECTS OF EXCESS PROTEIN AND PROTEIN HOMEOSTASIS

For economic reasons, generally, rations are formulated to meet only the minimum protein needs. The major problem with excess protein feeding is the added cost involved. Although contrary opinions exist, there is little research information to indicate that even very excessive consumption of true proteins is seriously detrimental to dairy cattle (NRC, 1978). The toxicity of excessive levels of nonprotein nitrogen sources will be discussed in Chapter 4.

Cattle have a remarkably effective homeostatic control mechanism for the excessive dietary protein intake (see Sections 2.6, 3.1, 3.2, and 4.1.b). True digestibility of the protein is not affected to a material degree by the amount of protein consumed. When more protein is eaten than needed by the animal, the nitrogen portion of the molecules is converted to urea in a series of steps. Excess urea is excreted in the urine, thus eliminating material from the body which otherwise would be harmful. The nonnitrogenous (carbonaceous) portions of the excess protein molecules are utilized for energy. Very high amounts of protein may decrease the overall efficiency of energy utilization due to the amount used in metabolizing and excreting the excess nitrogen.

3.8. PROTEIN RESERVES: DEPOSITION, MOBILIZATION, AND PRACTICAL IMPLICATIONS

In contrast to energy, some vitamins, and certain minerals, dairy cattle do not store large amounts of protein in forms which can be mobilized to meet the needs of the animal when the diet is deficient (Swick and Benevenga, 1977). Thus, the protein needs should be supplied on a daily basis.

A 1300-lb dairy cow has about 180 lb of protein in the body in addition to that in the digestive tract (Coppock, 1969). With steers (Biddle *et al.,* 1975) only about 6% of the protein is in a form which can be readily mobilized and used when the dietary intake is deficient (Coppock, 1969). For a 1300-lb cow, 6% of the 180 lb of body protein would only be enough to provide the protein needed to produce about 280 lb of milk. Thus, the easily mobilizable protein is not sufficient to provide the protein required for production of large amounts of milk. With very large weight losses the total amounts of protein in the total animal body are substantially reduced (Paquay *et al.,* 1972; Swick and Benevenga, 1977). Although the animal can survive, such major losses are quite inconsistent with economical milk production. When protein intake is excessive, the nitrogen is excreted in the urine as urea and the remainder of the molecules used for energy. Thus, there is little storage of protein above normal amounts (Swick and Benevenga, 1977). With fattening dairy cattle, a small part (about 15%) of the extra weight is protein. This extra protein in the fat tissue is removed with the fat when cattle are in a negative energy balance. Although the removed protein is available for meeting either the protein or the energy requirements of the animal, the ratio of the protein to energy obtained from the use of body fat is very low.

The very unbalanced source of nutrients obtained from using fat has major practical implications with high-producing cows in early lactation. Often cows

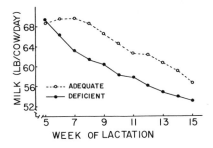

Fig. 3.4 Cows underfed on protein by about one-fourth, beginning in the fifth week of the lactation, produced considerably less milk than controls given adequate protein with all other conditions the same. (Courtesy of Darwin G. Braund, 1976, from Cooperative Research Farms Dairy Research Studies.)

with a high production potential cannot consume enough feed in early lactation to meet their energy needs (see Section 3.5). Thus, a cow in good condition at calving can make up the energy deficiency for a considerable period from body fat. However, if not fed protein to balance this energy from body fat, she may become protein deficient. Because of limited body reserves of protein, milk production declines rather quickly to a level supported by the current protein consumption (see Fig. 3.4). As a result, it is important to supply extra protein to the high-producing cow that is in negative energy balance (Braund, 1976; Coppock, 1969; Crowley, 1975; Hemken, 1975).

3.9. AMINO ACID DEFICIENCIES IN HIGH-PRODUCING DAIRY CATTLE; USE OF METHIONINE

The tissues of dairy cattle require at least 9 essential amino acids which are similar to those needed by rats and people (see Section 3.0.b) (Black *et al.*, 1957; Purser, 1970). These essential amino acids are absorbed from the small intestine and come from four sources: (1) digestion of rumen microorganisms (see Section 3.0.a); (2) digestion of feed protein escaping degradation and breakdown in the rumen; (3) amino acids bypassing the rumen; and (4) reabsorption of proteins endogenously secreted into the intestinal tract (Chalupa, 1975).

Substantial current thinking suggests that the amount of protein synthesized by the microbes in their cells usually is limited primarily by the amount of fermentable energy in the feed (Burroughs *et al.*, 1975b; Chalupa, 1975). If there is adequate fermentable protein or usable nonprotein nitrogen, essential minerals and other growth factors (Clark, 1975; Hungate, 1966) rumen microbial protein synthesis is estimated to be 10.4% of the TDN (total digestible nutrients) in the feed (Burroughs *et al.*, 1975b). This is based on three evaluations as follows: (1) that 52% of the ration TDN is digested in the rumen; (2) that 25% of the digested TDN is transformed into microbial crude proteins when adequate usable nitrogen is present; and (3) that 80% of the microbial protein is α-amino acids and therefore digestible (true digestibility—see Section 3.1).

The development of this formula is based on considerable research and represents an important step in a better understanding of protein nutrition for dairy cattle. Using a somewhat different approach, others have arrived at a similar answer (Conrad and Hibbs, 1968; Satter and Roffler, 1975). Whether the evaluations would be relatively constant under all feeding conditions, especially with widely different kinds of ingredients, is not clear. For example, there is some indication that the maximum microbial cell yield may vary perhaps from 10 to 20% of the feed fermented (Hungate, 1966; Purser, 1970). The yield of microbial protein appears to increase with more rapid rumen fermentation such as occurs with faster growth or higher-producing cows (Chalupa, 1978). With higher milk production, turnover rate of the microbes is faster with less of the energy from fermentation being used for maintenance of the microbes, leaving more for cell growth (Chalupa, 1978). Irrespective of the degree to which the 10.4% is variable rather than a fixed value, the formula and the concepts (Burroughs et al., 1975a,b) on which it is based have several important practical implications, as discussed in the next paragraph.

It is apparent that the amount of protein obtained from digestion of the rumen microbes is not sufficient to meet the protein needs of lactating cows producing at a high level (Chalupa, 1975) (see Section 3.10). Regardless of the percentage of protein in the diet, rumen microbes may degrade essentially all the readily fermentable protein. This is sometimes designated as degradable protein. Most of the nitrogen components are used by the microbes either for synthesis of their own protein or for conversion to products other than amino acids. Thus high-producing dairy cows must receive some protein which will bypass fermentation in the rumen (see Section 3.10).

The rumen microbes have remarkably similar amino acid composition with a relatively wide variety of rations (Chalupa, 1975). Although the composition of the rumen microbes is comparatively well balanced for all the essential amino acids relative to the needs of the animal tissues, the balance is not optimum for all functions (Chalupa, 1975). For sheep, amino acid composition of both casein and soybean meal appears to be better balanced than that of rumen microbes (Little and Mitchell, 1967). In contrast, zein (corn protein) is not nearly so well balanced. The protein of rumen protozoa has a better balanced amino acid composition than rumen bacteria (Chalupa, 1975).

Substantial research suggest that the total amount of the sulfur containing amino acids, methionine and cystine, are first limiting for sheep (McDonald, 1970). A major reason is the large amount of sulfur amino acids in wool. Whether one amino acid generally is more limiting in practical feeding of dairy cattle apparently has not been conclusively established.

The use of methionine in a form which may bypass degradation in the rumen has received considerable research attention. Methionine hydroxy analog (MHA) has been fed to lactating cows in several controlled experiments in University and Experiment Station farms and in field studies (Chandler et al., 1976). Generally the fat content of the milk has increased, but the effect on milk yield has been much less consistent. Chandler (1971) suggests that the greatest benefit appears to be exhibited by cows producing over 70 lb of milk per day. It was further suggested that it was desirable to feed the methionine hydroxy analog only in peak lactation and to discontinue its use when the milk production dropped to 50 or 60 lb per day (Chandler, 1971). In addition to the questionable effects on milk production, research information does not clearly establish that the methionine hydroxy analog bypasses ruminal degradation (Chalupa, 1975). In some studies the methionine hydroxy analog appears to have decreased performance (Muller and Rodriguez, 1975).

Other amino acids have been suggested as possibly deficient in some practical situations (Schwab *et al.*, 1976). However, the importance of this, if any, is not firmly established. Because of the interest in the subject among research workers, more definitive answers may be obtained within the next several years.

3.10. RUMEN BYPASS OF PROTEINS AND AMINO ACIDS: PRINCIPLES, TECHNOLOGY, AND APPLICATIONS

As discussed earlier (see Section 3.0.b) the tissues of dairy cattle must have essential amino acids comparable to nonruminants and these must be available at the absorption sites in the small intestine. If the necessary nutrients are provided for maximum synthesis of rumen-microbial protein, a substantial portion of the protein and essential amino acids can be obtained from digestion of the rumen microbes in the abomasum and small intestine (Chalupa, 1975). For instance, the rumen microbes can supply most of the protein needed for maintenance (Chalupa, 1975). However, the high-producing dairy cow cannot obtain sufficient protein and essential amino acids from digestion of microbes (see Section 3.9) (Burroughs *et al.*, 1975b; Chalupa, 1975). Some must come from feed protein not degraded in the rumen. The proportion of protein in natural feeds going through the rumen without degradation varies but under most conditions is between 20 and 60% (Table 3.3) (Chalupa, 1975; Mertens, 1977).

TABLE 3.3
Estimated Protein Solubility and Degradation in the Rumen of Selected Feeds[a]

			Ruminal degradation	
Feed	Crude protein (% of DM)	Protein solubility[b] (% of crude protein)	Measured[c] (%)	Burroughs estimates[d] (%)
Urea	280	100		100
Dried milk	27	93	90–97	85
Corn silage + 0.5% urea, 35% DM	12	59		
Alfalfa, fresh forage	20	(58)[e]		
Mixed grass + legume, fresh forage	18	(56)		
Corn silage, 35% DM	9	(54)		

TABLE 3.3
Estimated Protein Solubility and Degradation in the Rumen of Selected Feeds[a] (Continued)

| Feed | Crude protein (% of DM) | Protein solubility[b] (% of crude protein) | Ruminal degradation | |
			Measured[c] (%)	Burroughs estimates[d] (%)
Orchard grass, fresh forage	15	(53)		
Linseed meal	39	51		75
Mixed grass silage, 35% DM	13	48		
Rye grain	14	41		70
Peanut meal	52	40	63–78	75
Wheat middling	19	39		70
Rapeseed meal	44	39		75
Alfalfa hay	20	(39)	32–60	
Grass hay, 2nd cut	14	36	73–89	
Wheat bran	18	35		70
Corn gluten feed	28	32		75
Sunflower meal	50	30	75	75
Wheat grain	14	30		70
Distillers dried grains (corn)	30	26		75
Oat grain	13	26		70
Citrus pulp	7	26		75
Mixed hay, 1st cut	10	24		
Hominy feed	12	23		
Alfalfa meal	19	23		
Corn gluten meal	21	18		
Soybean meal (soluble)	52	18	39–55	
Barley grain	13	17	40–72	70
Green hay, 1st cut	10	17		
Meat & bone meal	54	16		75
Cottonseed meal (soluble)	45	15	60–80	
Soybean meal	52	13	39–55	75
Milo distillers grains	33	13		75
Corn grain	10	12		62
Sardine meal	69	10	10–50	
Corn gluten meal	47	7		75
Cottonseed meal	45	7	60–80	
Milo grain	12	(5)		52
Beet pulp	10	4		
Brewers grains	25	3		
Milo gluten meal	47	3		

[a] Adapted from Mertens (1977).
[b] Wohlt et al. (1973).
[c] Chalupa (1975).
[d] Burroughs et al. (1975a).
[e] Values in parentheses are estimates.

Considerable confusion exists in the use of terms in this area of protein nutrition and metabolism. Often soluble protein and degradable protein have been used interchangeably. Although they are related and correlated, they are not equivalent (Chalupa, 1978). Solubility refers to the crude protein which is soluble in some test solution. Appreciable amounts of proteins which are not soluble are fermented by the rumen microbes (Pichard and Van Soest, 1977). Nondegradable or bypass protein is that which is not fermented by the rumen microbes. Many proteins are partially degraded in the rumen. With these, a faster rate of passage increases the percentage bypassing fermentation.

Two important factors influencing the amount of protein bypassing degradation in the rumen are the length of time spent in the rumen and fermentability of the protein (Chalupa, 1975; Mertens, 1977). Protein degradability can be substantially modified by various treatments or procedures, including many routinely used in feed processing and preservation (Fig. 3.5). One of the most effective means of increasing the protein bypassing microbial degradation in the rumen is heating the feed. For example, much more protein of soybean meal which is heated during processing, goes through the rumen intact than occurs when raw soybeans are fed. Likewise, when forages are heated, protein degradability increases (Fig. 3.5) (Goering and Waldo, 1974).

The desirable proportion of nondegradable to degradable protein has not been well defined. Even so it is quite evident that the diet must contain sufficient

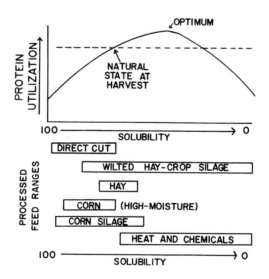

Fig. 3.5 Heating proteins in feeds decreases solubility but at the same time lowers digestibility. With limited heating the lower protein solubility can increase performance of dairy cattle by providing more bypass protein. However, with excessive heating, large amounts of protein become indigestible and useless. (Adapted from concepts by Goering and Waldo, 1974.)

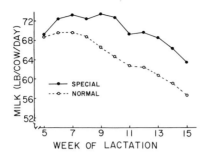

Fig. 3.6 The solubility of the protein can have a substantial effect on the milk yield from high-producing cows. This graph shows the effects of feeding the same amount of protein with one group receiving protein having apparently "normal solubility" and the other given protein with the solubility especially regulated to provide higher amounts of insoluble protein. (Courtesy of Darwin G. Braund, 1976, from Cooperative Research Farms Dairy Research Studies.)

degradable protein to permit efficient rumen microbial fermentation. Especially for high-producing cows, there must be appreciable bypass or nondegradable protein available to meet their protein needs. Figure 3.6 illustrates the beneficial effects which may be achieved by regulating protein degradability.

As illustrated in Fig. 3.5 and Table 3.1, excessive heating can greatly decrease protein digestibility (see Section 3.1). Thus, while some heating of proteins often is beneficial, overheating can make much of the protein useless (Chalupa, 1975; Goering *et al.,* 1972, 1973; Goering and Waldo, 1974). Likewise, sufficient degradable and usable nitrogen compounds must be available if the rumen microbes are to digest feed efficiently.

Other methods of processing feeds to reduce the degradability and increase the rumen "bypass" of proteins include a variety of chemical treatments (Chalupa, 1975). Of these, formaldehyde has received the most attention, but many others have potential practical uses (Chalupa, 1975; Ferguson *et al.,* 1967; Muller *et al.,* 1975). In each case, the desirability of these processing procedures must be evaluated by the effects on costs.

Various encapsulation processes, especially for amino acids, have been investigated in attempts to improve the performance of ruminants by more efficiently supplying the needed essential amino acids at the absorption site. Likewise, selecting nondegradable chemical forms of the amino acids is a promising way of improving the protein nutrition of dairy animals (Chalupa, 1975). Of the nondegradable forms of amino acids, methionine hydroxy analog (MHA) has been investigated most (see Section 3.9).

The amount of protein degraded increases with the length of time spent in the rumen (Ferguson *et al.,* 1967; Mertens, 1977). It is possible to physically bypass the rumen with liquids by evoking closure of the esophageal groove which appears to be a conditioned reflex associated with pleasure (Chalupa, 1975; Orskov, 1972).

Some of the differences observed in the performance of dairy cattle many years ago can be explained by the role of "rumen-bypass" proteins. For example, although the milk protein casein is very high quality, when fed to ruminants often their performance has been lower than that obtained with zein, the low quality corn protein. (In nutritional use, quality of protein usually refers to the balance of essential amino acids). Due to differences in degradability some 90% of the casein is degraded in the rumen compared to only about 35% for zein (Chalupa, 1975).

A considerable part of the advantage of ruminants in the food chain revolves around their ability to utilize, through rumen fermentation, feeds which man and other nonruminants cannot use. However, "nothing is for free" as conversion of feed energy and protein to microbial protein is far short of 100%. The efficiency of converting usable dietary protein into usable microbial protein has been estimated as only a little over 50% (Chalupa, 1975). When a portion of the dietary protein bypasses rumen digestion, both the energy and protein losses in the rumen are avoided.

Experimentally, proteins, such as casein, and/or amino acids have been infused directly into the abomasum or small intestine through surgically prepared fistula (Clark, 1975; Schwab et al., 1976). Such postruminal infusions have resulted in increased milk production. Although this procedure is not now practical, the beneficial response suggests that the feeding system involved may be less than optimum and might be modified to obtain the same benefits. The greatest value of the postruminal infusion studies is through increasing the basic understanding of protein and amino acid nutrition of dairy cattle.

3.11. EVALUATING PROTEINS FOR DAIRY CATTLE

Evaluation of proteins in feedstuffs for dairy cattle is a very different problem from that for nonruminants. Dairy cattle, other than baby calves, use protein both for the rumen microbes and for direct absorption of amino acids from the small intestine without microbial degradation. To be utilized by the microorganisms, the protein must have characteristics permitting degradation to component parts. Although the terms are often used interchangeably, solubility of a protein and the ability to be degraded in the rumen are related but not the same (Bull et al., 1977; Mertens, 1977) (Table 3.3). Often nonprotein nitrogen is the main component of soluble proteins. Better methods of measuring this important feed characteristic are needed.

Traditionally, one of the important ways to evaluate proteins is by apparent digestibility (see Section 3.1). Except where the protein has been considerably denatured with heat, chemicals, or other treatment, usually the apparent digestibility is closely correlated with total crude protein content (see Section 3.1). The nitrogen content of the acid detergent fiber can be used as an indication of indigestible protein (see Section 12.5) (Goering *et al.*, 1972, 1973; Mertens, 1977; NRC, 1978). Pepsin-insoluble nitrogen also can be used in the same way (Goering *et al.*, 1972; Mertens, 1977). Unfortunately, the amount of protein damaged beyond use by cattle is greater than the quantity measured as acid detergent-insoluble nitrogen or pepsin-insoluble nitrogen (Goering *et al.*, 1972). Thus, both methods underestimate the protein damage caused by heat or other factors.

Because of the effect on utilization after absorption, the amino acid composition of proteins which are not degraded by the rumen microbes is important in the same way as for nonruminants. This is known as the biological value, and often is given the term "quality." Generally, forage proteins have an amino acid composition similar to that of microbial protein, with the quality of both being comparatively good. The amino acid balance of soybean protein is relatively good and much better than corn grain protein.

No single measure adequately evaluates the usefulness of protein for dairy cattle. Burroughs *et al.* (1975a,b) developed a system of evaluating protein nutrition known as the metabolizable protein feeding standard. One of the important aspects of this system is its use in calculating the usefulness of nonprotein nitrogen. This is expressed as the urea fermentation potential (UFP) (see Section 4.5).

REFERENCES

Agricultural Research Council (ARC) (1965). "The Nutrient Requirements of Farm Livestock," No. 2. Ruminants. HM Stationery Office, London.

Biddle, G. N., J. L. Evans, and J. R. Trout (1975). *J. Nutr.* **105**, 1584–1591.

Black, A. L., M. Kleiber, A. H. Smith, and D. N. Stewart (1957). *Biochim. Biophys. Acta* **23**, 54–59.

Blackburn, T. H. (1965). *In* "Physiology of Digestion in the Ruminant" (R. W. Doughtery *et al.*, eds.), pp. 322–334. Butterworth, London.

Braund, D. G. (1976). *Feed Manage.* **27** (11), 10–12.

Brisson, G. J., H. M. Cunningham, and S. R. Haskell (1957). *Can. J. Anim. Sci.* **37**, 157–167.

Broster, W. H. (1972). *Handb. Tierernahr.* **2**, 293–322 (as cited by National Research Council, 1978).

Broster, W. H., V. J. Tuck, T. Smith, and V. W. Johnson (1969). *J. Agric. Sci. Camb.* **72**, 13–30 (as cited by National Research Council, 1978).

Bull, L. S., M. I. Poos, and R. C. Bull (1977). *Distill. Feed Res. Counc. Conf., Proc.*, **32**, 23–32.

Burroughs, W., P. Gerlaugh, B. H. Edington, and R. M. Bethke (1949). *J. Anim. Sci.* **8**, 9–18.

Burroughs, W., D. K. Nelson, and D. R. Mertens (1975a). *J. Dairy Sci.* **58**, 611–619.

Burroughs, W., D. K. Nelson, and D. R. Mertens (1975b). *J. Anim. Sci.* **41**, 933–944.

Chalupa, W. (1975). *J. Dairy Sci.* **58**, 1198–1218.

Chalupa, W. (1978). *Am. Chem. Soc. Symp. Nutr. Improve. Food Proteins*, 1977. Adv. in Expt. Med. and Biol. Vol. 105, pp. 473–496. Plenum Publ. Co., New York.

Chandler, P. T. (1971). *Hoard's Dairyman* **116,** 1279, 1310, and 1311.

Chandler, P. T., C. A. Brown, R. P. Johnston, Jr., G. K. Macleod, R. D. McCarthy, B. R. Moss, A. H. Rakes, and L. D. Satter (1976). *J. Dairy Sci.* **59,** 1897–1909.

Clark, J. H. (1975). *J. Dairy Sci.* **58,** 1178–1197.

Conrad, H. R., and J. W. Hibbs (1968). *J. Dairy Sci.* **51,** 276–285.

Coppock, C. E. (1969). *Hoard's Dairyman* **114,** 15.

Crowley, J. W. (1975). *Hoard's Dairyman* **120,** 464–465.

Downes, A. M. (1961). *Aust. J. Biol. Sci.* **14,** 254–259.

Ferguson, K. A., J. A. Hemsley, and P. J. Reis (1967). *Aust. J. Sci.* **30,** 215–217.

Forbes, E. B. (1924). *Natl. Res. Counc., Bull.* **42**

Gardner, R. W. (1968). *J. Dairy Sci.* **51,** 888–897.

Gardner, R. W., and R. L. Park (1973). *J. Dairy Sci.* **56,** 390–394.

Glover, J., D. W. Duthie, and M. H. French (1957). *J. Agric. Sci.* **48,** 373–378.

Goering, H. K., and D. R. Waldo (1974). *Proc. Cornell Nutr. Conf.* pp. 25–36.

Goering, H. K., C. H. Gordon, R. W. Hemken, D. R. Waldo, P. J. Van Soest, and L. W. Smith (1972). *J. Dairy Sci.* **55,** 1275–1280.

Goering, H. K., P. J. Van Soest, and R. W. Hemken (1973). *J. Dairy Sci.* **56,** 137–143.

Gordon, C. H., J. C. Derbyshire, H. G. Wiseman, E. A. Kane, and C. G. Melin (1961). *J. Dairy Sci.* **44,** 1299–1311.

Hemken, R. W. (1975). *Feed Manage.* **26** (6), 9 and 10.

Holter, J. A., and J. T. Reid (1959). *J. Anim. Sci.* **18,** 1339–1349.

Huber, J. T. (1975). *J. Anim. Sci.* **41,** 954–961.

Hungate, R. E. (1966). "The Rumen and Its Microbes." Academic Press, New York.

Jacobson, N. L. (1969). *J. Dairy Sci.* **52,** 1316–1321.

Jahn, E., and P. T. Chandler (1976). *J. Anim. Sci.* **42,** 724–735.

Jakobsen, P. E. (1957). Proteinbehov og proteinsynthese ved fosterdannelse hos drøvtyggere. *Bretn. Forsgslab.* p. 299 (as cited by National Research Council, 1978).

Knight, A. D., and L. E. Harris (1966). *J. Anim. Sci.* **25,** 593 (abstr.).

Little, C. O., and G. E. Mitchell, Jr. (1967). *J. Anim. Sci.* **26,** 411–413.

Lofgreen, G. P., J. C. Loosli, and L. A. Maynard (1951). *J. Anim. Sci.* **10,** 171–183.

Martin, Y. G., W. J. Miller, and D. M. Blackmon (1969). *Am. J. Vet. Res.* **30,** 355–364.

Maynard, L. A., and J. K. Loosli (1969). "Animal Nutrition." McGraw-Hill, New York.

Mertens, D. R. (1977). *Proc. Ga. Nutr. Conf. Feed Ind.* pp. 30–41.

Muller, L. D., and D. Rodriguez (1975). *J. Dairy Sci.* **58,** 190–195.

Muller, L. D., D. Rodriguez, and D. J. Schingoethe (1975). *J. Dairy Sci.* **58,** 1847–1855.

National Research Council (NRC) (1971). "Nutrient Requirements of Dairy Cattle." 4th rev. ed. Natl. Acad. Sci., Washington, D.C.

National Research Council (NRC) (1976). "Urea and Other Nitrogen Compounds as Protein Nitrogen Replacements in Animal Nutrition." Natl. Acad. Sci., Washington, D.C.

National Research Council (NRC) (1978). "Nutrient Requirements of Dairy Cattle," 5th rev. ed. Natl. Acad. Sci., Washington, D.C.

Orskov, E. R. (1972). *World Congr. Anim. Feeding, 2nd,* Vol. I, p. 627 (from Chalupa, 1975).

Paquay, R., R. DeBaere, and A. Lousse (1972). *Br. J. Nutr.* **27,** 27–37.

Paquay, R., J. M. Godeau, R. DeBaere, and A. Lousse (1973). *J. Dairy Res.* **40,** 93–103.

Pichard, G., and P. J. Van Soest (1977). *Proc. Cornell Nutr. Conf. Feed Ind.* pp. 91–98.

Preston, R. L. (1972). *Proc. Univ. Nottingham Nutr. Conf. Feed Manuf.* Vol. 6, pp. 22–37.

Purser, D. B. (1970). *J. Anim. Sci.* **30,** 988–1001.

Satter, L. D., and R. E. Roffler (1975). *J. Dairy Sci.* **58,** 1219–1237.

Schwab, C. G., L. D. Satter, and A. B. Clay (1976). *J. Dairy Sci.* **59,** 1254–1270.

Stobo, I. J. F., and J. H. B. Roy (1973). *Br. J. Nutr.* **30,** 113–125.

Stobo, I. J. F., H. H. B. Roy, and H. J. Gaston (1967). *Anim. Prod.* **9,** 7-33.
Sutton, A. L., and R. L. Vetter (1971). *J. Anim. Sci.* **32,** 1256-1261.
Swanson, E. W. (1977). *J. Dairy Sci.* **60,** 1583-1593.
Swick, R. W., and N. J. Benevenga (1977). *J. Dairy Sci.* **60,** 505-515.
Thomas, J. W. (1971). *J. Dairy Sci.* **54,** 1629-1636.
Van Horn, H. H., and D. R. Jacobson (1971). *J. Dairy Sci.* **54,** 379-382.
Wohlt, J. E., C. J. Sniffen, and W. H. Hoover (1973). *J. Dairy Sci.* **56,** 1052-1057.

4

The Use of Nonprotein Nitrogen (NPN) for Dairy Cattle

4.0. INTRODUCTION

Throughout much of the world, protein is one of the most limiting nutrients both for humans and farm animals. Among the major types of animals only ruminants are able to efficiently utilize nonprotein nitrogen (NPN) for conversion to proteins. Thus, using NPN as a substitute for some natural proteins in feeding dairy cattle, often lowers costs as well as increases the amount of protein for humans and other nonruminants.

Contrary to much popular opinion, most dairy cattle receive substantial amounts of NPN, irrespective of whether a special compound is used in formulating the feed. Many natural feeds such as pastures, hays, and especially silages contain an appreciable percentage of NPN (see Section 4.3). Present trends suggest that the use of special NPN sources in ruminant feeds will continue to increase. Because of this importance along with the complexities involved in efficient and safe use, an understanding of NPN utilization is needed in practical dairy cattle feeding and nutrition (Virtanen, 1969).

4.1. CONVERSION OF NPN TO PROTEIN IN THE RUMEN; AMMONIA, THE COMMON DENOMINATOR

In dairy cattle, the rumen microbes utilize NPN as a key ingredient in synthesizing their own protein. Although other nitrogen compounds are used, ammonia (NH_3) is a major source of nitrogen for these microbes (Chalupa, 1978; National Research Council, 1976). When dairy cattle are fed natural proteins, generally a substantial part is broken down to amino acids and from there to ammonia in the rumen by microbial action. Likewise, much of the various NPN compounds found in natural feeds or as special supplements also are converted to ammonia by the microbes.

In a very real sense ammonia is the common denominator of protein and nitrogen metabolism for the rumen microbes (Hungate, 1966; NRC, 1976). If a NPN compound is not converted to ammonia, little of it is used by the microbes. The reactions occurring when urea is utilized by dairy animals are summarized below (adapted from NRC, 1976).

$$\text{Urea} \xrightarrow[\text{microbes}]{\text{urease from}} \text{ammonia (NH}_3) + \text{carbon dioxide (CO}_2) \tag{1}$$

$$\text{Carbohydrates} \xrightarrow[\text{rumen microbes}]{\text{enzymes from}} \text{volatile fatty acids (VFA)} + \text{keto acids (carbon chains)} \tag{2}$$

$$\text{Ammonia} + \text{keto acid (carbon chains)} \xrightarrow[\text{from microbes}]{\text{enzymes}} \text{amino acids} \tag{3}$$

$$\text{Amino acids} \xrightarrow[\text{microbes}]{\text{enzymes from}} \text{microbial protein} \tag{4}$$

$$\text{Microbial protein} \xrightarrow[\text{and small intestine}]{\text{enzymes in abomasum}} \text{free amino acids} \tag{5}$$

$$\text{Free amino acids absorbed from the small intestine} \rightarrow \text{body of the cow} \tag{6}$$

Except for differences in the way and rates at which they are converted to ammonia, other NPN sources follow a similar scheme. The natural proteins that are degraded in the rumen are first digested to amino acids under the influence of proteolytic enzymes from the microbes.

Although the amino acids and peptides from natural protein can be used directly by the microbes (Nolan, 1975), a substantial portion is converted to ammonia plus a carbon chain. Just as with the ammonia from NPN, that from proteins can be used by the microbes to synthesize amino acids for their own protein; some of the ammonia is absorbed by the animal. (The fate of the absorbed ammonia is discussed in Section 4.1.b and rumen microbial synthesis of essential amino acids in Section 3.0.a.)

4.1.a. Metabolism of Ammonia in the Rumen

One key practical aspect of NPN utilization is maintaining an adequate but not highly excessive concentration of ammonia in the rumen. When the ammonia content in the rumen is too low, the growth rate of the rumen microbes is suboptimum. This is accompanied by reduced rates of forage and feed utilization and decreased animal performance. As the ammonia level in the rumen increases, ever higher quantities are absorbed through the rumen wall. If the amounts absorbed reach a sufficiently high rate, toxicity occurs (see Section 4.2).

The ammonia used by rumen microbes to synthesize amino acids provides only a small fraction of the total molecules. In addition, "carbon chains" are required. These carbon chains come largely from carbohydrates with lesser amounts from amino acids that have the amino group (NH_2) removed. Major differences exist in the efficiency with which various carbon chains are used in amino acid synthesis. Readily fermentable carbohydrates, especially starches from concentrates, are a preferred source.

In the synthesis of the sulfur-containing amino acids, methionine and cystine, the rumen microbes require sulfur (see Section 5.21). Likewise, the microbes require many other nutrients including the essential minerals (see Chapter 5). Most natural proteins supply substantial amounts of many of the essential mineral elements, but urea and other synthetic sources of NPN contain little or none. Thus, when urea is substituted for natural proteins, it is more important to be sure that adequate amounts of essential mineral elements are provided.

4.1.b. Ammonia Metabolism in the Body of Cattle

Although ammonia may be absorbed from the omasum, small intestine, and cecum, much of it is absorbed from the reiticulorumen (Chalupa, 1977; NRC, 1976). Normally, absorbed ammonia is converted to urea in the liver (NRC, 1976; Waldo, 1968). Urea thus formed may be excreted in the urine or recycled to the rumen by direct diffusion through the rumen wall, and in smaller amounts via the saliva.

The proportion of the urea recycled and reused in the rumen compared to that excreted in the urine is largely dependent on the amount of protein and NPN in the diet relative to the needs of the animal. When the protein and NPN intake are deficient, urea excretion in the urine may be very low. With intakes greatly above requirements, most of the excess urea is eliminated in the urine. The ability to either recycle and reuse the urea or to excrete it via urine is the key discrimination step in the homeostasis of protein and nitrogen compounds (see Section 3.7). When the capacity of the liver to convert ammonia to urea is exceeded, ammonia accumulates causing toxicity (NRC, 1976).

4.2. TOXICITY OF UREA AND OTHER NPN SOURCES; AMMONIA ALKALOSIS OR TOXICITY

Used correctly, urea and many other NPN compounds are very useful and safe nutrients in the feeding of dairy cattle. When improperly fed, urea can be a deadly poison. The toxicity of urea and other NPN compounds is caused by excessive levels of ammonia in the blood. Since ammonia is a normal constituent of the blood, it is helpful to understand how the toxicity occurs.

Unless some special treatment is used to reduce its solubility, the urea reaching the rumen is rapidly converted to ammonia. When high amounts of urea are consumed over short periods, the microbes cannot utilize the ammonia for synthesis of microbial protein as rapidly as it is produced. Ammonia levels in the rumen fluid will rise shortly after feeding with levels peaking in about 3 hours. More frequent feeding or continuous feeding will reduce or eliminate the wide variability.

If the other nutrients required for efficient growth of rumen microbes are present, the microbes use ammonia more rapidly than if any is deficient (see Section 4.4). When conditions are not conducive to its rapid utilization by the microbes, ammonia is more likely to build up in the rumen fluid. The quantity of ammonia absorbed by the dairy animal is greatly influenced by both the concentration in the rumen fluid and the pH of this fluid. Absorption is more rapid when the rumen fluid pH is high (NRC, 1976). Since ammonia is slightly alkaline, increasing concentrations of ammonia elevates the pH and increases its absorption across the rumen wall. Rumen fluid is not as well buffered against an increase in pH as against a decrease (NRC, 1976). When normal amounts are absorbed, ammonia is converted to urea in the liver (see Section 4.1.b). A blood ammonia content expressed as nitrogen of 1.0 mg per 100 ml will cause toxicity (NRC, 1976). Ammonia nitrogen in the rumen fluid is perhaps a better diagnostic tool with 80 mg per 100 ml associated with toxicity and is the point at which the ability of the liver to convert ammonia to urea is exceeded (NRC, 1976; Waldo, 1968). The ammonia level of rumen fluid needed for efficient rumen microbial synthesis is only 5–7 mg per 100 ml (see Section 4.4).

Acute ammonia toxicity in cattle (NRC, 1976) generally progresses in the following order. At first, the dairy animal becomes uneasy and nervous, secretes excessive amounts of saliva, and develops muscular tremors. Subsequently, incoordination, respiratory difficulties, frequent defecation and urination develop; the front legs begin stiffening, and the animal becomes prostrate. Most animals struggle violently, bellow, have terminal tetanic spasms, a marked jugular pulse, and often bloat. Within 30 minutes to two and one-half hours after the initial symptoms develop, death may occur (NRC, 1976).

Often before the tetanic spasms occur, the ammonia or urea toxicity can be successively treated either with cold water or acetic acid (NRC, 1976). Depending on the size of the animal, a large amount, 5 to 10 gallons, of cold water given orally will lower the temperature of the rumen fluid, thereby decreasing the rate of urea conversion to ammonia. In addition, cold water dilutes the ammonia, further reducing absorption through the rumen wall. About one gallon of dilute acetic acid or vinegar given with cold water is more effective than cold water alone because it combines with the ammonia producing ammonium acetate that lowers rumen fluid pH and decreases the absorption rate (NRC, 1976).

When dairy cattle die of urea toxicity, there are no characteristic neocropsy lesions; but often congestion, hemorrhages, and edema of the lungs are observed (NRC, 1976). Definite diagnoses of urea toxicity by routine necropsy examination of dead animals may not be possible.

The possible adverse effects of borderline urea toxicity on surviving dairy cattle has been the subject of considerable interest and speculation. Obviously performance is adversely affected for a short period, but abortions did not occur in cows severely affected before acetic acid treatments were administered (NRC, 1976). Likewise, reproductive performance was not affected in surviving cows.

Although dairy cattle often have been killed by urea toxicity, this need not be a serious problem with good management. Key factors which affect the probability of urea toxicity include the following (NRC, 1976):

(1) Animals unadapted to urea are more likely to develop toxicity. Thus, the amount of urea fed should be increased slowly over a period of at least 2 to 4 weeks.

(2) Diets composed primarily of poor quality roughage often do not provide enough readily fermentable carbohydrates for efficient urea utilization, thereby elevating the amount of free ammonia to be absorbed by the animal and increasing the probability of toxicity.

(3) Fasting or a relatively empty rumen at the time urea is fed increases the probability of toxicity because the microbes do not utilize the ammonia formed as rapidly.

(4) Any diet which increases the pH of the rumen fluid increases the possibility of urea toxicity by increasing ammonia absorption.

(5) A low water intake reduces rumen fluid volume and increases ammonia absorption by the animal.

4.3. SOURCES OF NPN FOR DAIRY CATTLE

Often urea is regarded, erroneously, as being almost synonomous with NPN. Mainly because of economics a major part of all supplemental NPN used is urea (Kertz and Everett, 1975). However, several other compounds are well-utilized by dairy cattle with some having important advantages over urea for specific situations.

Substantial quantities of NPN are fed to dairy cattle as a part of natural feeds. For instance, according to data cited by Waldo (1968), 10–30% of the crude protein in fresh forage is NPN. This increases to 25–50% when forage is dried for hay and to 60–75% in unwilted silage.

As mentioned earlier, too rapid conversion to ammonia is a major problem in the use of urea. Accordingly, many attempts have been made to develop products, combinations, and procedures to decrease the conversion rate. One approach is to "coat" urea with materials to reduce the solubility rate. Several urea–carbohydrate combinations have been developed with a few offering improved urea utilization (NRC, 1976). Some procedures involve treating combinations of grains and urea with heat, moisture and/or pressure. The situation is confusing because the beneficial effects of the combinations may be partially due to the processing on the grain and not just a slower conversion of urea to ammonia (NRC, 1976). One of the most widely studied urea carbohydrate combinations is "starea," produced by cooking and extruding mixtures of grain and urea (Bartley and Deyoe, 1975). The starch is gelatinized, apparently providing a more available source of energy for the microbes. Bartley and Deyoe (1975) indicate that feeding "starea" instead of urea improves feed intake and milk production.

In some instances, combinations of urea and forages such as alfalfa have given superior results (NRC, 1976). One of the best known of these is Dehy–100, a product containing 16% nitrogen or 100% crude protein equivalent (Conrad and Hibbs, 1968).

Ammoniated products and feeds are often used as sources of NPN for dairy cattle (NRC, 1976). Examples are ammoniated rice hulls, ammoniated beet pulp, and ammoniated citrus pulp. When properly used, these NPN sources can successfully provide a part of the nitrogen needs of dairy cattle, but special considerations are important. For instance, high levels of ammoniated molasses have produced nervous symptoms with serious problems such as causing cattle to run wild (NRC, 1976).

Many ammonium salts are well utilized by ruminants as sources of NPN (NRC, 1976). Among these, the ammonium polyphosphates, as well as monoammonium and diammonium phosphate have the added advantage of providing phosphate. Many of the ammonium salts of organic and inorganic acids appear to be well utilized sources of NPN (NRC, 1976). However, some ammonium salts are relatively toxic and feeds containing them are unpalatable (Conrad and Hibbs, 1968).

Biuret, a condensation product made from urea, is converted to ammonia at a slower rate than urea, and therefore is less toxic. As of this writing, this compound has not been approved by the United States Food and Drug Administration for use with dairy cows, apparently because of the possibility some might be secreted into milk (NRC, 1976). An important disadvantage of biuret, compared with urea, is its higher cost per unit of nitrogen. Likewise, an adaptation period is more critical before biuret can be utilized efficiently by rumen microbes (NRC, 1976). The microbes must develop a new enzyme for biuret. When added to forages during ensiling, biuret does not ferment to ammonia as does urea. Thus, it has the advantage of not neutralizing the organic acids needed for good silate preservation (NRC, 1976).

4.4. EFFICIENT UTILIZATION OF UREA (NPN): IMPORTANT FACTORS AND LIMITATIONS

Fortunately, the management and feeding practices that are essential for efficient urea utilization also prevent toxicity. Urea and other NPN compounds are useful to dairy cattle only when the nitrogen is utilized by the rumen microbes to synthesize microbial protein which is later digested and used by the animal. Accordingly, conditions needed for efficient microbial growth and fermentation must be present. An adequate supply of all the nutrients required by the microbes is needed, including most of the essential minerals, a suitable source of energy and "carbon chains" for synthesizing the amino acids. Usually, the greatest attention must be devoted to supplying readily fermentable carbohydrates used for energy and carbon chains. Generally, best results have been achieved when starch supplied a substantial portion of these carbohydrates (Conrad and Hibbs, 1968). For maximum utilization these authors suggest that for every pound of urea 10 lb of readily fermentable carbohydrates are needed, with two-thirds as starch.

Several research workers have estimated the percentage of the fermentable carbohydrates converted to microbial protein. Earlier this was thought to be a fixed value (Conrad and Hibbs, 1968). More recent evidence suggest that the yield of microbial protein can vary appreciably (see Section 3.9).

A continuous type of relatively even rumen fermentation appears to be most efficient. Probably this is much more important for high producing or rapidly growing dairy animals with high nutrient needs than for low producers having smaller nutrient requirements. The continuous type of relatively even and rapid rumen fermentation can be achieved by frequent or continuous feeding. Often dairy cows are given forages free choice but only fed concentrates twice per day. If the urea or another source of easily degradable nitrogen is fed in the concentrates only twice per day, peaks and valleys may occur in rumen ammonia levels. Present information suggests that urea is utilized somewhat more efficiently when it is either fed continuously, fairly frequently, or in a slowly released form (Huber, 1975). One way to make the urea or other NPN more evenly available is to feed it with corn silage.

Since urea and other NPN sources tend to increase the pH of silages, which adversely affects silage fermentation, best results have been obtained with corn silage. This material has a relatively low natural protein content plus large amounts of readily fermentable carbohydrates for use by rumen microbes. The corn silage also undergoes a good fermentation and develops sufficient acidity to insure good preservation. Results of adding urea to corn silages have been variable. Generally, it appears wise to restrict the amount of urea added to about 10 lb per ton of chopped corn and to only use urea when the dry matter content is between 30 and 40% (see Chapter 9).

Whereas natural proteins provide many other essential nutrients including a substantial amount of energy and minerals, urea supplies only nitrogen. Thus, urea is useful for dairy cattle only when the diet otherwise would be deficient in protein. Also, it is of value only to the extent that total digestible protein available in the abomasum and small intestine is increased.

When the ammonia level in the rumen fluid exceeds about 5–7 mg per 100 ml, there may be relatively little or no increase in the amount utilized for synthesis of microbial protein (Fig. 4.1) (Satter and Roffler, 1975; Waldo, 1968). At around 7 mg per 100 ml of rumen fluid the quantity of ammonia absorbed about equals the urea coming into the rumen from saliva and through the rumen wall (Waldo, 1968). As ammonia in the rumen fluid increases above this concentration, the amount absorbed and excreted in the urine also increases (Satter and Roffler, 1975).

If the natural proteins in the feed are being broken down to ammonia in sufficient amounts to fully meet the needs of the rumen microbes, adding urea may not be beneficial to the animal. Accordingly, even though all other nutrients are present, urea may not correct a protein deficiency under some conditions (Burroughs *et al.*, 1975; Satter and Roffler, 1975). Based on this type of informa-

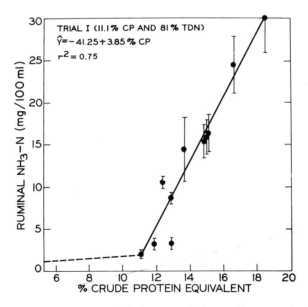

Fig. 4.1 Effect of increasing urea infusions on ruminal ammonia concentration in 1000-lb steers fed a basal diet containing 81% total digestible nutrients, (TDN, dry basis). Note the linear increase in ruminal ammonia with crude protein equivalent of more than about 11% (Roffler *et al.*, 1976). (Courtesy of L. D. Satter.)

tion, it was suggested that supplemental NPN may have little or no value in typical dairy rations containing more than 12–13% crude protein on a dry matter basis (Burroughs *et al.*, 1975; Satter and Roffler, 1975). These suggestions were based on several assumptions which probably are true under some conditions but not in others (see Section 3.9). For example, Goering and Waldo (1974) discussed the wide variations occurring in the amounts of protein not degraded or fermented in the rumen due to forage processing. Chalupa (1975) and Bull *et al.* (1977) suggest that it is feasible to utilize NPN to supply the ammonia needs of the rumen microbes and to supplement the diet with natural proteins or amino acids that are not degraded in the rumen to ammonia. Using this approach, it should be possible to efficiently utilize considerable urea, even in high protein diets when such levels are needed to meet protein requirements.

The concept that supplemental urea may have little or no value in rations containing above 12–13% crude protein appears to be inconsistent with practical observations with feeding high producing dairy cows in early lactation. Cows requiring a high percentage of dietary protein probably can efficiently use urea in certain diets having substantially more than the 12–13% crude protein. When a higher amount of protein goes through the rumen without being degraded, more urea could be utilized. Also the protein homeostatic control mechanisms may play a role. As discussed earlier (see Section 3.7) cattle have a remarkably good homeostatic control system for protein. A logical possibility is that when there is a small reduction in body reserves of protein, conservation mechanisms decrease the urinary excretion of urea resulting in a greater amount of recycling through the rumen. Possibly the amino acids absorbed from the small intestine may be used more efficiently by the animal with a lower proportion utilized for energy in the body tissues.

(In all things and among almost all groups fads exist.) It seems to this author that in our zeal to better understand the very important role of the unique digestive system of the ruminant animal, as scientists, often we have overlooked the equally important and interacting roles of happenings in the other body tissues. Contrary to much current thinking, considerable evidence indicates that events within the other tissues may have just as much influence on activities in the digestive system as the reciprocal relationship.

More definitive answers to the utilization efficiency of urea in high protein diets await further research and developments. Even so, it is evident that, generally, easier and more efficient use of urea is achieved on diets of low to medium protein content than with high protein rations. Likewise, it appears to be clearly possible to achieve efficient utilization of urea in fairly high protein diets. The degree to which this can be practical is not clear, but it is strongly influenced by the type of natural protein and the usable energy content of the diet.

For some NPN compounds, including urea, more efficient utilization is obtained after dairy cattle have had time to adapt to these substances (see Section 4.2). Part of the adaptation appears to be with the microbes. The liver also can adapt to more efficiently convert ammonia in blood to urea (Waldo, 1968) (see Section 4.3 for biuret).

Frequently when urea and other NPN compounds are fed, the amount of feed eaten is reduced (Kertz and Everett, 1975). The extent of this reduction associated with the adverse effects of urea on feed palatability and that caused by systemic effects is not clear. If feed consumption is reduced, animal performance is usually decreased.

4.5. OTHER PRACTICAL CONSIDERATIONS IN THE USE OF NPN

When urea and many other NPN compounds are fed, it is most essential to obtain thorough mixing in the feed. Often urea toxicity has been associated with improper feed mixing. Large numbers of dairy animals can be killed from one feeding with an unreasonably high amount of urea.

Some feeds, including raw soybeans, contain an enzyme, urease, which hydrolyzes urea to ammonia. Generally, urea should not be mixed with feeds containing free urease, as it will cause decomposition of urea in the feed. Likewise, when some ammonium type compounds are mixed with silages, ammonia may be released resulting in both loss of nitrogen and lower palatability of the feed. Relatively little ammonia is released in a silage with a low pH below 4.2.

Urea is an unpalatable feed ingredient, especially at high concentrations (Table 4.1) (Kertz and Everett, 1975). In addition, urea is hygroscopic and highly soluble often making it difficult to store high urea feeds in humid conditions. It is important that the mixed ration feed be sufficiently palatable to prevent a serious reduction in feed consumption. Generally 1% urea in rations for dairy cows does

TABLE 4.1

Effect of Moisture Content on the Palatability (Voluntary Consumption) of a Concentrate Ration Containing 1.15% Urea[a][b]

	Moisture	
Cow no.	11.2%	14.4%
	(lb/day)	(lb/day)
1	10.4	3.0
2	10.3	1.5
3	10.7	4.5
Average	10.5	3.0

[a] Adapted from Kertz and Everett (1975).

[b] Cows had a choice of the two concentrates simultaneously for 30 minutes per day.

not seriously affect feed consumption (Kertz and Everett, 1975). With higher concentrations, the effects are variable. One of the advantages of adding urea to corn silage, or in certain special combinations with grains, or with forages (Section 4.3), is elimination or reduction of the palatability problem.

Burroughs and associates (1975) developed the concept of urea fermentation potential of feeds (see Section 3.11). These values are based on calculations of the amounts of potential rumen energy available from a feed over and above that needed to utilize the ammonia produced from fermentation of the degradable crude protein of the feed (Chalupa, 1978). The two factors involved in the calculations are the amounts of crude protein degraded in the rumen and the quantity of digestible energy available for the microbes. Corn grain is an example of a feed with a high urea fermentation potential as it contains a high fermentable energy content and a low content of fermentable nitrogen. The urea fermentation potential value of feeds is a part of a proposed new system of determining protein requirements of dairy cattle based on metabolizable protein available to cattle at the intestinal absorption site (Burroughs et al., 1975). Although much additional information and improved evaluation procedures are needed before the new approach fully suffices in the practical feeding of dairy cattle, it promises to be a better way than is now available.

Feeding NPN to dairy cattle will make more natural protein available for humans and nonruminants. However, for the individual dairyman, generally, it is only desirable to feed urea or other special NPN sources when this is the most economical way to supply the needed nutrients.

Many years ago when urea feeding first began in the United States, some simple rules of thumb were developed. Although our knowledge has greatly expanded making it possible to more accurately tailor urea use to the situation, the old rules still provide quick reference points. Among the recommendations, the thumb rules suggested that the amount of supplemental urea should not exceed 1% of the total ration, or one-third of the total crude protein equivalent, or 3% of the concentrates.

REFERENCES

Bartley, E. E., and C. W. Deyoe (1975). *Feedstuffs* **47** (30), 42–44 and 51.
Bull, L. S., M. I. Poos, and R. C. Bull (1977). *Distill. Feed Res. Counc. Conf., Proc.* **32**, 23–32.
Burroughs, W., D. K. Nelson, and D. R. Mertens (1975). *J. Dairy Sci.* **58**, 611–619.
Chalupa, W. (1975). *J. Dairy Sci.* **58**, 1198–1218.
Chalupa, W. (1977). *J. Anim. Sci.* **46**, 585–599.
Chalupa, W. (1978). *Am. Chem. Soc. Symp. Nutr. Improve. Food Proteins,* 1977. Adv. in Expt. Med. and Biol., Vol. 105, pp. 473–496. Plenum Publ. Co., New York.
Conrad, H. R., and J. W. Hibbs (1968). *J. Dairy Sci.,* **51**, 276–285.
Goering, H. K., and D. R. Waldo (1974). *Proc. Cornell Nutr. Conf.* pp. 25–36.
Huber, J. T. (1975). *J. Anim. Sci.* **41**, 954–961.

Hungate, R. E. (1966). "The Rumen and Its Microbes." Academic Press, New York.

Kertz, A. F., and J. P. Everett, Jr. (1975). *J. Anim. Sci.* **41,** 945–953.

National Research Council (NRC) (1976). "Urea and Other Nonprotein Nitrogen Compounds in Animal Nutrition." Natl. Acad. Sci., Washington, D.C.

Nolan, J. V. (1975). *In* "Digestion and Metabolism in the Ruminant" (I. W. McDonald and A. C. I. Warner, eds.), pp. 416–431. University of New England, Armidale, N.S.W., Australia.

Roffler, R. E., C. G. Schwab, and L. D. Satter (1976). *J. Dairy Sci.* **59,** 80–84.

Satter, L. D., and R. E. Roffler (1975). *J. Dairy Sci.* **58,** 1219–1237.

Virtanen, A. I. (1969). *Fed. Proc. Fed. Am. Soc. Exp. Biol.* **28** (No. 1), 232–240.

Waldo, D. R. (1968). *J. Dairy Sci.* **51,** 265–275.

5

Mineral and Trace Element Nutrition of Dairy Cattle

5.0. INTRODUCTION

In both research and practical dairy cattle feeding and nutrition, the mineral and trace elements often have received little attention. However, this is an important and complex area with 21 elements established as essential or probably essential for animals. The amounts needed, their functions, the likelihood of a practical deficiency or toxicity, and many other important factors vary enormously among these elements. Each element has a unique chemical structure that is responsible for its total role in animal nutrition. While many similarities exist in functions, metabolism, supply, and practical importance between two or more mineral elements, the differences are sufficient to make each element a separate story requiring individual consideration.

5.1. THE ESSENTIAL MINERALS AND TRACE ELEMENTS

Calcium (Ca), phosphorus (P), magnesium (Mg), potassium (K), sodium (Na), chlorine (Cl), sulfur (S), iodine (I), iron (Fe), copper (Cu), cobalt (Co), manganese (Mn), zinc (Zn), selenium (Se), molybdenum (Mo), fluorine (F), chromium (Cr), silicon (Si), and vanadium (V) have been established as essential for one or more species of animals and undoubtedly are necessary for dairy cattle (Miller and Neathery, 1977). Likewise, there is sufficient evidence to suggest that tin (Sn) and nickel (Ni) are probably essential. Probably other elements will be added to the required list.

5.2. CLASSIFICATION OF MINERALS

Minerals are classified in a number of ways, with several classification schemes having a place in understanding their nutritional roles.

5.2.a. Major and Trace Elements

Because of the vast differences in amounts involved, one of the most convenient and widely used classifications of essential mineral elements is major minerals (or elements) and trace elements. This division is based on the amounts required in the diet and/or in the animal body.

The dietary requirements for all the major essential elements is substantially more than 100 ppm (parts per million) which is 0.01%, but the need for none of the essential trace elements exceeds this level. Comparably, the average tissue concentrations in the while body of animals greatly exceeds 100 ppm for all major elements but is less than this for all trace elements. The seven essential major elements are calcium, phosphorus, sodium, chlorine, potassium, magnesium, and sulfur (Table 5.1).

TABLE 5.1
Approximate Content of Essential Mineral Elements in the Earth's Crust and in the Animal Body[a]

Elements	Symbol	Earth's crust (%)	Earth's crust (ppm)	Whole cattle (%)	Whole cattle (ppm)
Major					
Calcium	Ca	3.4	(34,000)	1.2	(12,000)
Phosphorus	P	0.12	(1,200)	0.7	(7,000)
Magnesium	Mg	2.0	(20,000)	0.05	(500)
Sodium	Na	2.6	(26,000)	0.14	(1,400)
Potassium	K	2.4	(24,000)	0.17	(1,700)
Sulfur	S	0.05	(500)	0.15	(1,500)
Chlorine	Cl	0.2	(2,000)	0.10	(1,000)
Trace					
Iron	Fe		(50,000)		(50)
Zinc	Zn		(65)		(20)
Copper	Cu		(45)		(5)
Iodine	I		(0.3)		(0.43)
Cobalt	Co		(23)		(<0.04)
Selenium	Se		(0.09)		(Trace)
Manganese	Mn		(1,000)		(0.3)
Fluorine	F		(700)		(Trace)
Chromium	Cr		(200)		(<0.09)
Molybdenum	Mo		(1)		(<0.07)
Silicon	Si		(277,000)		(?)
Nickel	Ni		(80)		(<0.14)
Vanadium	V		(110)		(0.3)
Tin	Sn		(3)		(0.43)

[a] Adapted from several sources including Mortimer (1967); Schroeder (1965); Underwood (1971); Neathery et al. (1973b); NRC (1978). When no values available for cattle, estimate made from other species.

On the basis of present knowledge, at least 14 trace elements appear to be essential for cattle. Concentrations of major minerals are generally expressed as a percentage. Likewise, trace mineral contents are usually listed as ppm, but sometimes as ppb (parts per billion).

5.2.b. Other Classifications

Mineral elements are classified as cations, including calcium, magnesium, sodium, potassium, iron, zinc, and manganese, or anions (phosphorus, sulfur, chlorine, iodine, and fluorine). Likewise, they can be classified on the basis of valence number and on their group position in the periodic chart of the atoms. These classifications are useful because they describe physical and chemical attributes of importance in nutrition. For example, the monovalent cations, sodium and potassium, have a very high absorption percentage and major interrelationships exist between them. In contrast, the absorption percentage of the divalent cations (calcium, magnesium, and zinc) is much lower.

Some confusion exists in use of the terms minerals and elements in nutrition and feeding. In practical nutrition the term "minerals" is generally used to denote all the inorganic elements. However, not all the elements are minerals and not all minerals are single elements. Thus when strict accuracy is required, the term element is preferred. Since the purpose of this book is to present feeding and nutrition of dairy cattle in a way that is simple, accurate, and highly usable by those who are interested in practical applications and academic endeavors, the terms mineral, element, and mineral element, generally, are used interchangeably. Likewise, trace element, trace mineral, and trace mineral element are used interchangeably.

5.3. MINERAL CONTENT OF DAIRY CATTLE

In addition to the 21 essential elements, the body of a dairy animal contains many others. The possible importance for these other mineral elements has been studied relatively little. As shown in Table 5.1, an enormous range exists in typical amounts of the different minerals in an animal. For example, there is some 300,000 times as much calcium in a dairy animal as cobalt; yet, each is equally critical to the existence of every animal. Even within the classifications of major elements and trace elements, there is a wide range in typical levels of the whole animal body. For instance, a dairy animal's body has about 30 times as much calcium as magnesium and some 70 times more zinc then manganese.

Distribution of mineral elements in the animal body is very different and generally quite characteristic for each element. More than 98% of the calcium is in the skeleton and teeth, but most of the iodine concentrates in the thyroid gland

(Ellenberger *et al.*, 1950). Zinc, in contrast, is widely distributed throughout the body with the amounts in various tissues generally of a similar order of magnitude. Overall, about 80–85% of total mineral matter is in the skeletal tissues (Underwood, 1966).

Concentrations of some elements in tissues, such as the copper in the liver, vary tremendously with dietary intake. Tissue levels of elements such as manganese are affected only moderately by intake. The content of a mineral in certain tissues may vary greatly, whereas the level in other tissues will change little. For example, muscle zinc varies relatively little, whereas in calves, but not cows, liver zinc may increase manyfold when a high, but nontoxic, zinc diet is fed. In addition to the dietary level of the element, many other factors affect the tissue content of some minerals and trace elements. These factors include other dietary constituents, physiological, and hormonal variables.

The mineral content of certain tissues and fluids often is quite useful as biochemical diagnostic measurement. However, each effective measure must be based on sound principles and carefully developed details. Some of the most useful of these biochemical measurements are mentioned later with the discussion of each mineral (Miller and Stake, 1974b).

5.4. FUNCTIONS OF MINERALS AND TRACE ELEMENTS (MINERALS CANNOT BE SYNTHESIZED)

The essential mineral elements, unlike all other required nutrients, cannot be synthesized by living organisms. Thus, they must be present in the environment within an acceptable concentration range. One or more essential mineral element is involved in virtually every biochemical reaction and/or process of every animal. Yet the reactions of which a given element is capable are relatively limited and never change. As expressed by H. A. Schroeder (1965)

> A metal is a metal and has always been so; its reactivity is predictable and cannot change. It has always been present in the living environment. Although an almost infinite variety of proteins are possible, and new enzyme systems can evolve in response to metabolic needs, the reactions in which metals can take part are relatively limited and very basic. Therefore, metabolically active trace metals and other elements may have been fixed in the first Eocene . . . with virtually little change during the past many million years. Furthermore, the types of reactions which they catalyze (oxidation, reduction, hydrogenation, dehydrogenation, deamination, cyclization, and hydroxylation) were used by living matter since the beginning of life or when the need for the reactions evolved.

Underwood (1966, pp. 4–7) presents an excellent general discussion of the functions of minerals. Some elements play a critical role in the structure of tissues of the whole body. Calcium and phosphorus which form most of the mineral base of bones and teeth are the best examples. The bones provide the critical structure and shape of the animal. In addition to calcium and phosphorus, several other minerals and trace minerals have key functions in bones.

Phosphorus plays a crucial role in the metabolic reactions involving energy transfer in all cells of the animal. With most physiological changes there is either a gain or loss of energy involving the formation or breaking of high energy phosphate bonds (see Section 5.17). Several elements are essential for protein synthesis, including sulfur which is a part of certain amino acids. Phosphorus, iron, manganese, nickel, zinc, and chromium are components of ribonucleic acid (RNA), a vital component in protein synthesis.

Most of the essential trace elements, and some major elements such as calcium, function as a part of enzyme systems. Generally these systems are either metalloenzymes or metal enzyme complexes. Metalloenzymes are characterized by a firm association between the metal and the protein portion of the molecule. Likewise, there is a fixed number of atoms of a given element per molecule. In contrast, with metal enzyme complexes the association between the protein and the metal is weaker and there may be some substitution of different elements.

Some of the mineral elements including calcium, phosphorus, magnesium, sodium, potassium, and chlorine play key roles in such vital functions as the regulation of osmotic pressure, acid–base balance, pH, membrane permeability, and tissue irritability. For example, the plasma and extracellular fluid levels of calcium and magnesium are crucial to normal neuromuscular action. When calcium is too low, milk fever symptoms occur; whereas inadequate extracellular fluid magnesium results in grass tetany.

Several elements have very specific and well-defined roles. Apparently, the only essential function of iodine is a part of the hormone thyroxine which is responsible for the regulation of metabolic rate of the animal. Likewise, the only clearly established essential function of cobalt is as a component of vitamin B_{12}. Iron is needed for the transport of oxygen by blood hemoglobin. In addition, iron has many other roles (see Section 5.22).

The above are but a small portion of the known functions of mineral elements in a dairy animal. Further, it is quite evident that there are numerous additional functional roles of mineral elements still to be discovered.

5.5. METHODS OF DETERMINING MINERAL CONTENT OF FEEDS

Some knowledge of the problems, limitations, accuracies, and pitfalls of methods used to determine the mineral content of feeds is needed for a practical understanding of the mineral and trace element nutrition of dairy cattle. In determining the mineral content of feeds, every element must be considered individually as each presents a different set of problems. Likewise, the nature of the feed often has a major influence on the analytical problems involved for a given element.

Just as with the mineral content in the animal body, there is an enormous range in the amounts of various mineral elements typically found in feeds. For example, the potassium in a forage often is as much as 100,000 times the selenium concentration. Usually, it is easier to analyze elements occurring in substantial proportions in feeds than those present in very minute amounts. Thus, it is not surprising that throughout the history of mineral nutrition, much more information has been available on the amounts of the major elements in feeds than for trace elements (Table 5.2). In fact the name "trace mineral" came into use from very early analytical reports in which these elements were reported as a "trace."

The early quantitative methods for trace elements usually were crude, tedious, and laborious. For example, in 1935 a single cobalt analysis in duplicate required five days (Underwood, 1970). Because of the difficulties involved, relatively

TABLE 5.2
Mineral Content Data for Commonly Used Feed Ingredients[a]

Mineral	Feeds with mineral values (% of 228)
Calcium (Ca)	84
Phosphorus (P)	32
Iron (Fe)	73
Manganese (Mn)	48
Potassium (K)	43
Magnesium (Mg)	41
Copper (Cu)	38
Sodium (Na)	27
Cobalt (Co)	23
Chlorine (Cl)	11
Sulfur (S)	11
Zinc (Zn)	10
Fluorine (F)	1
Iodine (I)	1
Molybdenum (Mo)	1
Selenium (Se)	1

[a] From Miller et al. (1972b), as adapted from NRC (1970). The NRC publication provides a compilation of the nutrient composition of 228 commonly used feeds and ingredients. This table shows the percentage of the 288 feeds listed for which there were data on each of the essential minerals. For the other essential mineral elements, there were no data.

few samples were analyzed for trace elements by early methods and research progress was slow (Miller and Kincaid, 1975).

Since 1960, there has been a virtual revolution in analytical methodology for trace mineral analyses. Now, most essential trace mineral elements can be determined rapidly with reasonably good precision and at a moderate cost per sample. Some examples of the newer methodology for trace mineral analyses include atomic absorption spectroscopy, direct reading emission spectroscopy, neutron activation analysis, plasma emission spectroscopy, anodic stripping voltammetry, spark source mass spectrometry, and x-ray fluorescence. These are at various stages of development and use, with several already being "work horses" in research laboratories. No one method is suitable for all elements. Probably the "new" methodology which has been of greatest use in trace mineral research is atomic absorption spectroscopy. Generally, this can detect elements in the parts per million range, with even greater sensitivity for certain elements. Direct reading emission spectroscopy simultaneously determines many elements in a sample. A fairly recent breakthrough in methodology for detecting trace elements is neutron activation analysis which can detect many elements (e.g., vanadium, cobalt, arsenic, selenium, and molybdenum) at lower levels than is possible with most other methods. The vastly improved analytical methodologies have had several important ramifications in trace mineral nutrition. For example, they have been a crucial key to research which recently established the essentiality of several additional trace elements.

The currently available sophisticated methodology and instrumentation have not eliminated the importance of careful techniques, considerable skill, and a knowledge of problems associated with each mineral and type of material being analyzed in securing accurate values. Still today, in some instances, erroneous analytical results are being obtained and reported. Even greater problems are errors in values obtained in earlier years. Gross errors occasionally occur even in the most widely accepted tables. A number of National Research Council publications (1969, 1976) list the zinc content of beet pulp as 0.8 ppm of dry matter or only about 10% of that actually found (Miller *et al.,* 1972b). Instead of being a feed ingredient with an extremely low zinc content, beet pulp is just moderately low in this element.

5.6. SOURCE OF MINERALS

As indicated earlier, mineral elements cannot be synthesized by cattle as can many of the vitamins, amino acids, and other organic compounds. All the mineral requirements must be obtained from the environment, with most coming from the feed. Only small amounts are obtained from water, soil, or nonfeed contamination.

Feed sources of required mineral elements are sometimes divided into "normal" or "natural" feed ingredients and mineral supplements. In the practical feeding of dairy cattle, important authorities differ sharply concerning the best way to supply needed minerals. For example, Underwood (1966, p. 25), the great Australian mineral authority suggests that "Wherever possible or practicable, the mineral requirements of animals should be met by the proper selection or combination of available feeds alone." In contrast, McCullough (1973, p. 30) says "Vitamins and minerals no longer need be provided by natural feedstuffs. In fact, present nonfeed sources of vitamins and minerals provide these nutrients at such low cost that nothing would really be lost if natural feeds were devoid of these nutrients."

Obviously, least cost formulation of feeds would indicate that the most economical method of providing the required minerals for dairy cattle would be the most desirable. As a practical matter, sufficient information often is not readily available to indicate the amount of supplemental minerals needed. The great variation of minerals in feed ingredients, inexact knowledge of requirements, various interactions, and the wide range in availabilities of mineral elements can combine to cause unexpected reductions in animal performance. The problem has been aggravated by the apparent widespread attitude, especially in the United States, that providing the correct mineral nutrition of dairy cattle needs only minor attention or study.

5.6.a. Variation in Mineral Content of Feeds

Generally, tables of feed composition present only average values. However, in understanding practical dairy cattle nutrition, an appreciation of the enormous variability in the mineral and trace element content of feed ingredients is needed. Perhaps a few examples of actual data will make this clearer. In Table 5.3 the range in content of a few minerals from legume-grass samples submitted to the Pennsylvania State Forage Testing Service is shown (see also Table 5.15). The range for different elements varied from a factor of 10 for sulfur to more than 200 for iron and calcium. Some of the variability in mineral composition is due to differences between crops, but within the same crops, concentrations often vary greatly. The wide range for a few of the trace elements from corn silages submitted to the Georgia Forage Testing Service during a single year is shown in Table 5.4.

The mineral content of grains usually varies somewhat less than that of forages. Even so, the variability in grains often is still very large. The selenium content of corn can vary by 10- to 20-fold within a single state and as much as 200-fold when different areas of the country are involved (Table 5.5). It is not sufficient to know that the mineral content of feeds varies greatly. Rather it

TABLE 5.3
Range of Nutritive Content of Legume–Grass Forage[a]

Element	Range	Fold[b] difference
TDN (%)	51.0 – 71.0	1.4
Crude protein (%)	6.6 – 33.0	5.0
Calcium (%)	0.01– 2.60	261.0
Phosphorus (%)	0.07– 0.74	11.0
Magnesium (%)	0.07– 0.75	11.0
Sulfur (%)	0.04– 0.38	10.0
Manganese (ppm)	6.0 – 265.0	44.0
Iron (ppm)	10.0 –2599.0	260.0
Copper (ppm)	2.0 – 92.0	46.0
Zinc (ppm)	8.0 – 300.0	38.0

[a] Summary of Penn State Forage Testing Service (1969–1973), adapted from Adams (1975).
[b] Maximum divided by the minimum.

becomes important to understand the main factors affecting mineral and trace element composition of feeds.

5.6.b. Factors Affecting Mineral Content of Feeds

An understanding of the factors influencing the mineral concentrations in feeds has considerable importance beyond that of academic interest. A knowledge of what to expect relative to the mineral content of the feeds is of crucial importance in planning feeding programs and in diagnosing possible mineral related problems.

TABLE 5.4
Mineral Content in Corn Silages from Georgia Forage Testing Program 1973–1974[a]

Mineral	Average	Range		Fold difference
Manganese (ppm)[b]	88	15	– 287	19
Iron (ppm)	231	56	–4750	85
Zinc (ppm)	29	11	– 79	7
Copper (ppm)	7	1	– 28	28
Molybdenum (ppm)	2	0	– 11.5	115+
Sulfur (%)	0.07	0.005 –	0.186	37

[a] Courtesy of Georgia Cooperative Extension Service.
[b] Dry matter basis. 132 samples except for sulfur which was 27.

TABLE 5.5
Selenium Content of Corn Grain from Selected States[a]

State	No. samples	Average	Range
South Dakota	9	0.40	0.11–2.03
Nebraska	6	0.35	0.04–0.81
Iowa	26	0.06	0.02–0.32
Illinois	31	0.05	0.02–0.15
Missouri	4	0.05	0.02–0.09
Wisconsin	5	0.04	0.02–0.13
Indiana	17	0.04	0.01–0.15
Michigan	43	0.02	0.01–0.09

[a] Adapted from Ullrey (1974).

The main factors affecting the mineral content of plants include (1) the genetics of the plant, (2) the soil where the plants were grown, (3) climate and weather, (4) the stage of maturity (Underwood, 1966), and (5) the part of the plant. These factors are interrelated in many ways. For example, often the climate has a major effect on the soil characteristics. In turn the soil and climate often play decisive roles in the types of plants grown. Some of these interrelationships are illustrated with selenium. As shown in Table 5.5, the selenium content of corn grown in Nebraska and South Dakota is much higher than in that produced in Michigan, Indiana, and Illinois. This effect is caused by the much higher level of available selenium in Nebraska and South Dakota soils, which to a considerable extent, resulted from the drier climate as well as the type of rock formations. Over a long time period, the dry climate caused less leaching of the soil selenium and a higher soil pH. Selenium is more available for plant uptake in a soil with a higher pH.

The genetics of the plant often has a major effect on the mineral content. Legumes, in general, will have some two to five times more calcium than grasses (Table 5.15) (Underwood, 1966). For good growth, many legumes are more dependent on an adequate pH and calcium level in the soil. Likewise, legumes generally contain more magnesium, potassium, zinc, iron, copper, cobalt, and sulfur than grasses.

Substantial differences in content of many mineral elements also exist among species of legumes and grasses. These characteristic differences, although usually greater in the forage parts of plants, also apply to seeds. One of the best known differences in seeds is the far lower level of manganese found in corn relative to other grains. The characteristically high amounts of silica in a few grasses such as coastal bermuda grass is discussed elsewhere (see Section 12.5) because of the practical effect on the usable energy value of these grasses.

Even when grown under identical conditions different varieties and strains of the same species may have very dissimilar levels of one or more essential minerals. The characteristic mineral composition of plants can be changed greatly by plant breeding. A well-known example is the dramatic reduction of iodine in the short rotation ryegrass developed in New Zealand. As shown in Table 5.6, the short rotation ryegrass had only one-fifth to one-tenth as much iodine as the parent stock.

The role plant breeding may play in creating new mineral deficiency, toxicity, and imbalance problems in the feeding of dairy cattle, apparently has been given little attention. As plant breeders become more successful in developing plant varieties and strains with superior performance under specific growing conditions, one of the adaptations may be plants containing less of any element(s) that may be low in the soil of the area. Thus, changes can be expected in the characteristic mineral and trace mineral composition of new "superior" strains. At times this may result in unexpected mineral feeding problems for dairy cattle, especially when the changes are not known or considered.

The soil where the plant is grown often has a major role in determining the content of the various minerals. An influence of the total content of each mineral in the soil obviously is expected. Perhaps at least as important is the availability of the soil minerals to the plant. One of the major factors affecting mineral availability to plants is soil pH. Many minerals, including zinc, manganese, iron, nickel, and cobalt are more available to plants grown on low pH soils (Table 5.7). In contrast, others, especially molybdenum and selenium, are more available on higher pH soils (Tables 5.5 and 5.7).

Just as there are interactions in mineral metabolism within animals, the same is also true in growing plants. For example, higher potassium fertilization, while increasing the potassium content of the forage, also tends to reduce sodium and magnesium. The lower magnesium is sometimes associated with a higher instance of grass tetany in cattle when high potassium fertilization is used.

TABLE 5.6
Effect of Plant Breeding on Iodine Content in a Short Rotation Ryegrass Relative to Parent Stock[a]

	Soil iodine		
Ryegrass	1.8	3.0	8.5
Perennial	1.35	1.60	1.50
Italian	0.90	—	—
Short rotation	0.23	0.17	0.15

[a] Values in ppm of dry matter. Adapted from Johnson and Butler (1957), through Underwood (1966) p. 16.

TABLE 5.7
Influence of Soil pH on the Content of Some Trace Elements in Plants[a]

Species	Soil pH[b]	Molybdenum[c]	Cobalt[c]	Manganese[c]
Ryegrass	5.4	0.52	0.35	58
Ryegrass	6.4	1.23	0.12	40
Red clover	5.4	0.28	0.22	140
Red clover	6.4	1.53	0.12	133

[a] Adapted from Underwood (1966, p. 20).
[b] pH difference due to addition of limestone.
[c] Data given as ppm in dry matter.

In some areas of the world, notably in Australia and New Zealand, specific minerals, especially trace elements, are often applied to the soil as part of the fertilization for the express purpose of alleviating mineral deficiencies in the grazing animals. This practice is used much less in the United States. Quantities of minerals required to meet animal requirements by fertilization often are vastly larger than needed in the feed because of the low percentage going into the plant.

The crucial role of the stage of maturity of forages on their protein and usable energy composition was discussed in earlier chapters. The stage of plant maturity, likewise, has an important influence on the content of several minerals. One of the most important of these effects is the large decline in phosphorus with maturity. This, along with low phosphorus soils and the long periods when cattle depended on mature dry forage, was a key reason for the once very severe phosphorus deficiency over large areas of many countries. Several other minerals including potassium, sodium, chlorine, copper, cobalt, nickel, zinc, and molybdenum generally decline as plants mature (Underwood, 1966). Silica increases with maturity, whereas manganese, iodine, and iron apparently are not clearly related to the stage of maturity.

The mineral content of different parts of the same plant often vary appreciably. The grain or seeds generally have a characteristically different mineral content from the vegetative or forage portion. For example, typically grains contain much less calcium than forages (Table 5.15).

5.6.c. Availability of Minerals in Feeds of Dairy Cattle

It is well established that the availability to cattle of several minerals varies substantially in different feeds. Such information is much less complete than that concerning the total content. Even the practical importance, and the degree of the differences in availability are not well established. Nevertheless, some pertinent facts are known. A key aspect in grass tetany is the lower availability of magnesium in immature, highly succulent forage compared to the more mature.

There is a tremendous range in the characteristic amounts of the different elements which dairy cattle can absorb. Essentially, all the sodium, potassium, and chlorine are absorbed. Some elements have a characteristically very low percentage absorbed. Typically, manganese absorption is in the 3–4% range.

The availability of a mineral to a dairy animal is the amount the animal can absorb, usually expressed as a percentage of the total in the diet. With a few elements (such as sodium, potassium, and chlorine), this can be readily determined with a "digestion trial" that measures the amount eaten and the amount going into the feces (see Section 5.7.d). However, determining availability of most minerals is more difficult. With many minerals only a part of that in the feces comes from the feed. A substantial amount is endogenously (endogenous means "from within the body") secreted into the digestive tract and then excreted in the feces. Thus, the conventional "digestion trial" will not give a good measure of the amount the animal absorbed.

Radioactive mineral elements have been used to determine the availability of minerals to dairy cattle. This is possible because the radioactive isotope can be measured separate from the remainder of that element with which it comingles in the animal. Both the radioactive and the nonradioactive element are metabolized and utilized the same way. The methodology involved is relatively complex, and a detailed discussion is beyond the scope of this book.

Even with isotopes, determining the true availability of many essential elements is quite difficult. This will become more clear in Section 5.7.d on the homeostatic control of mineral elements. In essence, however, with several elements, including calcium, zinc, and iron, the percentage absorbed depends on the amount fed. When higher amounts are eaten, the absorbed percentage declines. While of great benefit to the dairy animal, this presents a major problem to scientists in determining the percentage available. Availability of minerals, on a relative basis, is sometimes determined in feeding experiments from the animal performance.

Many years ago, it was generally thought that natural sources of minerals were more available to cattle than inorganic supplements. Current information suggests that often this is not the case. The elements of many mineral supplements are highly available relative to natural resources.

5.6.d. Importance of Chemical Form of Elements

Although mineral elements do not change (see Section 5.4), they exist in different chemical forms, valences, combinations, and associations. Often the chemical form has a major influence on the way the element is metabolized by the animal. For example, iron oxide is much less available for dairy cattle than many other iron compounds. Sufficient amounts of the iron oxide (ferric oxide) will satisfy the iron needs of dairy cattle (Becker *et al.*, 1965). Cattle are able to tolerate substantially less fluoride from a soluble source, such as sodium or calcium fluoride, than from the less soluble fluorides in phosphate compounds (see Section 5.28).

In natural feedstuffs the mineral elements often occur in combination with organic compounds and associations. The nature of the specific organic compound can have a crucial effect on the availability of the element and/or the way a mineral element is metabolized and utilized after absorption. For example, the toxicity of selenium from natural feeds varies appreciably. In general, selenium from sodium selenate and sodium selenite appears to be less available and less toxic than that from many natural feeds. Similarly molybdenum from added inorganic sources seems to be less toxic than that from pastures. In contrast, cattle apparently can tolerate higher amounts of iron from natural feed sources than from ferrous sulfate (Standish *et al.*, 1969; National Research Council, 1978).

Chemical form not only influences absorption and utilization of minerals, but also metabolism after absorption may be drastically altered with different forms of the same element. One of the best known examples of this was discovered some years ago when 3,5-diiodosalicylic acid (DIS) was widely used as a source of iodine in salt blocks (J. K. Miller *et al.*, 1968). DIS was not readily leached from salt blocks and is well absorbed, at least by rats. When DIS was fed to cattle and sheep, iodine deficiencies developed. Subsequent research demonstrated that dairy cattle were unable to remove the iodine from the organic part of the DIS molecule. Thus, the iodine in DIS could not be utilized.

Major differences exist in the way cattle use some mineral compounds relative to monogastric animals such as rats, pigs, and chickens. Cattle readily utilize phosphorus in the phytate form, but chickens do not. This fact is of considerable practical importance since much of the phosphorus in feed grains occurs as the phytate.

Unfortunately, information on the effects of chemical form on metabolism and utilization of mineral elements is quite limited. Possibly most of such answers needed in feeding minerals to dairy cattle are yet to be obtained. This presents both a challenge to research workers and a justification for practical feed formulators to allow some margin of safety to cover unexpected problems.

5.6.e. Supplemental Minerals

Since energy, protein, and fiber are required by dairy cattle in large quantities, these nutrients are given primary consideration in selecting the major feed ingredients. The feeds used to satisfy these needs usually also supply appreciable amounts of essential minerals. Depending on the combination(s) of ingredients, there is a tremendous range in the amounts of the various elements provided.

Any minerals needed but not furnished by the major feed ingredients, should be provided as supplemental minerals. Most practical combinations of feeds chosen for energy, protein, and fiber will be inadequate in *sodium* which normally is supplied as common salt (Table 5.8). Likewise, such combinations frequently do not contain a sufficient concentration of one or more of the following: phosphorus, calcium, magnesium, cobalt, iodine, copper, sulfur, or potassium (Table 5.8). Milk is very low in iron; thus, calves given only milk for extended periods will become deficient. The frequency of selenium, zinc, and manganese inadequacies in dairy cattle rations, before supplemental minerals are added, is not clear. Current information suggests that borderline deficiencies occasionally occur. Deficiencies of several essential elements have never been observed in cattle (Table 5.8).

Supplemental minerals may be provided in a number of ways with the most desirable method depending on the feeding system(s) used. Frequently all the supplemental minerals for lactating cows are mixed with the concentrate or the complete feed. When concentrates are not fed, as is especially common with growing animals and dry cows, the supplemental minerals can be fed "free choice." If only supplemental trace elements are needed, they may be provided as trace-mineralized salt, either in granular form or in a block. A more complete mineral mixture should be given when some of the major minerals are needed. Another relatively widespread method of providing essential mineral elements is as part of a "liquid feed supplement" along with nonprotein nitrogen. This method presents special solubility and stability problems with a few of the elements including calcium and iodine.

TABLE 5.8

Classification of Essential Mineral Elements According to Probability of Deficiency in Practical Rations Formulated Primarily to Provide Energy, Protein, and Fiber Needs of Dairy Cattle

Generally inadequate	Usually or frequently inadequate	Sometimes inadequate	Not known to be deficient
Sodium	Phosphorus	Selenium[b]	Chlorine
	Calcium	Zinc	Molybdenum
	Magnesium	Manganese	Fluorine
	Cobalt		Chromium
	Iodine		Silicon
	Iron[a]		Vanadium
	Copper		Nickel
	Sulfur		Tin
	Potassium		

[a] Iron deficiency is likely in calves fed milk for an extended period.
[b] Whether selenium should be in this column or under "Frequently inadequate" is not clear.

Contrary to widespread popular belief, when given a choice of various individual minerals, dairy cattle will not consume the amounts of different minerals needed (Coppock *et al.*, 1976). The amounts eaten depend a great deal on the palatability of the particular source of the element. Thus when feasible, mixing the desired mineral with the feed is preferable to free choice feeding (Coppock *et al.*, 1976).

5.7. MINERAL METABOLISM IN DAIRY CATTLE

An appreciation of the general principles of mineral metabolism is needed for a practical understanding of dairy cattle nutrition. Too often fundamental principles have been overlooked in planning, conducting, or interpreting mineral research. Likewise, some practical recommendations have been in conflict with important principles of mineral metabolism. There are tremendous differences in the characteristic patterns of absorption, transport, excretion, tissue turnover, and homeostatic control among the mineral elements. Each element has its own specific pattern.

5.7.a. Absorption of Minerals

Most mineral absorption takes place from the digestive tract but other routes are important in a few situations. For example, inhaled cadmium (a toxic element) is absorbed far more efficiently than that consumed. Likewise, some absorption can occur through the skin or from various types of injections.

The part of the digestive tract where absorption occurs most actively varies among the minerals. Several divalent cations, including calcium, zinc, manganese, and iron appear to be absorbed primarily from the small intestine. Studies with sheep indicate that the reticulorumen is a major site of magnesium absorption with variable amounts absorbed from the small intestine and possibly some from the omasum (Tomas and Potter, 1976).

Mechanisms of absorption vary widely among the elements. Several monovalent ions such as sodium, potassium, and chlorine are absorbed by simple diffusion. Other minerals, especially some of the divalent cations including calcium, zinc, and iron appear to be absorbed by facilitated diffusion which is a form of active transport (Table 5.9, Stake, 1974). In recent years the mechanisms of calcium absorption have been under active investigation. A special calcium binding protein plays a major role in calcium absorption (Taylor, 1973). Although information is much less detailed, binding proteins play a key role in the absorption of iron, copper, and zinc. Likewise, some of the mineral elements may be absorbed in association with various organic compounds. In this respect various chelates often have a significant role. Several elements are absorbed by

TABLE 5.9
Mode of Absorption of Essential Mineral Elements[a]

| | | Facilitated diffusion | |
| | Simple | | Binding |
Element	diffusion	Chelates	proteins
Sodium, potassium, chlorine	+ +	—	—
Phosphorus	+ +	?	?
Sulfur	—	—	—
Calcium	+ +	+ +	+ +
Magnesium	—	—	—
Iron	?	+ +	+ +
Copper, zinc	?	+ +	+ +
Manganese, cobalt, nickel	—	—	—
Fluorine, chromium, vanadium,			
Molybdenum, tin, silica,	—	—	—
Selenium, iodine			

[a] From Stake (1974). Key to table: + +, Definite supporting data; ?, inadequate or conflicting data; —, little or no information.

more than one mechanism (Table 5.9). Calcium is absorbed by both facilitated diffusion and simple diffusion. As suggested in Table 5.8, current information on absorption mechanisms is very incomplete for most of the mineral elements.

Perhaps the absorption process will be better understood by reviewing an example in somewhat more detail. Zinc, as well as many other trace elements, occurs in the animal almost entirely combined with organic compounds (Miller, 1971b), and often may be transported in a tight combination with an organic compound. In absorption and transport, zinc apparently is passed from one binding protein to another. The absorption can be visualized as a two-step process, although more than two steps are involved. These steps are absorption by the intestinal mucosa and subsequent transfer into the blood plasma followed by transport to other tissues. The protein which binds the zinc in the intestinal mucosa is quite different from that in the blood plasma with still different binding proteins in other tissues. The zinc may or may not be combined with other compounds such as chelates in addition to the protein. Thus, there may be a zinc–protein complex or a zinc–chelate–protein complex. The nature of the chelate can affect the metabolism and utilization of the zinc.

In discussing absorption the terms net absorption and true absorption are often used. True absorption is the total amount going into the tissue. In some situations an appreciable part of that which is absorbed is reexcreted into the digestive tract within a very short time. Net absorption is the true absorption less the amount reexcreted. Normally, it is much easier experimentally to determine net absorption than true absorption.

5.7.b. Transport and Turnover of Minerals within Tissues

Following absorption, mineral elements in widely varying chemical forms are transported throughout the body via the plasma. Some elements are transported in combination with organic compounds such as proteins or amino acids (see Section 5.7.a). Others, including sodium, potassium, and chlorine are transported as ions or as parts of ions such as phosphates.

The combination of mineral elements within the many tissues of the body are extremely complex and varied, covering innumerable compounds. In an attempt to better understand the processes and practical applications involved, scientists have developed some general concepts. One of these is turnover rate. Turnover rate simply describes the length of time an element remains in a tissue organ or other body unit after it enters and before it is removed.

There is an enormous difference in the turnover rates among elements, chemical combinations, and tissues. It is only a moderate exaggeration to suggest that some turnover rates proceed almost at the speed of light while others change as slowly as geological time periods. Much of the sodium in the body moves about very rapidly, with equilibrium being reached within minutes after it enters. In contrast, some of the calcium in teeth almost never exchanges. Between these extremes are a great range of exchange or turnover rates such as the calcium in the blood plasma which turns over several times per day. Many of the differences in turnover rate have practical importance: the much smaller amount of bone having readily exchangeable calcium appears to be the primary reason mature cows have milk fever more frequently than do young cows.

5.7.c. Excretion of Mineral Elements

Excretion of mineral elements, for convenience, can be divided into that not absorbed and that which has been absorbed and subsequently excreted. The portion which is absorbed and subsequently excreted is called endogenous excretion (see Section 5.6.c).

All of the unabsorbed minerals are excreted in the feces. Most of the endogenous excretion of mineral elements is either through the feces and/or the urine. However, some elements are eliminated in the sweat, and in certain situations, such as a very high selenium intake, a limited amount is lost via the breath or vapor. These endogenous excretions are considered in a different context from the amounts deposited in tissue and secreted into milk. Some writers have called minerals going into milk an excretion. This is not only an inaccurate term, but is

TABLE 5.10
Endogenous Excretion of Zinc and Manganese by Holstein Calves and Cadmium by Goats over a Period of 13 Days Following Intravenous Injection[a]

	% of dose		
Element	Feces	Urine[b]	Retained
Zinc	15	0.3	85
Manganese	69	0.2	31
Cadmium	7	0.03	93

[a] Determined using the radioactive element. (Data courtesy of W. J. Miller and Associates, University of Georgia).
[b] Urine values may be somewhat too high due to fecal contamination.

poor esthetics. Mineral elements are secreted into milk. Most of the endogenous excretion of sodium, chlorine, potassium, and magnesium is through the urine, whereas feces is the main endogenous excretion route of iron, zinc, manganese, and cadmium (Table 5.10).

5.7.d. Homeostatic Control of Tissue Levels of Minerals

(Much of the following discussion is adapted from earlier publications (Miller, 1973, 1975a,b).

The amounts of essential mineral elements in cattle feeds vary enormously (Section 5.6.a), but for normal performance, concentrations of functional forms in animal tissues must be maintained within relatively narrow limits (Underwood, 1977). Understanding how ruminants adapt to this dual situation of highly variable mineral intake and the necessity of keeping functional tissue levels fairly uniform is of considerable practical importance. In discussing these adaptations it is helpful to use the physiological term homeostasis or homeostatic control. Homeostasis is defined as "the condition of relative uniformity which results from the adjustments of living things to changes in their environment" (see Section 2.6).

Homeostatic control for every nutrient is essential for survival. The remarkable adaptability of ruminants to a great variety of feeds is possible because of good homeostatic control mechanisms for most nutrients. Whereas energy balance or homeostasis is achieved primarily by varying feed intake, homeostasis of minerals and other nutrients is not attained this way. During evolution, cattle developed adaptation (homeostatic control) mechanisms enabling them to perform normally with huge differences in consumption of most essential trace mineral elements. The range of intakes consistent with normal performance is determined by the homeostatic control of the animal. On a practical basis one must consider the possibility and serious consequences of both deficiencies and toxicities when intakes are outside the range of the animal's ability to maintain functional forms of the elements within acceptable tolerances.

Both the degree of mineral homeostatsis and the mechanisms involved vary tremendously with different minerals. Rejecting excesses of minerals, including essential ones, is just as necessary as retaining those needed. For example, the manganese content in most good forages is very high relative to that needed in cattle tissues. Thus, appreciable discrimination against manganese must be exerted.

The huge differences existing among minerals in the characteristic amounts absorbed, retained, and excreted must have evolved as a means of survival. The concept can be illustrated as follows. Although quite variable, the earth's crust, including the soil, has some 15 times more manganese than zinc. Forages usually contain only a little more manganese than zinc, yet the animal body has about 70 times as much zinc as manganese (Table 5.1). Plants discriminate against manganese relative to zinc with a further major discrimination by the animal.

Mineral homeostasis can be considered from two viewpoints. First (as discussed in Sections 5.7, 5.7.a, 5.7.b, and 5.7.c) the characteristically different ways animals metabolize individual minerals are important. Another view concerns how animals adapt to the highly variable intake of a given element. The major routes by which cattle adapt to the large differences in dietary mineral intakes are changes in (1) absorption, (2) excretion in urine, (3) tissue deposition in harmless and/or readily usable reserve forms, (4) secretion into milk, and (5) endogenous excretion via feces. Other usually minor ways include exhalation, skin sloughing, hair loss, and perspiration. A summary of the mechanisms ruminants use to maintain homeostatic control of selected elements is presented in Table 5.11.

Absorption changes. Changes in the percentage absorbed is a major adaptive control route for differences in consumption of some essential divalent cations, such as calcium, iron, zinc, and manganese. For instance, zinc absorption varies from less than 10% to about 80% in young calves and lactating cows with changes in dietary zinc intake. The total amount of zinc absorbed, in grams/day, increases with higher dietary intake, but much less than the increase in consumption. Change in absorption percentage apparently contributes little to the regulation of sodium, potassium, chlorine, iodine, cadmium, and magnesium. Increasing dietary cadmium for goats by 400-fold caused little change in percentage absorbed and retained.

Urinary excretion. The main way cattle adapt to dietary magnesium dif-

TABLE 5.11

Routes Animal Use in Adapting to Varying Intake Levels of Certain Elements[a]

	Degree of importance in adaptation			
Route	Of major importance	Fairly important	Minor contribution	Little or no importance
Absorption (changes in percentage)	Calcium Iron Zinc Manganese		Magnesium Selenium	Sodium Chlorine Potassium Iodine Cadmium
Endogenous fecal loss variations	Manganese	Iodine	Zinc	Selenium Cadmium
Urinary loss changes	Magnesium Sodium Chlorine Potassium Iodine Selenium Fluorine	Calcium	Nickel Cadmium	Zinc Copper Manganese
Tissue deposition in harmless or readily usable reserve forms	Calcium Iron Copper Molybdenum Fluorine	Iodine Cadmium	Magnesium Sodium Zinc Cobalt Manganese Selenium	
Secretion of variable amounts in milk	Iodine Molybdenum	Zinc	Copper Cobalt Manganese Selenium Fluorine	Calcium Sodium Chlorine Potassium Iron Nickel Cadmium

[a] From Miller (1975b); see also Miller (1975a). For many elements, information is insufficient to determine the influence of some of the possible adaptation routes. For every element, the capacity of the animal to adapt can be overcome with sufficient deviation from normal intake resulting in deficiency or toxicity effects.

ferences is by excretion of the excess in the urine. Urine is also a major homeostatic control route for sodium, potassium, chlorine, iodine, fluorine, and selenium.

Tissue trace element composition changes. Changes in tissue concentration of elements as an adaptation mechanism varies greatly among elements. When an excess of an element is fed and absorbed, some may be deposited either as a readily usable reserve and/or in a harmless, but not readily mobilizable, form. Substantial calcium can be stored in bone and withdrawn when needed. Cattle can accumulate considerable copper in the liver and withdraw it as needed. Likewise, there is appreciable storage of iron in liver. Soon after absorption, animals combine cadmium with specifically induced cadmium-binding proteins thus greatly decreasing turnover and toxicity.

Variable endogenous fecal excretion. Of the important adaptation routes, information is most inadequate on endogenous fecal excretion due partially to difficulties in conducting such studies. Fecal losses of absorbed minerals have considerable significance in understanding mineral nutrition. Such losses may equal or even exceed true absorption leaving little or no net retention. A net absorption near zero often occurs in normal, adult, nonlactating animals for minerals whose predominant endogenous excretion route is feces. Examples include manganese and calcium. Since endogenous fecal excretion of manganese is a major control route, a higher dietary intake is quickly reflected in an elevated endogenous fecal loss.

Secretion into milk. Iodine is unique among essential elements in that secretion into milk is closely related to intake (Table 5.12). In fact, milk iodine is regarded as a good biochemical measure of the iodine status of lactating cows.

The concentration of many elements in milk, including calcium, chlorine, iron, nickel, and cadmium, is affected remarkably little by dietary intake. Milk sodium and potassium are not influenced by a dietary excess and decrease only slightly with deficient intakes. With other minerals, including copper, cobalt, zinc, manganese, and molybdenum, dietary effects are small but may be of some practical importance. The magnitude can be illustrated with zinc. Reducing zinc in the ration from a typical 40 ppm to 17 ppm decreased milk zinc from 4.2 to 3.3 ppm, whereas elevating dietary zinc to 700 ppm increased milk zinc to a plateau of about 8 ppm (Table 5.23). The mammary gland is a barrier against large increases in milk zinc content. Milk is remarkably well protected against the toxic elements, arsenic, cadmium, and mercury, but not as well protected from lead.

TABLE 5.12
Iodine in Milk with Varying Iodine Intakes[a]

Intake (mg/cow/day)	Milk (ppb)[b]
1.6	28
12.7	78
20.0	267

[a] Adapted from Underwood (1966) p. 118. Data of Kirchgessner (1959).
[b] ppb = parts per billion.

5.8. DETERMINATION OF MINERAL
REQUIREMENTS AND TOLERANCES

To most effectively use the mineral requirement and tolerance level data, some knowledge of how the values are determined is helpful. The two basic approaches are the factorial method and the feeding experiment method (see Section 1.5). Insofar as possible, it is desirable to compare results by both methods.

With the factorial method, the requirements are measured in two stages. First, the net amounts of the element deposited in tissues during growth and/or reproduction and that secreted into milk are estimated. To this is added the "inevitable" losses of the mineral from the body. This is the maintenance or endogenous loss. Second, the percentage of the mineral in the feed which is available to the animal is determined by metabolism studies. The total net requirement is divided by the percentage availability to arrive at the total amount needed in the diet. The factorial method is described in greater detail in the British publication on the nutrient requirements of ruminants (Agricultural Research Council, 1965).

The greatest weaknesses of the factorial method are the difficulties involved in determining the maintenance or endogenous losses and for several elements, the percentage which is available to the animal. Since current information is inadequate to calculate requirements for many of the essential mineral elements by the factorial method, results from feeding experiments alone must be used.

The feeding experiment method consists of feeding known amounts of the mineral under study and observing the response and performance of the animal. To obtain an accurate estimate of the requirement, satisfactory criteria of adequacy are needed. For some elements, the criteria in use have poor sensitivity. These are discussed more fully in some of the sections concerned with individual elements.

In developing the estimated requirements for the minerals (see Appendix, Tables 1, 2, and 3), information from both the factorial and the feeding experiment method were used if available (NRC, 1978). When results of feeding experiments are not in agreement with those of the factorial method, a judgment must be made. Understanding certain general aspects is helpful in making such decisions. For example, a ration which does not reduce milk production or growth over relatively short periods may be quite deficient if tissue levels of the element are being depleted. This situation would occur with calcium or iron-deficient diets. The following quotation from the ARC (1965, p. 15) concerning interpretation of mineral requirement data is pertinent.

> In general, results which indicate no change in animal performance when less than the estimated requirement is given do not necessarily mean that the estimate of requirement is too high because the trials may have employed relatively insensitive criteria of dietary lack. If, however, practical trials indicate that factorial estimates of requirement are too low, then it is

essential to reexamine the bases of these requirements and to accept as requirements values based on practical tests.

The tolerance or maximum safe levels are determined almost entirely in feeding experiments.

5.9. EFFECTS OF MINERAL DEFICIENCIES IN ANIMALS

The effects of deficiencies vary widely, as will be discussed later for each mineral. There is also considerable overlapping of symptoms with many different deficiencies causing the same symptoms and having similar effects. Thus diagnosis of individual mineral element deficiencies is not easy. A deficiency of any required element ultimately has an adverse effect on the health and performance of a dairy animal. However, appreciable depletion of some required elements is possible before performance is impaired.

5.10. TOXICITY AND TOLERANCES

Every nutrient, including each essential mineral element, is potentially toxic if consumed in sufficient amounts. Generally, the mineral levels which dairy cattle

TABLE 5.13
Estimated Maximum Safe Levels of Several Minerals for Dairy Cattle Relative to Minimum Requirements[a]

Class of cattle	Mineral	Minimum requirement	Maximum safe level	Safe range (max/min)
Lactating	Sodium chloride (%)	0.46	5	11
Nonlactating	Sodium chloride (%)	0.25	5	20
Calf	Iron (ppm)	100	1000	10
Older cattle	Iron (ppm)	50	1000	20
All	Cobalt (ppm)	0.1	20	200
All	Copper (ppm)	10	80	8
All	Manganese (ppm)	40	1000	25
Young	Zinc (ppm)	40	500	12
Mature	Zinc (ppm)	40	1000	25
Cows	Iodine (ppm)	0.5	50	100
Non-cows	Iodine (ppm)	0.25	50	200
All	Selenium (ppm)	0.1	5	50

[a] Both the minimum requirements and maximum safe levels often are influenced by several variables. See text and Appendix, Table 3 for more details. From NRC (1978) and Neathery and Miller (1978).

can tolerate without adverse effects on performance or health are far higher than
the minimum requirement. Thus the range between minimum needs and toxic
levels is far greater than that of most other nutrients. For example, dairy cattle
can safely tolerate 30 to 50 times the minimum required level of selenium which
sometimes is cited erroneously as having a relatively narrow range between the
deficient and toxic (Table 5.13). Usually the range between a deficiency and a
toxicity is relatively greater for trace elements than for major minerals.

Toxicity of a few of the essential mineral elements including fluorine,
selenium, molybdenum, iodine, and copper can be a practical problem in dairy
cattle under some feeding conditions. Some mineral elements which have not
been shown to be essential for animals are of interest because of their toxicity, or
potential toxicity (see Section 5.31). These include cadmium, lead, and mercury.

5.11. COST OF ESSENTIAL MINERALS

The total amount of essential mineral elements required by dairy animals is
less than 3% of the dry diet. Of course, the compounds used to supply the
requirements usually contain other elements. Except for phosphorus, the cost of
most of the major minerals which need to be added to dairy cattle diets is
relatively low per pound.

The total amount of trace elements needed is only about 0.02% of the feed.
Most combinations of practical feed ingredients will supply a substantial portion
of the essential mineral elements. Accordingly, it is apparent that the total cost of
supplementing the needed minerals in dairy cattle feed is very small. The really
important cost is paid when performance is adversely affected by a deficiency,
toxicity, or imbalance of an essential mineral element.

It is generally assumed that acute mineral problems do not exist or are extremely rare in practical feeding situations. The lessons of history do not convey
such confidence. Often problems which reduced performance by a large percentage existed for many years before they were discovered to be caused by minerals.
Thus some of the unexplained health and performance problems of dairy cattle
may be manifestations of mineral deficiencies, excesses, or imbalances.

5.12. IMPORTANCE OF AVOIDING BORDERLINE
DEFICIENCIES AND TOXICITIES

An acute mineral or trace mineral deficiency or toxicity problem can be
devastating to the profits of a dairy farm. Borderline problems, however, are
vastly more costly overall because they are so much more widespread. The gross

deficiencies are like the tip of an iceberg with most of the problem hidden as a borderline or marginal malady. Some estimates have indicated that for every animal exhibiting a gross or acute mineral deficiency problem, the performance of a thousand may be adversely affected by a borderline deficiency.

A marginal deficiency (or a toxicity) of an essential mineral is likely to reduce milk production, growth rate, resistance to disease or infection, and/or reproduction by a small percentage; usually all animals are not equally affected. Because of the variation normally encountered in these measures, performance of a herd can be reduced enough to materially lower the net income without the problem being either suspected or easily diagnosable by the best of specialists. Even in carefully controlled experimental conditions, differences of 5% or more are usually required before a problem can be detected.

Although there appears to be general agreement among authorities that this interpretation of the effects of borderline deficiencies is correct, many hesitate to voice it. This hesitation is due to the belief that such statements would be used by the unscrupulous to sell large quantities of unneeded minerals or services. One of the major dangers of the unneeded minerals and trace minerals is that borderline toxicity and imbalance problems may be created which will have a similar undetected effect on performance and thereby on net profits.

5.13. MINERAL INTERACTIONS

The metabolism, amounts required, and maximum safe dietary levels of essential mineral elements are affected in a major way by the level of other minerals in the diet. A wheel of some of the better known interactions (or interrelationship) is shown in Figure 5.1. Although, space does not permit a full discussion of the interactions between minerals or even specific mention of some, several interrelationships must be considered in many practical feeding situations.

Often a mineral interaction involves some form of antagonism between two or more elements. An example is copper and molybdenum (see Section 5.24). One of the important and best known interrelationships among minerals is between calcium and phosphorus. Frequently, the ratio between calcium and phosphorus is used in studying the nutrition of these two elements. Most of the interactions between minerals have not been studied sufficiently with dairy cattle, if indeed such is possible, to arrive at quantitative data to accurately define the safe dietary ratio limits.

High calcium levels can increase the requirements of some of the trace elements including zinc and manganese for some species. In cattle, these relationships are not well defined. Many of the interactions listed in Figure 5.1 are taken from other species as there are few or no data for dairy cattle. It is recognized that

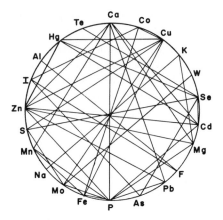

Fig. 5.1   The nutrition and metabolism of many minerals is greatly influenced by the amounts of other nutrients in the diets. This chart illustrates some of the interactions which have been noted between mineral elements. Organic constituents of the diet also strongly affect the nutrition of mineral elements. (Courtesy of W. J. Miller, Univ. of Georgia.)

this is presumptive as interactions between mineral elements do not have the same effects in different species of animals.

The mechanisms by which one element affects the nutrition of another vary considerably with different pairs of elements. Likewise, for many of the interactions, the mechanisms have not been defined. One of the important ways some elements affect others is by changing the percentage absorbed. Most often a higher level of the interacting element reduces that of the other. Another mechanism is through more rapid excretion, whereas a third type is through effects at the tissue or functional level. Often increasing the dietary level of one element will materially change the distribution of another among the body tissues or fluids. Many of the more important interactions among mineral elements are discussed under the various elements in later sections.

5.14. INTERACTIONS BETWEEN MINERALS AND ORGANIC DIETARY CONSTITUENTS

Although more attention is usually given to the interactions between or among mineral elements, interrelationship among mineral elements and organic compounds of the diet may be equally important. One of the best known of these interactions is that between vitamin D and calcium. Vitamin D plays a crucial role in the absorption of calcium through its relationship to the synthesis of calcium-binding protein (see Sections 6.3.a and 6.3.b). This vitamin also affects the metabolism of several other minerals including magnesium, zinc, cadmium, and lead.

The sparing effect of vitamin E (tocopherols) on the amount of selenium required is well known. A high level of tocopherols in southern forages may explain the apparently low incidence of selenium deficiencies in grazing cattle.

As discussed earlier (see Section 5.6.d), much of the phosphorus in grains is present as phytate. Phytate phosphorus is poorly available to some species of monogastric animals. Since the young calf is a monogastric animal before a functional rumen develops, it is reasonable to expect the phytate phosphorus might not be readily available to the very young calf. Phytate also adversely affects the utilization of some trace elements, such as zinc in monogastric animals, but apparently has much less influence in ruminants.

The much higher tolerance of dairy cattle for molybdenum from the added inorganic element compared to that in pasture may infer a relationship between some organic constituent(s) of the pasture and molybdenum (see Section 5.24). The metabolism of several minerals is controlled or greatly affected by certain hormones. Especially noteworthy are the effects of (1) the parathyroid hormone on calcium and phosphorus, (2) aldosterone on sodium and potassium, and (3) thyrocalcitonin and estrogen on calcium. Although information is quite meager, it seems probable that there are many yet to be defined interrelationships of importance in feeding dairy cattle between organic compounds and essential minerals.

5.15. THE ESSENTIAL MINERAL ELEMENTS

Previously in this chapter, many of the principles of mineral nutrition for dairy cattle have been considered. In the following sections other important aspects of the essential mineral elements are discussed. The main aspects considered are functions, deficiency effects, requirements, tolerance or maximum safe dietary levels, diagnosing deficiencies and toxicities, metabolism, homeostatic control mechanisms, sources of the mineral elements, practical importance, and other items of special or practical interest. The minimum requirements (see Appendix, Tables 1, 2 and 3) were formulated with the objective of selecting the lowest amount of each element that would adequately fulfill the needs of the dairy animals under most typical farm conditions. These values are somewhat higher than the minimum amounts needed to prevent deficiency effects under ideal conditions. Accordingly, if these levels of minerals are fed, performance should be unimpaired and there should be no borderline deficiency effects. Because of substantial interactions which greatly affect the requirements of some elements, the footnotes to the tables and the discussion for each element should be consulted for possible changes with special situations.

5.16. CALCIUM

Calcium (Ca) is a critical nutrient in the practical feeding of dairy cattle. Even though 98% of the body Ca is in the bones and teeth (Ellenberger *et al.*, 1950), Ca performs numerous key functions in the soft tissues of cattle. These include blood clotting and normal tissue excitability. The functional roles of Ca in the soft tissues of the body are so critical that the Ca level of the blood plasma is one of the most closely regulated of substances (Miller, 1970a). Perhaps this can be illustrated best by noting that milk fever occurs when Ca in the blood plasma drops to a level substantially below normal. The dairy cow can be fed a Ca-deficient diet for months or even years without a major decrease in the Ca level of the blood plasma.

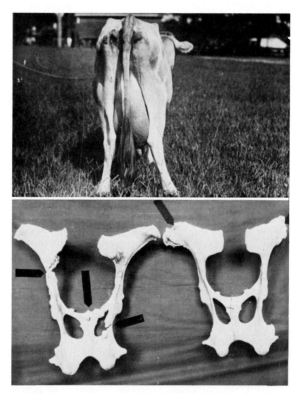

Fig. 5.2   This cow had been fed a calcium-deficient diet for years resulting in considerable depletion of the mineral from the bone. The pelvis of the cow shown above is pictured at the lower right. Note that both hip bones were broken (knocked down). At lower left is the pelvis of another cow fed a calcium-deficient diet for an extended period. Note the breaks indicated by arrows. (Courtesy of R. B. Becker, University of Florida.) (See also Becker *et al.*, 1933a, and 1953.)

TABLE 5.14
Average Breaking Strength of Leg Bones from Jersey Cows Fed at University of Florida[a]

Calcium and phosphorus supplement	No. of cows	Breaking strength (lb)
None	1[b]	335
Bone meal at 2% of concentrate	7	3140
1% bone meal and 1% marble dust	6	3232

[a] Adapted from Becker *et al.* (1953).
[b] Cow shown in Fig. 5.2.

Regulation of plasma Ca content is quite complex with many practical implications (Miller, 1970a). The regulation involves several hormones including the parathyroid hormone, calcitonin, estrogens, and others. When plasma Ca starts to decline, Ca is mobilized from the bone bringing plasma Ca back to normal. Bones provide substantial reserves of Ca permitting the cow to produce milk of the same Ca content when dietary intake is deficient. The Ca reserves of the bones serve a very vital function and are exhaustible. When the Ca is removed the ash content decreases proportionally so there is less total bone and therefore weaker bones (Fig. 5.2). If Ca depletion is sufficiently advanced, the bones become very soft and easily broken (Becker *et al.*, 1933a, 1953). Calcium withdrawal from bones over relatively short periods with subsequent replacement is not harmful. Some depletion during early lactation compensated for in late lactation or the dry period does not appear to be detrimental.

When dietary Ca is low, cattle absorb a higher percentage of that ingested (ARC, 1965; Van't Klooster, 1976; NRC, 1978). Likewise, with excess dietary Ca, the percentage absorbed declines (Fig. 5.3). These changes in absorption occur fairly quickly enabling dairy animals to meet their Ca needs over a wide range of intakes without accumulating excess Ca in the body (Figs. 5.4 and 5.5).

The fine regulation of plasma Ca, along with the ability to mobilize bone reserves and to make large changes in the percentage absorbed, enable the dairy animal to perform normally with long periods of inadequate dietary Ca (ARC, 1965; NRC, 1978). These regulatory mechanisms greatly complicate the work of the researcher attempting to determine the dietary requirements of Ca (Miller, 1975a; NRC, 1978). Thus, in spite of considerable research on Ca requirements, the minimum Ca needs of dairy cattle have not been established with a high

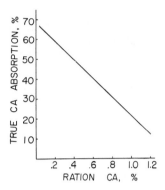

Fig. 5.3 Change in the percentage dietary calcium absorbed is a major factor in the homeostatic control of calcium. In this illustration for a lactating cow, note the decline in true calcium absorption as the calcium content of the diet increases. (Developed from results of many sources and research studies in several countries; courtesy of W. J. Miller, Univ. of Georgia.)

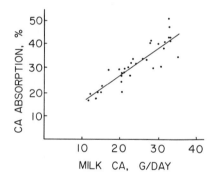

Fig. 5.4 Effect of level of milk production on the percentage of dietary calcium absorbed when all cows fed 72 gm of calcium per day. In this experiment the increased absorption of calcium was equal to 80% of the calcium secreted into the milk. (Adapted from Van't Klooster, 1976.)

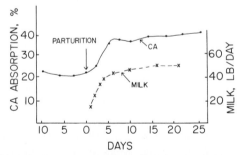

Fig. 5.5 Effect of lactation on percentage of radioactive calcium absorbed. Note the large and fairly rapid increase in absorption following parturition. The same amount of calcium was fed throughout the study. (Adapted from Van't Klooster, 1976.)

degree of accuracy. On the basis of all evidence available the estimated Ca requirements presently in Tables 1, 2, and 3 of the Appendix should be sufficient for maximum performance of dairy cattle under all typical farm conditions. Feeding less than these amounts for short periods often will not affect performance (Belyea *et al.*, 1976).

The estimated Ca requirements were calculated by the factorial method (see Section 5.8). Available Ca needed for maintenance, growth, pregnancy, and/or lactation were totaled and the dietary requirements calculated from these data with a factor for the percentage of Ca available in the feed (ARC, 1965; NRC, 1978). The maintenance requirement (endogenous or metabolic losses) and the true availability of the dietary Ca are the two aspects in the factorial procedure for which information is most inadequate (ARC, 1965; NRC, 1978).

Briefly, the values used in determining the Ca needs by the factorial method were based largely on the following information. The maintenance requirement for Ca was calculated from isotope data, using 0.73 gm per 100 lb (or 1.6 gm/100 kg) of body weight per day, with adjustments for differences in metabolic size of mature animals (NRC, 1978). An 88 lb calf contains about 541 gm of total body Ca (Ellenberger *et al.*, 1950; Hogan and Nierman, 1927). Calcium deposited during gestation is low until the last 2 months when 75% of the total fetal Ca is deposited (Fig. 5.6). Each pound of milk with 4% fat contains an average of 0.56 gm (1.23 gm/kg) of Ca. Assuming an availability of 45%, the requirements for lactation is 1.23 gm of Ca per pound of milk (2.7 gm per kg of milk). Dairy cattle tend to maintain Ca reserves by increasing or decreasing the percentage absorbed (Figs. 5.3, 5.4, and 5.5). Thus, determination of the true availability in feeds is the weak link in this information.

The effect of age on Ca availability was considered in developing the (NRC, 1978) Ca requirement values. Some studies indicated that calves absorbed 90% of the Ca in milk. In older cattle, true absorption was quite variable, ranging from 22 to 55%, but averaging about 45%. Calcium in milk fed to young calves is more efficiently absorbed and retained than that in rations consisting of forages

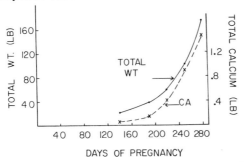

Fig. 5.6 Total weight and calcium deposited in fetus and products of conception during pregnancy for a calf weighing 100 lb at birth. (Adapted from ARC, 1965.)

and concentrates. Young animals apparently absorb Ca more efficiently (Hansard *et al.*, 1957). Some evidence suggests Ca from inorganic sources is more available than from organic sources (Hansard *et al.*, 1957). Differences due to age, however, were greater than those associated with feed sources (Hansard *et al.*, 1954, 1957). Calcium absorption also may be influenced by other factors including vitamin D, phosphorus intake, and acid–base balance. High fat diets reduce Ca availability by increasing fecal Ca losses through the formation of Ca soaps (Oltjen, 1975).

These Ca requirement values (see Appendix, Tables 1, 2, and 3) obtained by the factorial method are consistent with results from many feeding experiments (ARC, 1965). When interpreting results of feeding trials, it is essential to remember that cattle are able to reduce reserves of skeletal Ca over extended time periods before milk production or growth are reduced. Some authorities feel that Ca balance studies can be used to accurately measure the Ca requirements of dairy cattle. However, many factors other than the dietary Ca affect Ca balances. Alone, balance studies do not provide a definitive measure of the Ca requirement.

A deficiency of Ca, unlike some of the essential mineral elements, usually does not result in many easily recognizable symptoms (NRC, 1978). The blood plasma levels are so well regulated that this does not provide a good biochemical measure (Netherlands Committee on Mineral Nutrition, 1973). A reduction in Ca content of the bone is the major change occurring when dietary Ca is inadequate. The Ca to P ratio in bone usually does not change materially when Ca is depleted from the bone (see also Section 5.17). Thus, diagnosis of a Ca deficiency generally depends on a measure of bone depletion (Miller and Stake, 1974b). Measures frequently used are bone ash, bone density, or breaking strength (Miller and Stake, 1974b). Bone demineralization is not a definitive indication of Ca deficiency as inadequate phosphorus or vitamin D can cause the same effect. On a practical basis often the best approach in establishing whether or not Ca is

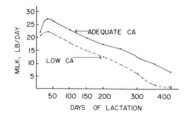

Fig. 5.7 Average milk production of a herd of Jersey cows which had been fed a very low calcium diet increased about 46% after the addition of supplemental calcium (steamed bonemeal). When inadequate calcium was fed, broken and weak bones were a problem (see Fig. 5.2). (Adapted from Arnold and Becker, 1936; Becker *et al.*, 1933a.)

adequate, is to compare the Ca content of feeds with the Ca requirements (Miller and Stake, 1974a,b).

Usually, only an extreme or long extended deficiency of Ca will decrease growth or milk production. However, such extreme deficiencies can occur if attention is not given to providing adequate Ca (Fig. 5.7).

Milk fever (parturient paresis) in cows is caused by disturbance in Ca metabolism. A marked drop in blood plasma Ca occurs due to a failure of Ca homeostasis (Miller, 1970a). This very complex problem is influenced by the dietary intake of Ca and phosphorus. High Ca intake during the dry period (over 100–125 gm per day) tends to increase the incidence of milk fever (see Section 18.1) (Jorgensen, 1974). Feeding a low-Ca diet for a few weeks prior to calving can reduce the incidence of milk fever (see Section 18.1). Milk fever as a metabolic problem is further discussed in Section 18.1.

Information on the maximum safe levels of dietary Ca is much less adequate than for the minimum requirements (NRC, 1978). Over a relatively wide range, cattle are able to avoid toxicity from high dietary Ca by excreting the excess in the feces. Even so, excessive Ca is antagonistic to the metabolism of several other elements including phosphorus, manganese, and possibly zinc (Fig. 5.1). Many of these relationships have been studied in laboratory animals but have not been quantitatively measured in cattle. Nevertheless, it is believed that Ca levels greatly above those presented herein should be avoided. Except just prior to calving, feeding moderate excesses of Ca to lactating cows has not caused obvious problems. When bulls were fed three to five times the amount of Ca recommended, a high incidence of bone and joint abnormalities, including stiffness, excess calcification and degenerative joints (osteopetrosis, vertebral ankylosis, and degenerative osteoarthritis) were observed (Krook et al., 1969, 1971).

Forages generally contain more Ca than do grains (Underwood, 1966). However, the range is tremendous in the content both within grains and forages and even in the same species (Table 5.15). Legumes, on an average, have 2 to 5 times as much Ca as grasses (see Section 5.6.b). High levels of potassium and of magnesium in the soil will reduce Ca content of the forage (Netherlands Committee on Mineral Nutrition, 1973).

Although dairy cattle require more Ca than any other mineral element except potassium, it is relatively easy and economical to add supplemental Ca to meet the requirements of dairy cattle. Ground limestone is a good economical source of Ca. Often excess lime has been added to dairy feeds because the cost per pound was much lower than that of the average ingredient used. Most phosphorus supplements contain more Ca than phosphorus. However, when only liquid supplements are used to provide added minerals, adequate Ca may not be provided (Miller and Stake, 1972). Because of the low solubility of Ca, it is difficult to include a substantial amount in liquid supplements.

TABLE 5.15
Content of Major Minerals in Feeds from Northeastern United States[a,b]

Feed		Calcium	Phosphorus	Sodium	Magnesium	Potassium	Sulfur
Legume forage	Average	1.18	0.30	0.024	0.24	2.55	0.26
(992)[c]	Range	(0.03–2.23)	(0.14–0.56)	(0.001–0.100)	(0.10–0.58)	(0.21–4.93)	(0.14–0.43)
Grass forage	Average	0.49	0.22	0.014	0.16	1.68	0.20
(352)[c]	Range	(0.10–1.58)	(0.09–0.56)	(0.000–0.110)	(0.04–0.42)	(0.24–4.04)	(0.14–0.29)
Corn silage	Average	0.27	0.23	0.005	0.18	1.07	0.14
(7179)[c]	Range	(0.01–1.88)	(0.01–0.93)	(0.000–0.350)	(0.01–0.55)	(0.02–3.28)	(0.04–0.22)
Corn grain	Average	0.03	0.31	0.003	0.12	0.42	0.14
(221)[c]	Range	(0.01–0.72)	(0.19–0.59)	(0.000–0.100)	(0.06–0.20)	(0.24–1.41)	(0.10–0.19)

[a] Samples analyzed by the Penn State Forage Testing Service (1969–1973), with most samples from Pennsylvania and New York. Data given as % in dry matter.
[b] Adapted from Adams (1975).
[c] Denotes number of samples except for sulfur analyses which were 39, 4, 249, and 8 for the feeds in descending order.

5.17. PHOSPHORUS

The 85% of the total phosphorus (P) (Ellenberger *et al.*, 1950) in the skeleton of dairy cattle along with calcium (Ca) are the major elements forming the mineral base of bones and teeth. In addition, (see also Section 5.4) P is an extremely important element in the other tissues of cattle. It is the key mineral element in the biochemical energy transformations of all the body cells. Biochemical energy utilization in animal cells involves the formation and breaking of high energy P bonds.

Although P is intimately related with Ca there are major differences in their nutrition and metabolism. For example, low dietary P reduces inorganic P in blood plasma much more than low Ca intake affects plasma Ca. Thus, decreased plasma inorganic P can be used as a biochemical aid in diagnosing a P deficiency. Normal values of plasma inorganic P are 4–6 mg/100 ml for cows and 6–8 mg/100 ml for calves under 1 year of age (NRC, 1978). Likewise, dairy

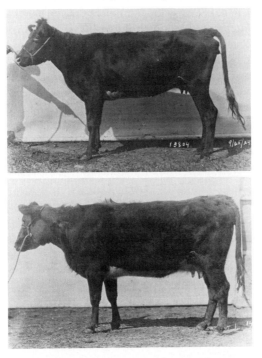

Fig. 5.8   This Minnesota cow (top photo) had a depraved appetite and had produced an average of only 6.4 lb of milk daily and gained only 0.15 pounds per day for 285 days on a phosphorus-deficient ration. After phosphorus (monobasic sodium phosphate) was added to the same ration (bottom photo), she had gained 2.18 lb per day for 90 days. (Courtesy of R. B. Becker, Univ. of Florida.) (See Eckles *et al.*, 1926.)

animals do not have hormones to readily mobilize P from the bone in response to the animal's needs. In fact, P is withdrawn along with Ca from bone in response to Ca needs.

The inability to finely regulate inorganic P in blood plasma results in more deficiency signs at an earlier stage than occurs with insufficient Ca. When too little dietary P is given to dairy animals, appetite declines, growth rate is reduced, milk production is lowered, and the efficiency of feed utilization decreases (Eckles *et al.*, 1932) (Fig. 5.8). Often feed utilization efficiency declines before appetite is affected (Beeson *et al.*, 1941). Decreased appetitie is not of definitive diagnostic value as it also is associated with several other deficiencies. Depraved appetite manifested by chewing of such things as wood, bones, and hair is often observed (Figs. 5.9, 5.10, 5.11, and 5.12). However, cows may suffer from severe P deficiency without exhibiting a depraved appetite. The clinical symptoms of P and cobalt deficiencies are similar but usually can be differentiated by hemoglobin and plasma values.

When P is deficient, just as with insufficient Ca, the bones lose minerals and become fragile. In chronic P deficiency, the animal sometimes develops joint stiffness (Figs. 5.9 and 5.10). Anestrus, failure to exhibit estrus or heat, and low conception rates may occur in females of breeding age with inadequate P intakes, but the P content of the milk is not affected. The most effective means of diagnosing a P deficiency appears to vary somewhat in different circumstances. For example, Underwood (1966) suggests that the most sensitive and earliest biochemical measure of a P deficiency is reduced serum inorganic P. In contrast, the Netherlands Committee on Mineral Nutrition (1973) did not consider serum inorganic P to be sufficiently sensitive to recommend it in diagnosing problems with cattle. As with Ca, analyses of the feeds for P content and a comparison of

Fig. 5.9 A phosphorus-deficient (sweeny) cow with a depraved appetite and showing typical pose while chewing objects such as bones, oyster shells, old rubber, etc. (Courtesy of R. B. Becker, Univ. of Florida.) (See Becker *et al.*, 1933b.)

Fig. 5.10 This Florida cow suffered from an advanced phosphorus deficiency (sweeny) and became permanently crippled. The prominent appearing shoulder was caused by distortion of the scapula. This cow had a depraved appetite and continually chewed a great variety of objects. Her blood-phosphorus level was only 2.0 mg/100 ml. The bottom photo from the same cow shows the severely eroded and weakened scapula and humerous bones. Note the hole "A" worn into the articular surface to the bone marrow cavity. (Courtesy of R. B. Becker, Univ. of Florida.) (See Becker *et al.*, 1933b.)

intake values with established requirements is probably the most useful way of determining P status of dairy cattle in practical situations.

The P requirements are shown in Tables 1, 2, and 3 of the Appendix (NRC, 1978). The procedures used in estimating the minimum amount of P needed were similar to those for Ca. This includes calculating the requirements by the factorial method plus comparing these values with results from feeding experiments.

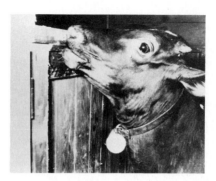

Fig. 5.11   A phosphorus-deficient calf chewing wood. Such a depraved appetite is a frequent manifestation of this deficiency. (Courtesy of S. E. Smith, Cornell Agricultural Experiment Station.) (National Research Council, 1971a.)

In Cornell experiments, Wise *et al.* (1958) demonstrated that 0.22% P in the air dry ration was sufficient for maximum growth rate in young dairy animals, but bone ash was higher when 0.30% P was fed. Less P is needed for maximum growth rate in dairy animals than for maximum bone ash or strength. It is more important to be sure P is fully adequate for animals being kept for breeding and lactation purposes than for those going to slaughter within a reasonable time period.

Generally, as a percentage of intake, absorption of P in dairy cattle is higher than Ca. Apparently, P absorption declines as animal become larger (ARC, 1965; NRC, 1978). The estimated requirements (see Appendix, Tables 1, 2, and 3) are based on P availability declining from 90% for young calves to 55% in cattle over 880 lb. Additional research is needed to more accurately define the true availability of P by dairy cattle. Considerable attention has been given to the Ca to P ratio. In bones this ratio is about 2:1 in older cattle but is somewhat lower in young cattle (NRC, 1978). In milk the ratio is about 1.3 to 1.

Except for prepartum feeds, ruminants are able to tolerate a much higher ratio of Ca to P than are laboratory animals. Calcium to phosphorus ratios below 1 to 1 probably should be avoided in ruminants (Smith and St. Laurent, 1970). Research at Cornell University demonstrated that growth rate and feed utilization of calves were satisfactory with Ca to P ratios ranging from 1:1 to 7:1 (Wise *et al.*, 1963). Decreased performance and nutrient conversions were noted at ratios above and below this range. In University of Missouri research a ratio of 8:1 reduced growth and feed utilization of Holstein steers (Ricketts *et al.*, 1970). No significant differences were found in lactating cows fed rations containing Ca and P at ratios of 1:1, 4:1, and 8:1 (Smith *et al.*, 1966). Long term experiments with pregnant heifers fed dietary ratios of 2:1 and 1:1 showed better absorption of both elements from the 2:1 diet (Manston, 1967).

If P is in short supply, feeding 10% less than indicated requirements for a

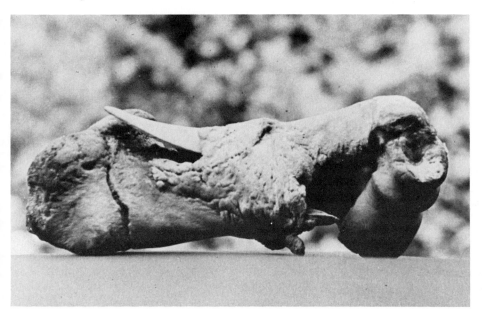

Fig. 5.12 Leg bone that had been broken and pushed together from a phosphorus-deficient cow grazed on muck soil near Belle Glade, Florida. It is believed that the high molybdenum in the forage ties up the phosphorus, thus increasing the dietary requirement. (Courtesy of R. B. Becker, Univ. of Florida.)

short time should not adversely affect performance (National Research Council, 1974b). Under these conditions, the levels of Ca and Vitamin D become more critical. Excess Ca may increase the P requirement and efficient use of P depends on adequate vitamin D.

Except for the limited research indicating adverse effects on performance when the diet contains more P than Ca, information on excess or toxic levels of P and on the maximum safe level is meager. One reason is the relatively high cost of P. Since large amounts of P are required and supplemental P has a high cost per pound, the cost of providing P is substantially greater than for any other mineral element.

Generally, feed grains and especially oil meal protein supplements, are relatively rich in P. Most of the P in grains is in the form of phytates (often called phytin). Unlike monogastric animals, cattle with a functional rumen effectively utilize the phytate P (see Section 5.14). The P content of forages is extremely variable. On an average, legumes contain more P than grasses (Table 5.15) (Underwood, 1966). Likewise, P is much higher in immature forages and declines sharply with maturity. As the protein falls, so does the P content (Underwood, 1966).

Without supplementation, except for salt, P deficiency is the most widespread

TABLE 5.16
Sources of Supplemental Phosphorus and Calcium for Dairy Cattle Feeds with Typical Contents of P, Ca, F, and Certain Other Minerals[a]

Source	P (%)	Ca (%)	F (%)	Mg (%)	Na (%)	K (%)	Fe (ppm)	Mn (ppm)	Zn (ppm)	Cu (ppm)
Dicalcium phosphate	18.5	20.0	0.18	0.6	0.08	0.07	10,000	300	220	80
Defluorinated phosphate	18.0	32.0	0.2	—	4.0	0.09	9,200	220	44	22
Bone meal (steamed)	12.0	24.0	—	0.64	0.46	—	840	30	424	16.3
Phosphoric acid (75%)	23.8	—	—	—	—	—	5	—	—	—
Diammonium phosphate (18% nitrogen)	20.0	0.5	0.2	0.45	0.04	—	15,000	500	300	80
Monammonium phosphate (11% nitrogen)	24.0	0.5	0.24	0.45	0.06	—	12,000	500	300	80
Monodicalcium phosphate	21.0	16.0	0.15	0.5	0.05	0.06	7,000	220	210	70
Sodium phosphate, monobasic	21.8	—	—	—	32.3	—	—	—	—	—
Sodium tripolyphosphate	25.0	—	0.02	—	31.0	—	42	—	—	—
Curacao phosphate	14.0	36.0	—	—	—	—	—	—	—	—
Calcium carbonate	—[b]	38.0	—	0.5	0.06	0.06	336	279	—	24

[a] Adapted from Allen (1977).
[b] Dash indicates data not available or not pertinent.

and economically important mineral problem in grazing cattle or in those fed predominantly on forages (Underwood, 1966). Thus, in feeding dairy cattle, it is most important to be sure that the diet contains adequate P.

Two of the most used P supplements are dicalcium phosphates and defluorinated rock phosphate. Smaller amounts of steamed bone meal are utilized. In manufactured feeds and liquid feeds supplements, phosphoric acid and/or ammonium polyphosphates are frequently used (Miller and Stake, 1972). Likewise, diammonium phosphate, disodium phosphate, monosodium phosphate, monoammonium phosphate, and sodium tripolyphosphate are sometimes utilized as sources of feed P. The non-Ca compounds are useful when there is reason to avoid addition of Ca to the ration. Although research information is far from complete, apparently the P in the above supplemental sources is relatively well utilized by dairy cattle.

The amounts of other minerals supplied by P supplements varies greatly. Several P supplements contain substantial Ca; often more Ca than the P (Table 5.16). Likewise, many P supplements also have appreciable amounts of trace minerals and may provide a major part of the requirements for some elements, including zinc. Excess fluorine in P supplements can present an important practical problem (Table 5.29). Rock phosphates which have not been defluorinated should be avoided in feeding dairy cattle (see Section 5.28).

5.18. MAGNESIUM

Magnesium (Mg) is of considerable practical interest in dairy cattle nutrition because of its central relationship to "grass tetany" which often results in extremely high mortality (see Section 18.2). Chemically and nutritionally Mg is closely related to calcium (Ca). These two elements are in the same group, IIA, of the periodic chart of the atoms, with Mg next to Ca in the series. However, there are major differences between Ca and Mg including much higher amounts of Ca in the body.

About 50–70% of the total Mg is in the bones of cattle, with the remainder about equally distributed between muscle and other soft tissues (Miller et al., 1972a). Normally the total Mg content of an animal is closely related to the Ca and nitrogen contents. Total Mg in growing animals can be estimated by adding 1/45th of total Ca to 1/140th of the total nitrogen (Blaxter and Rook, 1954).

Magnesium plays many important functional roles in dairy animals. In the bone, it contributes to the structural integrity of the skeleton. Magnesium is the second most plentiful cation (after potassium) of intracellular fluids. Likewise, it plays a key role as the activator of numerous enzyme systems especially those involving energy transfer and utilization (Krehl, 1967). Among the numerous enzyme systems and reactions activated by Mg are those required in synthesis of fat, protein, nucleic acids, and coenzymes; muscle contractions; and numerous other essential biochemical reactions (Miller *et al.*, 1972a; Wacker and Parisi, 1968).

Although only about 1% of the total Mg is in the extracellular fluid (blood plasma and interstitial fluid), this Mg bathes the body cells and is of great importance. If absorption or mobilization from tissues is inadequate, this fraction can drop very rapidly. When the Mg in the extracellular fluid declines substantially below normal, the consequences are quite serious. Grass tetany is a result of too little Mg in this fluid.

In dairy cattle two types of Mg deficiencies occur (Miller *et al.*, 1972a). The least prevalent is that in calves given an all-milk diet or in animals otherwise fed insufficient Mg for an extended time until body reserves are depleted. The more prevalent Mg deficiency, "grass tetany," usually develops before there is any material depletion of body reserves. Grass tetany is often a major problem in lactating cows grazing lush, highly fertilized pastures during cool seasons. The time of year varies in different sections of the world.

The Mg reserves are primarily in the bones. As animals become older, the amount of reserves which can be easily mobilized decreases. This is believed to be the reason for the much higher incidence of grass tetany in older cows (Fig. 5.13). A parallel is the higher incidence of milk fever in older cows, also greatly influenced by the lower readily mobilizable Ca in the bones of older animals (see Sections 5.16 and 18.1).

Symptoms or signs of Mg deficiency in calves caused by long-time feeding of a deficient diet included reduced appetite, greatly increased excitability, calcification of soft tissue, and hyperemia (increased blood supply) (Blaxter *et al.*, 1954; Moore *et al.*, 1938). The calf also is susceptible to convulsions that are erroneously called tetany, falling on its side with legs alternately relaxed and rigidly extended. Death often occurs during the convulsions. Profuse salivation and frothing at the mouth may be evident. Symptoms of grass tetany are similar.

The much higher incidence of grass tetany in lactating cows relative to nonlac-

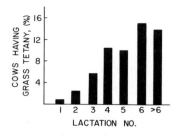

Fig. 5.13 The incidence of grass tetany is higher in older cows. (Adapted from Blaxter and McGill, 1956.)

tating animals is caused by the relatively high amount of Mg in milk (about 0.013%). Thus the requirement for Mg, expressed as a percentage of the diet, increases with the level of milk production. Likewise, because of tissue deposition, in young animals the Mg requirement is increased by more rapid growth.

The availability of Mg to dairy cattle varies greatly and is an important factor in the amount required in the diet. Availability values up to 70% have been observed in young, milk-fed calves, but in older calves these decline to 30–50% (Peeler, 1972; Rook and Storry, 1962). Because of experimental problems in determining "true availability" of Mg, most research data represent apparent availabilities (net absorption which is true absorption less endogenous fecal excretion) (Miller *et al.*, 1972a). In this discussion, and in most of the scientific literature, when availability of Mg in feeds is discussed it means apparent availability.

Usually Mg in grains and concentrates is more available to cattle (about 30–40%) than that in forages (Miller *et al.*, 1972a, Peeler, 1972). This along with the substantial amounts of grain usually fed to lactating dairy cows is one of the major reasons why grass tetany is less prevalent among dairy cows than beef cows in the United States. Usually Mg in preserved forages is more available than in pastures. In sharp contrast to most nutrients, Mg absorption is lowest in young, succulent, highly digestible pasture forage and increases with maturity (Table 5.17). This is contrary to the situation with most nutrients. The apparent availability of Mg in pasture forage in British and Dutch studies ranged from 7–33% with an average of 17% (Blaxter and McGill, 1956; Kemp, 1963).

While the reasons for the low availability of Mg in lush, young forages are not completely clear, several factors seem to contribute; including high potassium, elevated concentrations of certain long-chain fatty acids, high nitrogen (or protein), and possible other organic acids such as citric or transaconitic (Miller *et*

TABLE 5.17
Effect of Growth Stage on Magnesium Metabolism[a]

Forage composition			
Crude protein (%):	26	18	14
Magnesium (%):	0.15	0.12	0.11
Magnesium			
Intake (gm/day)	15.9	14.8	13.7
In feces (gm/day)	14.3	12.5	11.0
In milk (gm/day)	2.2	2.0	1.9
In urine (gm/day)	0.57	0.69	0.98
Available (%)	10	16	20

[a] Adapted from Kemp *et al.* (1961).

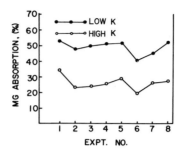

Fig. 5.14 High levels of dietary potassium lowers the apparent absorption of magnesium. (Adapted from Fontenot, 1972; Newton *et al.*, 1972.)

al., 1972a). Of these factors the importance of high potassium probably is most clearly established (Fontenot, 1972) Fig. 5.14).

Within practical intake ranges, absorption of Mg is not materially affected by the level of intake (Miller, 1975a). This is in sharp contrast to its sister element, Ca. A moderate excess of dietary Mg does not accumulate in tissues but is excreted in the urine (Fig. 5.15). This excretion along with the ability to mobilize

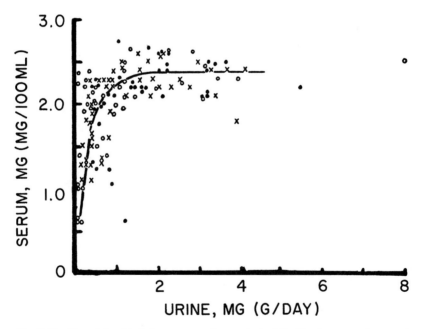

Fig. 5.15 The relationship between magnesium content of the blood serum and magnesium excretion in the urine. ×, ○, and ● represent data from different sources. (Adapted from Kemp *et al.*, 1961.)

some reserves from the bone provide the primary homeostatic control routes for Mg. The ability to withdraw Mg from bone stores is limited both in amounts and rate. Over an extended period of borderline deficiency, appreciable Mg depletion from bone can occur (Miller *et al.*, 1972a).

Because of the many factors which influence the Mg requirement of dairy cattle under practical conditions, it is not possible to select a dietary level that is adequate for some major practical situations without frequently having more than needed. Feeding a moderate excess of Mg is more desirable than too little, because of the ability to eliminate via urine some excess Mg, and the less effective homeostatic control with low intakes (Miller, 1975a,b).

The suggested dietary magnesium requirement is 0.07% in the diet of young calves, increasing to 0.20 percent in the diet of lactating cows fed substantial amounts of preserved forages and/or concentrates (see Appendix, Table 3) (NRC, 1978). Under conditions conducive to grass tetany (i.e., most of the nutrients from lush, highly fertilized pastures in cool seasons) for high producing lactating cows, 0.25% or more dietary Mg is suggested. In these situations it is generally wise to provide some supplemental Mg in a readily available form such as magnesium oxide.

Toxicity of Mg is not known to be a practical problem with dairy cattle. Apparently, dairy cattle can tolerate comparatively high intakes of Mg for extended periods without harm. For instance, 0.6% added magnesium, as magnesium oxide, has been fed with low-roughage rations to alleviate low milk-fat problems, with no adverse effects (Miller *et al.*, 1972a). Recent studies in our laboratory indicate that 1.3% Mg substantially reduces the growth rate of calves. One of the major effects of excess Mg is diarrhea (Fig. 5.16). The adverse effects of high amounts of Mg on the palatability of feeds decreases the probability of actue toxicity. There is need for further research to establish the maximum safe levels of Mg and the effects of toxicity in dairy cattle.

Probably magnesium oxide is used more than other sources of supplemental Mg for dairy cows. Although the data are not unequivocal, magnesium sulfate (Epsom salts) may be less satisfactory due to its laxative effect. Recent research in our laboratory has shown that excess magnesium oxide also is very laxative. The Mg of dolomitic limestone appears to have a low availability (Fontenot, 1972). The Mg in the carbonate and chloride forms apparently has a good availability.

In evaluating the availability of Mg compounds, it is important to utilize measures that will give definitive answers. To do this, an understanding of the homeostatic control mechanisms is essential. In some earlier research, blood plasma Mg levels have been used as an index of availability. Because much of the excess Mg is excreted via urine, feeding a more available Mg source may not increase plasma levels substantially. Thus, total urinary excretion of Mg is a very useful measure in determining the true availability of supplemental Mg sources.

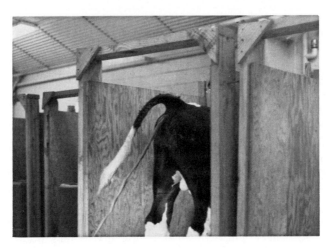

Fig. 5.16 One of the first indications of excessive amounts of magnesium in diarrhea. This shows a calf fed a diet containing a high level of magnesium as magnesium oxide. (Courtesy of R. P. Gentry, Univ. of Georgia.)

Unlike common salt (sodium chloride), most Mg salts are quite unpalatable. An important practical aspect in feeding supplemental Mg is combining it with other palatable ingredients. Various combinations of magnesium oxide with salt, protein supplements, molasses, other concentrate ingredients, and other feeds have been used or studied extensively.

In diagnosing the Mg status of dairy cattle, blood plasma Mg and urinary Mg are key biochemical measures (Miller and Stake, 1974b). Although the clinical signs of grass tetany are caused by inadequate Mg in plasma and other extracellular fluids, often low plasma Mg does not result in tetany. Even though the clinical signs of grass tetany are fairly characteristic, plasma Mg is very useful in confirming the diagnosis. In cattle, the normal range of plasma (or serum—the two are essentially interchangeable for this use) Mg is about 1.8–3.2 mg/100 ml (Underwood, 1966). Most lactating cows having tetany will have plasma levels below 1.0 mg per 100 ml (Underwood, 1966). However, often cows with less than this level will not have tetany.

One of the most important uses of plasma and urine Mg determinations is in diagnosing whether grass tetany is likely to occur (Miller and Stake, 1974b). Low plasma Mg for a high percentage of the lactating animals suggests the possibility of an outbreak of grass tetany and has been used as the basis of an early warning system in Europe. Urinary excretion of Mg is a threshold phenomenon with very little Mg eliminated this way until intake is above requirements. Most of excess intake is excreted in the urine. Accordingly, a low Mg content in the urine, indicates intake is deficient or barely adequate (Miller and Stake, 1974b). Even with grass tetany, Mg levels in bone, most soft tissues, and milk remain within normal ranges.

5.19. SODIUM AND CHLORINE (COMMON SALT)

The importance of common salt (sodium chloride, NaCl) in the diet of cattle has been recognized for thousands of years and is included in biblical references. Homer called salt "divine"; "Today salt is the basis for huge industrial complexes..." (Krehl, 1966). Certainly, in the practical feeding of dairy cattle, "salt" has been recognized much longer than any other mineral as having a crucial role. Even with this long history, there are still major areas of salt nutrition which are not generally understood. For example, it is widely, but erroneously, believed that cattle will only eat sufficient salt to meet their requirement. Since salt is very palatable, cattle will voluntarily consume far more than enough to meet their minimum nutritional needs.

Sodium and chlorine have major functional roles in animals. Sodium, in ionic form, is the predominant cation of the extracellular fluid including blood plasma, whereas the Cl ion fills this role for the anions. The levels of these electrolytes are well regulated in extracellular fluids. Correct amounts of Na and Cl in these fluids are crucial to the maintenance of normal fluid volume, pH, acid–base balance, and osmotic relationships. Sodium has a major role in the transmission of nerve impulses, and in maintaining proper muscle and heart contractions. Likewise, many enzyme systems function in a Na environment. The substantial amount of Na in bones contributes to hardening of the outer layers, but bones contain very little Cl. Chlorine is a key element in gastric juice and an essential component of the chloride–bicarbonate shift.

Most dairy cattle are given supplemental salt. Without added salt, a substantial percentage of the dairy cattle would not receive sufficient Na. Even with the long history of feeding salt, it was not until fairly recently that Aines and Smith (1957) established that Na was the deficient elements when no salt was fed. Although Cl has many essential functions, a deficiency has not been observed or experimentally produced in ruminants. One of the main reasons Na is often deficient in unsupplemented dairy cattle feeds is the low content of Na in most natural feeds (Tables 5.15 and 5.18). The forage in many areas, especially where rainfall is low, often is quite high in salt. In such areas, including parts of the Western United States and much of Australia, supplemental dietary salt is not needed.

The clinical symptoms of a Na deficiency are well described in research publications with a substantial portion of the information obtained by Babcock (1905) at the University of Wisconsin more than 70 years ago. The earliest signs

TABLE 5.18
Typical Sodium, Chlorine, and Potassium Contents of Forages[a]

Type	Sodium	Chlorine	Potassium
Legumes	0.08	0.4	2.4
Grasses	0.14	0.5	2.0

[a] Adapted from Underwood (1966) p. 15. Data given as % in dry matter.

observed when Na is deficient is an intense craving (pica) for salt evidenced by chewing and licking of various materials such as wood, urine, and soil (Fig. 5.17) (Smith and Aines, 1959). This extreme carving is observed in as little as 2 to 3 weeks after salt is removed from the feed (Babcock, 1905; Underwood, 1966). Other signs may not develop for many months. The long delay is due to the remarkable ability of cattle to conserve salt by reducing NaCl excretion in urine, sweat, and feces to an extremely low level. The length of time before additional symptoms of Na deficiency occur is influenced by the degree of the deficiency and the level of milk production. Since milk contains a relatively high Na content, high-producing cows become deficient much sooner.

Signs of Na (salt) deficiency which develop after a long depletion time include a decrease or loss of appetite, an unthrifty haggard appearance, lusterless eyes, rough hair coat, decreased milk production, and a rapid weight loss, or lower growth rate in young animals, (Figs. 5.18, 5.19). Terminal symptoms include shivering, incoordination, general weakness, cardiac arrythmia, and death (NRC, 1978). If salt is provided before the terminal stage of collapse, recovery is rapid and dramatic.

Fig. 5.17 A sodium (salt)-deficient cow licking run-off from a manure pile (Smith and Aines, 1959). (Courtesy of S. E. Smith, Cornell Univ.)

Fig. 5.18 The cow in the top photo prior to being fed a salt (sodium)-deficient diet. Bottom photo shows the same cow after being fed a practical-type diet without supplemental salt for one year. (Smith and Aines, 1959.) (Courtesy of S. E. Smith, Cornell Univ.)

Fig. 5.19 A cow with an extreme sodium (salt) deficiency showing a very gaunt appearance after a large loss of weight. This cow was fed silage, hay, and concentrates without supplemental salt. (Smith and Aines, 1959.) (Courtesy of S. E. Smith, Cornell Univ.)

Since a deficiency of Cl has never been observed in cattle, the signs of such a deficiency are not known. As is the case with so many of the essential mineral elements, the Na and Cl content of the milk are not materially affected by the amount in the feed. The Na content of milk is quite variable (ARC, 1965) and apparently is genetically controlled (Kemp, 1964; Schellner et al., 1971). Average Na values in milk from several studies varied from 0.042 to 0.105% (overall average of 0.06%). (ARC, 1965). Similarly the average Cl content of milk is about 0.115%.

Requirements for sodium (Na) are increased in hot climates where substantial amounts of sweat are produced. Likewise, Na needs are higher with lactation because of the amount going into milk, and in rapidly growing young animals due to deposition in tissues. About 0.135% of body gain in cattle is Na and some 0.094% is Cl (ARC, 1965). Bone contains about 0.4% Na; body fluids, 0.35%; muscle, 0.07%; and fat, very little (ARC, 1965). The estimated Na requirement for nonlactating dairy cattle is 0.10%, equivalent to 0.25% salt, of the dry matter in the total diet (see Appendix, Table 3). Because of the large amount of Na in milk, the requirement for lactating cows is about 0.18%, equivalent to 0.46% salt, of the dry matter in the feed. The requirements for Cl are not known but are less than that in the salt necessary to meet the Na requirements (less than 0.28% Cl for lactating cows and 0.15% for other dairy cattle) (NRC, 1978).

The Na content of forages varies widely and is substantially influenced by the level in the soil (Underwood, 1966). Generally the Na content is much higher in soils having a low rainfall. Tropical forages often are very low in Na. Most feeds contain far less Na than potassium, especially in humid regions (Table 5.15). Na content tends to be reduced with high levels of potassium fertilization, especially on sandy soils. Although the addition of Na to soils will increase Na in forages, it is much easier to supply Na by feeding salt. Most cereal grains are very low in Na (Underwood, 1966).

Although cattle develop a craving for salt and a depraved appetite for abnormal objects, comparable signs are caused by other abnormalities. Thus they are not reliable tools for diagnosing a Na deficiency. Procedures have been developed in the Netherlands for diagnosing a Na deficiency (Table 5.19) (Netherlands Committee on Mineral Nutrition, 1973). This procedure is based on the sharp decrease in the Na content of the saliva and the accompanying increase in potassium when dietary Na is inadequate. Saliva samples are easily collected with small sponges.

With inadequate dietary Na, the amount excreted in urine declines sharply, but because of the huge fluctuations in urine output, total collections for a number of days are needed to provide a good diagnosis (Netherlands Committee on Mineral Nutrition, 1973). This approach requires considerable effort.

Any supplemental Na needed by dairy cattle can be readily and economically

TABLE 5.19
Sodium Status of Cattle from the Sodium and Potassium Content of the Saliva[a]

Sodium (%)	Potassium (%)	Diagnosis
Over 0.3	Less than 0.05	Adequate, normal
0.2–0.3	0.05–0.15	Borderline—from inadequate to normal. No clinical signs
0.1–0.2	0.15–0.25	Inadequate but usually no gross clinical deficiency signs
Below 0.1	Above 0.25	Marked deficiency with possible other severe clinical symptoms

[a] Adapted from Netherlands Committee on Mineral Nutrition (1973).

supplied by common salt in any one of several convenient ways. Often salt is included in the concentrate. Likewise, it can be offered free choice either as granular (loose) salt or as block salt. Although cattle will consume more granular salt, they will eat sufficient block salt to meet their requirements (Smith *et al.,* 1953). Because salt is much more palatable than many other mineral elements, often other minerals are included with it in a mixture fed free choice. Likewise, frequently the trace minerals are mixed with salt and the combination fed as trace mineralized salt or the salt may be included in a liquid feed supplement. Although the usual way of providing supplemental sodium is as common salt, Na in other Na compounds also will supply the needs of dairy cattle.

Most or all of the Na and Cl of the usual feeds appears to be absorbed by cattle. Unlike many of the other mineral elements, the availability of these elements does not present a problem. In some materials such as bones, a part of the Na is in unavailable chemical combinations. Cattle have remarkably good homeostatic control mechanism(s) for conserving Na and Cl when dietary intake is low, and for eliminating any excess. This is mainly by reducing urinary excretions to extremely low levels. Likewise, the recycling of blood Na back to the rumen via saliva is an important conservation mechanism. The reserve supply in the rumen is carefully conserved. Cattle dispose of excess Na and Cl via urinary excretion.

Relative to the amount required, cattle tolerate a comparatively high amount of salt in the diet, especially when water is readily available. Nine percent salt in the total ration did not adversely affect steers (Meyer *et al.,* 1955). However, much less salt in the water will produce toxicity. With heifers, 1% salt in the water caused no adverse effects but 1.2% was harmful (Weeth and Haverland, 1961). Although the maximum amount of salt which can be safely consumed by dairy cattle has not been clearly established, it is suggested that salt (NaCl) not exceed 5% of the total dry matter intake. This is more than 10 times the amount needed in any practical diet to meet the needs for optimum performance.

5.20. POTASSIUM

Until quite recently, it was generally thought that there was relatively little possibility of a potassium (K) deficiency in practical dairy cattle feeding (ARC, 1965; Ward, 1966). Present information indicates that dairy cattle require more K than any other mineral element. Potassium is closely related chemically and nutritionally to sodium (Na) which occupies the adjoining position in the periodic chart of the elements. Even so, there are major differences between K and Na. Where Na is the dominant cation of the extracellular (outside and between the cells) fluids, K is the major cation within the cells (intracellular fluids) (Krehl, 1966; Ward, 1966). Potassium performs many of the functions inside the cells that Na does outside, including contributing to proper acid–base relationships, osmotic balances, and activation of various intracellular enzymes. Correct amounts of K in cells and fluids are essential for normal contraction and relaxation of the heart and other muscles.

The metabolism of K is similar to that of Na. Essentially all the dietary K is absorbed and most of the excess is excreted in the urine. The body does not maintain very much K reserve. High K concentrations occur in muscle and liver but far lower levels are found in tissue fluids and bones (ARC, 1965). Because nearly all the K is found in the lean tissue of animals, the total lean body mass can be estimated from the K content by determining the amount of the radioactive isotope ^{40}K naturally present in the body (Ward, 1966). At birth the calf contains an average of about 0.20% K with subsequent gain having about 0.16% (ARC, 1965; Hogan and Nierman, 1927). Milk contains an average of 0.15% K (ARC, 1965; Hemken, 1975a). Although the K of milk may vary by as much as 50% between cows, it remains quite constant for the same cow (Ward, 1966).

Current information suggests that severe K deficiencies do not occur frequently in dairy cattle fed most practical rations (Hemken, 1975a). It seems possible that borderline deficiencies could be fairly prevalent, but a borderline K deficiency is difficult to diagnose (Hemken, 1975a). In lactating cows, the signs of a relatively severe K deficiency include a marked decrease in feed intake, reduced weight gains, decreased milk production, pica (depraved appetite), loss of hair glossiness, decreased pliability of hides, lower plasma potassium, slightly reduced milk K, and higher hematocrit values (Fig. 5.20) (Pradham and Hemken, 1968). Severe deficiencies have occurred on diets containing 0.06–0.15% K in the dry matter. With a borderline K deficiency (observed with 0.45–0.55% K in dry diet), the most noticeable sign is decreased voluntary consumption of feed

Fig. 5.20 Potassium-deficient cow. The symptoms resemble, somewhat, those of an underfed cow. (Courtesy of R. W. Hemken, Univ. of Kentucky.)

(Dennis *et al.,* 1976). In other animal species, K deficiency has been associated with overall muscular weakness and poor intestinal tone.

The K requirement of lactating dairy cows is about 0.8% of the dry diet (Dennis *et al.,* 1976; Pradham and Hemken, 1968; NRC, 1978; Ward, 1966). Although adequate data are not available to reach a firm conclusion, the requirement probably is similar in other dairy cattle (NRC, 1978). The indicated requirement of 0.8% is higher than the 0.5% suggested as being adequate some years ago (Ward, 1966).

Most forages contain a much higher percentage of K than needed by dairy cattle (Tables 5.15, 5.18), but in many concentrates, the content is below the suggested requirement levels. Thus, rations consisting primarily of concentrates may not have sufficient K to meet the requirement (Hemken, 1975a). The K content decreases with advancing maturity of forages and apparently can be reduced substantially by leaching of mature standing forages in humid areas. Potassium is highly soluble in water and occurs in an uncombined form in many biological materials.

Young, very lush forages grown on highly fertilized, especially with K, soils in cool weather may have an extremely high K content frequently above 3% of the dry matter. The high K content in such forage appears to be somewhat antagonistic to magnesium absorption and/or utilization (Fontenot, 1972). Thus, high-K forage is a factor in grass tetany of lactating cows (Kemp, 1960; Ward, 1966; Miller *et al.,* 1972a). In situations where grass tetany is not a problem, the maximum level of K which can be safely fed to dairy cattle has not been determined (Ward, 1966).

5.21. SULFUR

Sulfur (S) is a component of the essential amino acids, methionine and cystine; and thus is closely associated with protein nutrition. Likewise, S is a part of the vitamins thiamin and biotin and of sulfated polysaccharides including chondroitin. Chondroitin is a key component of cartilage, bone, tendons, and blood vessel walls. Also, S is a constituent of other important compounds including the anticoagulant, herparin; glutathione, which functions as a coenzyme and in detoxification reactions; thiocyanates; and taurine, that is necessary in fat digestion. The total body tissues of cattle contain about 0.15% S and milk, 0.03% (NRC, 1978).

Development in recent years have increased the need for S additions to practical dairy rations. Two of the most important changes are increased feeding of corn silage, which often is very low in S, and the expanded use of nonprotein nitrogen. Natural proteins contain appreciable S but urea and other nonprotein nitrogen compounds do not. Although the impact is not as clear, other changes in agricultural practices are tending to reduce S in some feeds. One such change is the use of higher purity phosphate fertilizers (Allaway, 1969). Shifting from the use of ordinary superphosphate to triple superphosphate eliminates a source of available S (Allaway, 1969). Likewise, efforts to decrease air pollution reduce the amount of S being added to some soils. In certain areas of the northwestern United States and of Australia, soils are especially low in S and often produce low S feeds.

Symptoms of a sulfur deficiency in dairy cattle are not specific (Elam, 1975; NRC, 1978). Accordingly, it is difficult to diagnose a deficiency, especially a borderline one. Some of the effects associated with inadequate S intake include decreased feed consumption, lower weight gains, depressed milk production, lower rumen-microbial protein synthesis, reduced numbers of rumen microorganisms, and decreased rumen digestion of cellulose and/or fiber (Elam, 1975; NRC, 1978). Likewise, several blood changes have been noted in S-deficient ruminants including reduced serum sulfate, increased blood urea, and accumulation of lactate in the rumen (Elam, 1975).

Apparently, the most critical S requirement in dairy cattle is for optimal rumen microbial growth. Among the essential mineral elements, S appears to be relatively unique in this respect: generally the requirements of rumen microbes for minerals are lower than those of dairy cattle for other functions. In this respect, S nutrition and metabolism resemble nitrogen more than it does most mineral elements. The rumen microbes require S in the synthesis of amino acids and proteins.

Requirements of dairy cattle for S are not well defined. The S requirement of dairy cattle can be approached from a consideration of the nitrogen to S ratio and/or from performance results in feeding experiments. Since most natural protein sources generally have been thought to supply adequate S with protein, often S supplementation has been related to the use of nonprotein nitrogen. In earlier years, supplying one part S for each 15 of nitrogen (N) was generally recommended (Elam, 1975). The ratio of N to S in rumen protein is about 15 to 1 (Pope, 1971). However, in work with sheep, improved nitrogen retention was observed when dietary S was increased to lower the N to S ratio from 12:1 to 9.5:1 (Moir *et al.*, 1968). Accordingly, a N to S ratio of about 10 to 1 frequently has been recommended (Elam, 1975; NRC, 1978; Pope, 1971). In lactating dairy cows, a N to S ratio of 12 to 1 gave sufficient S to attain maximum feed intake (Bouchard and Conrad, 1973a). Reasons suggested for a ratio of N to S narrower than 15:1 include less recycling of S than of nitrogen in the ruminant body and the wastage of S associated with dissimilatory bacteria in the rumen (Pope, 1971).

A variety of different types of feeding experiments have been conducted with the results often not entirely definitive (Elam, 1975; NRC, 1978). For example, in Cornell University work (Grieve *et al.*, 1973) adding supplemental S to corn silage rations containing 0.11 and 0.13% S in dry matter did not significantly increase milk production or feed intake. However, with one treatment added S increased daily milk production by 4.4 lb per cow. In University of Kentucky research, adding supplemental S to a corn silage and concentrate ration containing 0.10% S in dry matter increased milk production and feed consumption (Jacobson *et al.*, 1967).

Since the S requirement for optimum microbial action appears to be the highest need for S of dairy cattle, often researchers have studied the effects of S on rumen function. In one such study, the optimum S level for cellulose digestion *in vitro* (in glass containers in the laboratory) was 0.16–0.24% of the dry matter (Bull and Vandersall, 1973). The estimated S requirement for lactating dairy cows is 0.20% in the average diet (NRC, 1978). The requirement for nonlactating dairy cattle was calculated from the minimum protein requirement with a N to S ratio of 12:1 (NRC, 1978).

In corn silage, often the S content is far lower than the required level. The amount in corn silage samples analyzed in the Georgia Forage Testing Laboratory in 1973–1974 averaged 0.07% in dry matter with a range of 0.005%–0.19% (Table 5.4). More than 80% of the corn silages had less than half (0.10%) the estimated needed level of 0.20% for lactating cows.

Excessive sulfur may overload the urinary excretion system and reduce feed intake and other measures of performance such as weight gains and milk production (Elam, 1975; NRC, 1978). Likewise, too much S interferes with the metabolism of certain other minerals, especially selenium (NRC, 1978; Pope, 1971). Very excessive dietary S levels may cause acute toxicity resulting in symptoms such as abdominal pain, muscle twitching, diarrhea, severe dehydration, strong odor of sulfide on the breath, congested lungs, and acute enteritis (inflammation of the digestion tract).

The maximum amount of S which can be safely fed is even less well defined than the maximum requirement (Bouchard and Conrad, 1974; NRC, 1978). Maximum safe levels are affected by several factors, including the chemical form of sulfur. Thus it appears that until more complete information is available, generally the maximum amount of S should be about 0.35% of the diet with no more than 0.20% from added sulfate S (NRC, 1978). However, dairy cattle can tolerate more S from natural feed ingredients than from added sulfate. Thus levels higher than 0.35% do not invariably cause adverse effects. Chalupa et al. (1971) fed Holstein calves 1.7% S as elemental S and observed no adverse effects.

Several compounds can be utilized as sources of supplemental S (Table 5.20). The ability of ruminants to utilize both organic and inorganic S is well established (Block and Stekol, 1950; Elam, 1975; NRC, 1978; Thomas et al., 1951). Inorganic sources usually are much lower in cost. Often some form of sulfate S is used (NRC, 1978). The S in sodium sulfate, calcium sulfate, potassium and magnesium sulfate, and ammonium sulfate appears to be readily available and roughly comparable (Elam, 1975; Bouchard and Conrad, 1973a,b,c; Bull and Vandersall, 1973; Johnson et al., 1971). Sulfur as the highly insoluble elemental sulfur (S) or lignin sulfonate is much less available (Albert et al., 1956; Elam, 1975; Bouchard and Conrad, 1973b). It has been suggested that elemental S, flowers of sulfur, is utilized about one-third as efficiently as the S in sulfate and methionine (Elam, 1975; Albert et al., 1956). Methionine and methionine analog S is readily utilized by dairy cattle, but the cost of S in this form is prohibitively expensive (Elam, 1975). Special aspects of methionine as an amino acid are discussed in section 3.9.

Several interactions of S with other nutrients are important. For instance, adding dietary sulfate has resulted in a higher requirement for selenium in lambs (Elam, 1975; Pope, 1971). The additional sulfate increases the selenium excretion in the urine (Elam, 1975; Ganther and Baumann, 1962). Interference of S and molybdenum with copper utilization is discussed in Section 5.24. Other interactions of S with calcium, phosphorus and potassium have been observed (Elam, 1975). The vast literature on S nutrition has been reviewed by several authors including many in a symposium (Muth and Oldfield, 1970; Garrigus, 1970; Moir, 1970; Whanger and Matrone, 1970).

5.22. IRON

The importance of iron in the formation of blood has been known since the seventeenth century (Underwood, 1971). Thus, there is a vast scientific literature

TABLE 5.20
Some Sources of Mg, K, S, and Trace Elements Often Used in Formulating Feeds[a]

Compound	Mg (%)	K (%)	S (%)	Fe (%)	Zn (%)	Cu (%)	Mn (%)	Co (%)	Se (%)	Na (%)
Magnesium sulfate (MgSO$_2$·7H$_2$O)	10.0	—	13.0	—	—	—	—	—	—	—
Magnesium sulfate (MgSO$_4$)	20.0	—	26.6	—	—	—	—	—	—	—
Potassium sulfate (K$_2$SO$_4$)[c]	—	44.8	18.3	—	—	—	—	—	—	—
Ferrous sulfate (FeSO$_4$·7H$_2$O)	0.05	—[b]	11.0	21.0	0.01	0.01	0.12	—	—	—
Ferrous carbonate (FeCO$_3$)	0.31	—	0.4	43.0	—	0.3	0.35	—	—	—
Copper sulfate (CuSO$_4$·5H$_2$O)	—	—	—	—	—	25.0	—	—	—	—
Cupric carbonate (CuCO$_3$)	—	—	—	—	—	53.0	—	—	—	—
Cupric oxide (CuO)	—	—	—	—	—	75.0	—	—	—	—
Cobalt sulfate (CoSO$_4$·7H$_2$O)	0.04	—	—	0.001	—	0.001	0.002	21.0	—	—
Cobalt carbonate (CoCO$_3$)	—	—	—	—	—	—	—	45.0	—	—
Cobalt sulfate (CoSO$_4$·H$_2$O)	0.06	—	—	0.001	—	0.001	0.003	33.0	—	—
Zinc sulfate (ZnSO$_4$·H$_2$O)	—	—	—	—	36.0	—	—	—	—	—
Zinc oxide (ZnO)	0.5	—	1.0	0.8	73.0	0.07	0.01	—	—	—
Manganese sulfate (MnSO$_4$·H$_2$O)	0.3	—	19.0	0.04	—	—	25.0	—	—	—
Manganous oxide (MnO)	2.4	—	—	3.4	0.42	0.2	60.0	—	—	—
Sodium selenite (Na$_2$SeO$_3$)	—	—	—	—	—	—	—	—	45.6	26.6
Sodium selenate (Na$_2$SeO$_4$)	—	—	—	—	—	—	—	—	41.8	24.3

[a] Adapted from Allen (1977).
[b] Dash indicates data not available or not pertinent.
[c] Other sources of sulfur used include sodium sulfate, ammonium sulfate, and calcium sulfate.

on this essential trace element. Perhaps, a key factor contributing to the early discovery of the importance of iron is its role in giving blood the red color. A major function of iron is as part of the hemoglobin, an iron–protein complex which carries oxygen from the lungs to tissue and carbon dioxide on the return trip.

Although a major portion of the total body iron is in hemoglobin, iron also plays a key role in other enzymes involved in oxygen transport and the oxidative processes including catalase, peroxidases, flavoprotein enzymes, and cytochromes. Likewise, iron is a part of myoglobin which transports oxygen in muscles. Iron makes up almost 5.0% of the earth's crust (Table 5.1) but the content in the whole body of cattle is only about 50–60 ppm (fresh weight basis) (ARC, 1965; Blaxter et al., 1957). The iron level in cow's milk is very low and does not change materially with increased dietary iron (Underwood, 1977). In one study (Blaxter et al., 1957) milk as it comes from the udder varied from 0.18 to 0.31 ppm iron. In contrast, values after the contamination of usual handling may be substantially higher with an average of about 0.5 ppm (Underwood, 1977).

The main problem involving iron in the practical feeding of dairy cattle is the deficiency which occurs when calves are fed only milk for a substantial period (NRC, 1978; Blaxter et al., 1957; Möllerberg, 1974; Underwood, 1971). While the iron reserves of the newborn calf are quite variable, generally they are sufficient to prevent serious iron-deficiency anemia, if dry feeds are fed beginning within a few weeks after birth (NRC, 1978).

Iron deficiency seldom occurs in older cattle (Fig. 5.21) unless there is considerable blood loss from some nonnutritional cause such as parasitic infestations or disease (NRC, 1978; Underwood, 1977). Most feeds as well as dirt, especially clays, contain considerable iron with many having tremendously more than minimum needs. Likewise, when adequate iron is fed, there is considerable storage reserve of iron in an animal. Feeding an iron-deficient diet for an extended period is required to deplete these reserves before a deficiency develops. During this period iron absorption increases, enabling animals to better use the dietary iron.

Anemia is the best known symptom of an iron deficiency in calves. However, reserves are depleted before serious anemia develops. Other nutritional deficiencies including inadequate copper or cobalt, or excess molybdenum, zinc, selenium, lead, or cadmium also can cause anemia. Symptoms of iron deficiency, in addition to anemia and related blood changes, include lower weight gains, listlessness, inability to withstand circulatory strain, labored breathing after mild exercise, reduced appetite, decreased resistance to infection, atrophy of the papillae of the tongue, blanching of visible mucous membranes, and a pale color of the muscle meat (NRC, 1978). Pale color, the traditional "trademark" of good veal (MacDougall et al., 1973; NRC, 1978; Niedermeier et al., 1959) is evidence of anemia. Whereas, this pale color per se has no beneficial effect to the

Fig. 5.21 This weak (iron-deficient) 12-year-old cow, which was grazing on a Blanton fine sand (yellow) soil in Florida, had to be helped up. Her hemoglobin was only 4.8 gm/100 ml of whole blood. After being given supplemental iron (ferric ammonium citrate), the hemoglobin value increased to 12.6 gm/100 ml, and she regained body condition and strength. (Courtesy of R. B. Becker, Univ. of Florida.) (See Becker *et al.*, 1965.)

consumer, it is not easily attained on a practical basis except with an all-milk or milk-replacer diet (see Chapter 14). Adequate supplemental iron will eliminate the pale color. In veal calves some supplemental iron increases rate of gain and reduces the incidence of diarrhea (Möllerberg, 1974).

Iron requirements of dairy cattle are not well defined (NRC, 1978; Underwood, 1977). One of the major reasons is the difficulty in determining the amounts needed. Whereas, the sharp changes in iron absorption and the substantial amounts of body stores are tremendously beneficial in permitting cattle to avoid a deficiency, these homeostatic control mechanisms greatly complicate establishing the amounts needed. The change in absorption percentage is illustrated by data from North Carolina (Matrone *et al.*, 1957): when 30 mg of iron was fed, calves utilized 60%; but with 60 mg, only 30%. Frequently, in research on iron requirements and deficiency effects, relatively insensitive measures such as growth rate, anemia, hemoglobin, and packed cell volume have been used to determine iron status. A measure of "adequate" iron reserves, such as percent saturation of transferrin, is a much more sensitive way of detecting an iron deficiency (Miller and Stake, 1974a,b).

In one experiment with veal calves, 40 ppm iron in the dry diet prevented severe anemia but was inadequate to avoid an initial drop in blood hemoglobin and a low transferrin saturation (Bremner and Dalgarno, 1973b). Apparently synthesis of hemoglobin takes priority over needs for myoglobin and some iron-dependent enzymes (Bremner and Dalgarno, 1973b; MacDougall et al., 1973). An iron level of 100 ppm in the dry diet is sufficient to meet all the needs of the calves to 3 months of age with 50 ppm adequate for other dairy cattle (NRC, 1978). If a pale-colored veal is desired, the mild deficiency resulting from 30–40 ppm dietary iron may make this a more desirable level (NRC, 1978).

The exact requirement would vary with age of cattle, growth rate, and availability of the iron source as well as with the criteria used to measure adequacy. The higher requirement in the young calf is caused by the greater amount of muscle deposited per pound gain and the higher feed efficiency (NRC, 1978). Iron absorption or availability varies with level of intake. Generally, the iron in the soluble compounds, ferrous sulfate, and ferric citrate are more available than that in ferrous carbonate and much more available than iron in ferric oxide and iron phytate (Ammerman et al., 1967; Bremner and Dalgarno, 1973a).

Iron toxicity is not a common problem in feeding dairy cattle, but, as with all other nutrients, a sufficiently high iron level in feed or water is detrimental (NRC, 1978). Among the first effects in cattle are lower feed intake, weight gains, and milk production. The amount of iron needed to produce undesirable changes depends on the source of iron and whether the level of other nutrients which might be adversely affected are already near the deficiency borderline. Apparently, cattle can safely consume substantially more iron from natural feed sources than from soluble sources such as ferrous sulfate (NRC, 1978; Standish et al., 1968, 1969). Under most conditions, cattle can tolerate 1000 ppm dietary iron, especially if the iron is from natural feed sources and adequate amounts of other minerals are fed (NRC, 1978).

5.23. ZINC

Except for iron, the average concentration of zinc (Zn) at about 20 ppm on a fresh basis in the body of dairy animals is much higher than that of any other trace element (Table 5.1). In contrast to many trace elements, zinc is fairly evenly distributed among the tissues and organs. Although some tissues have materially more zinc than others, the relative differences are much smaller than for most essential trace elements. Generally zinc is closely associated with protein and skeletal tissues, with relatively little in the fat and lipids. As with most trace mineral elements, zinc rarely occurs alone in the animal but is combined with protein or other organic compounds. Zinc functions largely or entirely in enzyme systems including both metalloenzymes and metal–enzyme complexes (see Section 5.4) (Miller, 1970b). As an essential part of these enzyme systems, zinc plays a key role in carbohydrate metabolism, protein synthesis, nucleic acid metabolism, and in many other biochemical reactions in the body.

Although a severe zinc deficiency causes definite pathological effects in cattle, relatively little is known of the specific enzymatic impairments responsible for these pathological changes (Miller, 1970b). Lowered feed intake is one of the first changes observed when a dairy animal is given a very zinc-deficient ration (Miller, 1970b). Zinc-deficient cattle grow more slowly due both to the decreased feed consumption and less efficient feed utilization (Miller, 1970b; Miller et al., 1965b). The reduced feed efficiency results from less efficient utilization of digested nutrients and not from lower digestibility (Table 5.21) (Miller, 1970b; Miller et al., 1966; Somers and Underwood, 1969). Other changes occurring in severely zinc-deficient cattle include retarded testicular growth and development (Fig. 5.22); skin parakeratosis that generally is most severe on the legs, neck, head, and around the nostrils; loss of hair; general debility; lethargy; increased susceptibility to infections; unthrifty appearance; stiffness of joints; gnashing of teeth; and excessive salivation (Figs. 5.23, 5.24, and 5.25) (Kirschgessner and Schwarz, 1975; J. K. Miller and Miller, 1962; W. J. Miller, 1970b).

When a zinc-deficient animal is given supplemental zinc, there is a very rapid and dramatic recovery (Fig. 5.23) (Miller, 1970b). A person familiar with the animals can easily observe major improvements in the appearance including decreased lethargy of a very zinc-deficient calf within 24 hours after supplemental zinc is fed. Feed consumption increases materially in 2 or 3 days (Miller and

TABLE 5.21
Dry Matter Digestibility, Nitrogen Balances, and Growth in Normal and Zinc Deficient Young Ruminants

Animal	Normal	Zinc deficient
Ram lambs[a]		
Dry matter intake (gm/day)	489	489
Dry matter digestibility (%)	66.8	64.5
Protein intake (gm/day)	77.5	77.5
Nitrogen in feces (gm/day)	4.5	4.6
Nitrogen in urine (gm/day)	3.7	5.8
Nitrogen balance (gm/day)	4.2	2.0
Weight gain (gm/day)	47	26
Holstein calves[b]		
Dry matter digestibility (%)	87.6	86.5
Goat kids[b]		
Dry matter digestibility (%)	83.4	84.4

[a] Adapted from Somers and Underwood (1969).
[b] Adapted from Miller et al. (1966).

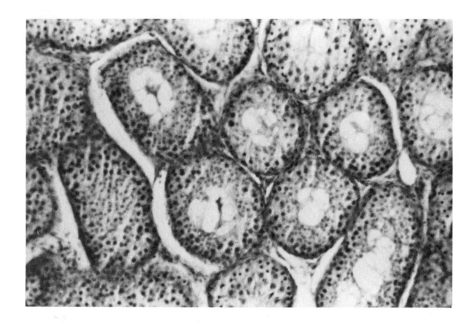

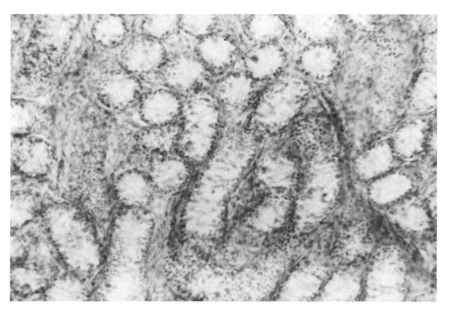

Fig. 5.22 Top photo shows testes of a normal goat. In bottom photo, note degeneration associated with a zinc deficiency. (Courtesy of W. J. Miller, Univ. of Georgia.) (See Neathery *et al.*, 1973a.)

Fig. 5.23   Top photo shows a zinc-deficient calf which had received a diet containing 3.6 ppm zinc. In bottom photo is the same calf after being fed supplemental zinc for 5 weeks. (J. K. Miller and Miller, 1962.) (Courtesy of W. J. Miller, Univ. of Georgia.)

Miller, 1962). In zinc deficient-rats, feed intake increased within 4 hours after zinc was fed, indicating a major biochemical defect had been quickly corrected.

An hereditary zinc deficiency caused by a simple recessive gene has been observed in a small percentage of Dutch Friesian cattle fed normal rations (Andresen *et al.,* 1970; Kroneman *et al.,* 1975). Clinically and biochemically this deficiency is almost identical to that produced experimentally with a low-zinc diet (Fig. 5.25) (Miller, 1970b, 1971b). In genetically normal dairy cattle, a severe zinc deficiency appears to be relatively rare under practical conditions.

The extent to which a borderline deficiency may occur in dairy cattle fed practical diets has not been clearly established. One reason is the absence of effective biochemical measures for determining a borderline deficiency (Miller

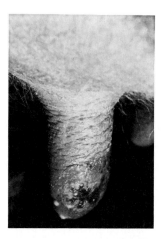

Fig. 5.24   Teat of zinc-deficient cow. Note the skin lesions similar to those on other parts of the body of zinc-deficient calves. (Courtesy of M. Kirchgessner, Univ. Munich, West Germany.) (Kirchgessner and Schwartz, 1975.)

Fig. 5.25   A Dutch Fresian calf showing typical skin lesions and other signs of a zinc deficiency. This calf was fed adequate zinc. The deficiency was due to a genetic defect. (Courtesy of J. Kroneman, Univ. of Utrecht, Netherlands.)

and Stake, 1974a,b). In contrast to the very drastic clinical symptoms associated with a severe deficiency, no comparable effects are evident in a borderline deficiency. When zinc is marginally inadequate, effects such as small reductions in feed intake, growth rate, resistance to infections, and possibly a decreased reproductive efficiency may be expected (Miller, 1970b).

Under experimental conditions, many biochemical changes have been observed in severely zinc-deficient cattle (Miller and Stake, 1974b). However, no good biochemical measures for diagnosing deficiencies under field conditions have been developed (Miller and Stake, 1974a,b). Those with some promise include plasma zinc, hair zinc, bone zinc, and alkaline phosphatase in plasma or other tissues (Kirchgessner and Schwarz, 1975; Miller and Stake, 1974a,b). These measures have considerable individual variability and are affected by factors other than zinc status. Another measure which varies with zinc status in cattle is uptake of radioactive zinc (^{65}Zn) by red blood cells or other tissues *in vitro* (Chesters and Will, 1978; Rosas and Bell, 1969; Miller and Stake, 1974a,b).

When cattle are fed a low-zinc diet, the percentages of the dietary *zinc absorbed* increases, with values as high as 80% sometimes observed (Kirchgessner and Schwarz, 1976; Miller, 1970b, 1975a; Schwarz and Kirchgessner, 1975). Likewise, with a high-zinc diet, there is a reduction in the percentage absorbed (sometimes to less than 10%). However, the absolute amounts of zinc absorbed (grams per day) increase or decrease with changing intake. As with calcium, these changes in absorption percentage take place rather quickly (within a few days) and enable the animal to meet its zinc needs with wide differences in amounts consumed. With a severe deficiency, further increases in absorption percentage may continue for several weeks (Kirchgessner and Schwarz, 1976). Although detailed mechanisms have not been established, the changes in zinc absorption from the intestine appear to be controlled by events in other parts of the body (Miller, 1975b).

Zinc absorption percentage is related rather closely to the needs of the animal (Kirchgessner et al., 1977; Miller, 1975b). A rapidly growing animal or one with high milk production absorbs more zinc than those having lower needs. This explains the decreasing absorption usually observed in older cattle (Kirchgessner and Schwarz, 1976; W. J. Miller et al., 1968; Miller, 1975a).

Decreases in *endogenous zinc excretion* also contribute to zinc homeostasis in cattle. Likewise, with a low-zinc diet, zinc in a few tissues including liver and kidney will decline to a small extent. There is a large decline in hair zinc taking place over long time periods. In contrast, with a very deficient diet, zinc in plasma may decline within a few days.

In calves, feeding a high-zinc diet results in a huge (several hundred percent) increase in the zinc content of a few tissues including the liver, pancreas and kidney (Miller et al., 1970). In contrast, no such increase occurs in mature cows

(Table 5.22) (Kincaid *et al.*, 1976; Miller *et al.*, 1977). Likewise, the zinc content of a few tissues including muscle and heart are not affected by dietary intake (Miller, 1973). In contrast to calcium and iron, cattle cannot obtain most of their zinc needs from body stores for an extended period of time. Thus, regardless of the amount of zinc fed earlier, with an extremely deficient diet, cattle may develop a deficiency within a very few weeks.

Average zinc content in the milk is about 4 ppm, on a fresh basis (Miller, 1970b), but there are material differences between cows (Neathery *et al.*, 1973c). The role of genetics in this cow difference has not been established. With a high-zinc ration, milk zinc increases to about twice normal but apparently plateaus at about 700 ppm dietary zinc with further increases having little effect (Miller *et al.*, 1965a). When cows were fed a low-zinc diet (17 ppm compared to 40 ppm for controls), the zinc content of milk decreased by about 25% (Table 5.23). With a zinc-deficient diet (6 ppm), the reduction in milk zinc is greater (Kirchgessner and Schwarz, 1975). Adjustments in milk zinc content and absorption percentage take place rapidly when dietary zinc is changed, with much of the change within one week (Neathery *et al.*, 1973c).

Most of the zinc in milk is associated with the protein, especially the casein, and very little with the fat (Parkash and Jenness, 1967). Apparently, the udder is a major control organ for zinc entering milk (Miller, 1970b; Miller *et al.*, 1965a).

Because of the large changes occurring in the percentage of zinc absorbed, the factorial method is not very effective in determining zinc requirements of dairy cattle. Most of the information on zinc requirements has been obtained in feeding experiments, but results have not been completely consistent. With young Holstein bulls, no benefit was obtained by increasing zinc beyond 9–10 ppm (Miller *et al.*, 1963). In some experiments, but not all, increased weight gains and feed efficiency were observed when supplemental zinc was added to rations containing 18–29 ppm zinc (Perry *et al.*, 1968; Zurcher, 1970). Similarly, in University of Georgia studies, lactating dairy cows fed 17 ppm zinc performed fully as well

TABLE 5.22

Liver Zinc when 600 ppm Added Zinc Was Fed for 21 Days to Different Animals[a]

	Zn (ppm) fresh	
	Control	+ 600 (ppm) Zn
Calf	32	150
Cow	57	45
Rat	25	29
Chick	23	21

[a] Adapted from Miller *et al.* (1978).

TABLE 5.23
Effect of Ration Zinc on Zinc in Milk[a]

Main ration ingredients	Ration zinc (ppm)	Milk zinc (ppm)	% dietary zinc in milk
Experiment 1			
Beet pulp	17	3.3	18
Beet pulp	39	4.2	9
Experiment 2			
Silage, grain	44	4.2	12
Silage, grain	372	6.7	2.2
Silage, grain	692	8.0	1.6
Silage, grain	1279	8.4	0.8

[a] Adapted from W. J. Miller (1975a).

as those given 40 ppm (Neathery *et al.*, 1973c). In contrast, adding zinc to a diet containing 25 ppm increased milk production in a South Dakota study (Voelker *et al.*, 1969). An entirely satisfactory explanation for the differences observed in the suggested zinc requirements in various experiments is not evident. However, it seems desirable to supply enough zinc to meet the highest indicated needs. Using this approach, the estimated zinc requirement of dairy cattle is about 40 ppm in the dry matter of the diet (NRC, 1978).

The maximum safe level of zinc for dairy cattle is far higher than the minimum requirement. With lactating dairy cattle, 1279 ppm zinc, dry matter basis, did not reduce performance (Miller *et al.*, 1965a). In growing cattle, 500 ppm zinc did not adversely affect performance but in one experiment, 900 ppm zinc decreased weight gains and feed efficiency (Ott *et al.*, 1966). Growing cattle developed a depraved appetite characterized by excessive consumption of salt and other minerals, and wood chewing on a 1700 ppm zinc diet (Ott *et al.*, 1966). Diarrhea, drowsiness, and paresis (paralysis) may result from excessive zinc intake (Van Ulsen, 1973). Based on the above and other similar information, the estimated maximum safe level of zinc is at least 500 ppm in young cattle and 1000 ppm in older cattle (NRC, 1978). The maximum safe level probably is affected by the amount of copper and iron in the diet as excessive zinc may aggravate borderline deficiencies of these elements.

Zinc is widely distributed in feeds with relatively few grains or forages consistently deficient in zinc. However, there are a few exceptions. As with most mineral elements, urea contains little or no zinc. Gelatin and egg whites are very low in zinc, but are rarely fed to cattle, except experimentally. Most other high-protein feeds contain substantial amounts of zinc. In areas such as part of Western Australia, where the soil is very low in zinc, some widely used feeds may be very deficient.

On a practical basis, several changes taking place in agricultural practices are increasing the probability of a borderline zinc deficiency when no supplemental zinc is fed. The use of galvanized pipe and equipment which can be a significant source of zinc, has greatly decreased. Likewise, the increased use of nonprotein nitrogen lowers the zinc content of feeds compared to most natural proteins. Higher rates of animal performance increase the zinc requirement. Also, liming the land decreases the zinc content of feed produced.

Although there is considerable research showing major interrelationships between zinc and other nutrients in nonruminants, the data for ruminants are much more limited. For example, the major decrease in zinc absorption when soybean meal is fed to swine does not occur with ruminants (J. K. Miller, 1967).

5.24. COPPER AND MOLYBDENUM

Because of their close interdependence in the practical feeding of dairy cattle, copper (Cu), and molybdenum (Mo) are discussed together. Deficiencies appear to be the major practical problem with Cu. In contrast, the only known practical problem with Mo for dairy cattle is a toxicity. A Cu deficiency can result from too little Cu per se or from interfering substances, particularly Mo and/or sulfate. Although the Cu requirement is affected by several factors, it is influenced most by the Mo content of the feed. Similarly, the amount of Mo which can be tolerated is greatly influenced by the Cu intake.

The importance of Cu in the practical nutrition of ruminants and other animals has been recognized since the 1930's. Accordingly, this trace element has been the subject of considerable research. Initially, Cu was shown to be essentially for hemoglobin (in red blood cells) formation and therefore necessary to prevent anemia (Underwood, 1966, 1977). Subsequent research demonstrated that copper is crucial in the pigmentation process of hair, bone formation, synthesis of myoglobin (the protein of muscles), reproduction, normal heart function, formation of connective tissues, and myelination of the spinal cord (Underwood, 1977). At the cellular level, Cu is a highly versatile element. It is essential in numerous oxidative enzymes including tyrosinase, uricase, ascorbic acid oxidase, cytochrome oxidase, and lysyl oxidase (Underwood, 1966, 1977).

In contrast to the many key roles played by Cu, the only established function of Mo is as an indispensible component of the flavorprotein enzyme, xanthine oxidase. Although the absolute requirement of dairy cattle for Mo has not been determined, it is believed to be very low; perhaps no more than 1 ppm. Adequate Mo is of considerable practical importance in plant growth. At least part of this importance is due to its role in nitrate reductase, which is involved in nitrogen metabolism of legumes.

The clinical effects of a Cu deficiency vary materially with different species of animals and to some degree within the same species (Underwood, 1977). In cattle many of the symptoms of a Cu deficiency are nonspecific including reduced growth or loss in weight; decreased milk production, and unthriftiness (Allcroft and Lewis, 1957; Becker *et al.*, 1953; Netherlands Committee on Mineral Nutrition, 1973; NRC, 1978; Suttle and Angus, 1976; Underwood, 1977). When Cu is deficient, the performance of dairy cattle may be substantially below normal even though there are no obvious deficiency signs other than nonspecific unthriftiness. Following geochemical surveys which revealed Mo enriched areas in certain parts of England, Cu supplementation increased growth rate by 10–70% in young dairy animals (Thornton *et al.*, 1972). Apparently no problem had been suspected.

Fig. 5.26 The heifer in the upper photograph was copper deficient. Although receiving iron, she was anemic. The bottom photo shows the same animal after receiving supplemental copper. (Becker *et al.*, 1953, 1965.) (Courtesy of R. B. Becker, Univ. of Florida.)

With an extreme Cu deficiency, often there is severe diarrhea followed by rapid loss of weight; cessation of growth; a change in hair coat color which may be dull, faded, bleached, graying, and in which white hair may turn dirty-yellowish, or black hair become brownish; a change in hair texture and appearance (tending to be open and rough over the withers); swelling of the leg bones at the ends, especially above the pasterns; fragile bones which may result in multiple fracture of ribs, femur or humerus; stiff joints that may cause a "pacing gait" in older cattle; delayed or subdued estrus and decreased reproduction; calving difficulties and retained placenta; calves born with congenital rickets; sudden death, or falling disease, caused by acute heart failure; and anemia (Figs. 5.26, 5.27, and 5.28). Owing to losses of pigment, the black hair around a cow's eye sometimes may develop a gray-spectacled appearance that apparently is specific for copper deficiency (Netherlands Committee on Mineral Nutrition, 1973). A deficiency of Mo has not been observed or experimentally produced in cattle (NRC, 1978).

Generally, most of the Cu deficiencies and/or Mo toxicities occur in grazing animals, especially heifers, or in calves fed only milk for long periods (Ammerman, 1970; Netherlands Committee on Mineral Nutrition, 1973; Underwood, 1966). Copper deficiencies are much less likely in cattle eating dried forages and occur even more rarely when concentrates make up a substantial part of the ration. As with iron, cattle are able to store large reserves of Cu in the liver.

Fig. 5.27 A dry cow showing a pacing gait characterized by an apparent stiffness in the hocks and springiness in the pasterns typical of cows grazing certain Florida muck soil (marsh ranges) containing high molybdenum levels. In advanced cases animals have difficulty breathing after minor exertion. The condition can be prevented or corrected with proper amounts of copper to offset the high molybdenum in the forage. (Becker *et al.*, 1965.) (Courtesy of R. B. Becker, Univ. of Florida.)

Fig. 5.28 This cow which was extremely short of breath after mild exercise, had grazed on Okeechobee muck soil marsh where the forage typically contains excessive molybdenum. The condition can be prevented by suitable amounts of copper. (Courtesy, R. B. Becker, Univ. of Florida.) (See Becker *et al.*, 1965.)

Cow's milk is quite low in Cu; but if the dam's intake of copper is adequate, the newborn calf will have a substantial store of Cu (400–500 ppm) in the liver (Netherlands Committee on Mineral Nutrition, 1973).

In sharp contrast to zinc, manganese, and some other elements, good biochemical procedures are available for diagnosing a Cu deficiency in dairy cattle (Miller and Stake, 1974a,b). Liver Cu is the best criterion of Cu status (Miller and Stake, 1974a,b; Netherlands Committee on Mineral Nutrition, 1973). The standards established for liver in the Netherlands are given in Table 5.24. Because of the greater ease in taking samples, blood plasma Cu is often used. Plasma Cu can be used to indicate a deficiency but does not reflect higher Cu stores in the liver (Hartmans, 1974). The normal range of plasma Cu for cattle is 0.5 to 1.5 ppm with values consistently below 0.5 indicating a deficiency

TABLE 5.24
Determining Copper Status from Liver Copper[a]

Newborn calf	Yearling	Heifer or cow	Assessment
Cu in liver (ppm in dry matter)			
400	200	150	Normal values in spring, unlikely to lead to severe deficiency in autumn
	50	25	Marginal, acceptable only at the end of the grazing season
<50	<20	<10	Severe deficiency, usually clinical signs

[a] Adapted from Netherlands Committee on Animal Nutrition (1973).

(Miller and Stake, 1974a,b; Underwood, 1971). The values are often listed as mg/liter or μg/ml which are the same numbers as ppm. These standards may be further refined for specific areas as done in the Netherlands (Table 5.25).

As discussed above, Cu and most other essential trace mineral elements function mainly through enzyme systems. In the diagnosis of trace element deficiencies, the use of enzymes can, in some situations, offer important advantages including (1) less sensitivity to contamination or artifacts thus requiring less care in sample handling, (2) more rapid and/or less expensive analyses, and/or (3) lower detection limits or greater precision.

The Cu status of cattle has been ascertained by measuring serum ceruloplasmin activity. Ceruloplasmin is a true oxidase enzyme, ferroxidase, synthesized in the liver. In cattle a high percentage of the copper in plasma or in serum is in the form of ceruloplasmin; thus there is a high correlation between ceruloplasmin and serum copper (Miller and Stake, 1974a,b; Todd, 1970; Underwood, 1977).

The minimum Cu requirement of dairy cattle is quite variable because of the great influence of other substances including Mo. Another major factor affecting the Cu requirement and/or the maximum safe level of Mo is the amount of sulfate or substances which can be converted to sulfate. The interaction between Cu, Mo, and sulfate is complex and not fully understood. As Underwood (1977, p. 125) has said, "sulfate can either aggravate or ameliorate the toxic effects of Mo, depending on the Cu status of the animal."

In addition to Mo and sulfate, other factors can influence the Cu requirement of cattle. For instance, more Cu is required from pasture than when dry forage or concentrates are fed. Presumably this indicates lower availability of pasture Cu. Copper in silage appears to be intermediate in availability. A level of 10 ppm Cu in the feed dry matter is estimated to be the minimum requirement under usual conditions (NRC, 1978). Even though 4 ppm Cu will meet the requirement under some conditions, more than 10 ppm Cu may be needed if pastures contain high concentrations of Mo or other interfering substances (Hartmans, 1974; Underwood, 1977).

Although not a major practical problem in dairy cattle, acute Cu toxicity can occur if excessive amounts of supplemental Cu or feeds contaminated with Cu

TABLE 5.25
Evaluation of Copper Status from Blood Plasma Copper Taken in Autumn[a]

Cu in blood plasma (ppm fresh)			
Calf	Yearling	Heifer or cow	Assessment
0.70–1.00	0.65–1.00	0.75–1.20	Adequate
	0.50–0.65	0.60–0.75	Slight deficiency
<0.50	0.30–0.50	0.40–0.60	Clear deficiency
	<0.30	<0.40	Severe deficiency

[a] Adapted from Netherlands Committee on Animal Nutrition (1973).

compounds are consumed (NCR, 1978; Underwood, 1977). Copper is widely used for many agricultural and industrial purposes. When cattle consume excessive Cu, large amounts accumulate in the liver before obvious toxicity symptoms become evident. The toxicity symptoms are due to the sudden liberation of considerable Cu from the liver to the blood causing what is termed the "hemolytic crisis." This condition is characterized by considerable hemolysis, jaundice, methemoglobinemia, hemoglobinuria, widespread necrosis, and often death (Allcroft and Lewis, 1957; Todd and Thompason, 1965; Underwood, 1977). A substantial increase in certain serum enzymes, including glutamic oxaloacetic transaminase (SGOT) and lactic dehydrogenase (LDH), precedes the "hemolytic crisis" by several weeks (Underwood, 1977) (see Section 5.27).

The maximum Cu which cattle can safely tolerate is not well defined (NRC, 1978). The amount also is influenced by the method of administration. For example, 4.8 gm of Cu (as cupric sulfate, $CuSO_4$) given to steers dry in gelatin capsules daily for more than one year caused no toxicity symptoms. When administered as a liquid drench, the same amount caused death within 60 days (Chapman et al., 1962). Just as the minimum requirement of Cu is increased by high dietary Mo, the tolerance level also is greater with more Mo. Cattle safely tolerate about 70–100 ppm Cu for long periods of time and even higher levels for short periods, such as a few weeks (NRC, 1978).

The most sensitive aspect of feeding high levels of Cu to dairy cows may be the effect on oxidized flavor in milk (Haase and Dunkley, 1970; Underwood, 1971). It is well known that adding Cu to milk accelerates the development of oxidized flavor. The Cu content of milk of dairy cows generally varies below 0.1 ppm with many values between 0.04 and 0.07 ppm but amounts as high as 0.2 ppm are not unusual soon after calving. Inadequate Cu intake reduces milk Cu (Underwood, 1977).

Published literature is somewhat conflicting relative to the effects of feeding high levels of Cu on the Cu content of milk and the development of oxidative flavor (Underwood, 1971; Haase and Dunkley, 1970; King and Dunkley, 1959). Drenching cows with large doses of copper sufate increased both the Cu and oxidized flavor in milk (King and Dunkley, 1959). Feeding cows 1.28 gm of supplemental Cu per day (as copper sulfate or copper EDTA), for a total copper content of about 93 ppm in the dry matter, increased Cu content of milk, but did not affect the oxidative stability of the milk (Dunkley et al., 1968). Generally, milk Cu does not appear to be greatly increased by feeding moderate, additional Cu (Underwood, 1977). Although rations containing 80 ppm Cu would not appear to increase the susceptibility of milk to oxidized flavor (NRC, 1978), until more definitive data are available, probably it is wise to avoid feeding substantially more Cu than required to prevent a borderline deficiency.

Among farm animals, cattle are the least tolerant of high Mo. Thus, in many areas of the world Mo toxicity is an important practical problem (Underwood, 1977). Forage grown on poorly drained soils, especially granite alluviums or black shales, and on high organic soils such as peats and mucks are likely to contain more Mo (Fig. 5.29) (Davis *et al.*, 1974; Underwood, 1977). Likewise, Mo availability to plants increases with a higher soil pH, while Cu uptake decreases with increasing soil pH.

In the animal body, Mo and Cu are antagonistic to each other. High Mo levels interfere with Cu metabolism. The relative amounts of Mo and Cu in feed are crucial in determining the occurrence of Mo toxicity. If copper is high, the tolerance for Mo is higher, but as Cu decreases, so does the tolerance to Mo. To a considerable extent, the major symptoms of Mo toxicity are those of Cu deficiency. However, diarrhea may be somewhat more prominent. Thus, the names of "teartness" or "peat scours" have been used to describe the Mo toxicity in various parts of Australia, Europe, New Zealand, and the United States.

Whereas, 3 to 5 ppm or less Mo is regarded as normal, typical "teart" pastures may contain from 20 to 100 ppm on a dry basis (Underwood, 1977; Cunningham, 1960). With Mo toxicity, over a period of time there may be a disturbance of phosphorus metabolism, with lameness, joint abnormalities, and osteoporosis (Underwood, 1977). Other dietary constituents modify the degree of Mo toxicity. A normal level of sulfate provides some protection against high Mo by increasing excretion (Cunningham *et al.*, 1959; Underwood, 1977; Vanderveen and Keener, 1964). Also, natural protein and substances capable of oxi-

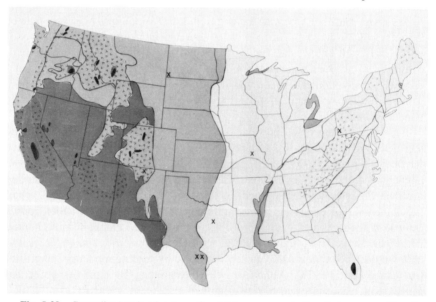

Fig. 5.29 Generalized regional pattern of molybdenum concentrations in legumes and its relationship to distribution of molybdenum-toxic areas in the United States. ■ , Areas with background levels of 6 to 8 ppm of Mo; ▧ , areas with background levels of 2 to 4 ppm of Mo; □ , areas with background levels of 1 ppm or less of Mo; ✔ , general location of naturally occurring molybdenumtoxic areas; x , general location of industrial molybdenosis. (Courtesy, J. Kubota, SCS, United States Plant, Soil and Nutrition Laboratory, Ithaca, New York.)

dation to sulfate appear to have a similar effect. The chemical form of Mo may have an important effect of its toxicity. For example, Mo in pasture may be much more toxic than a similar amount experimentally fed (Cunningham, 1950; Vanderveen and Keener, 1964).

5.25. COBALT

Cobalt, although having only one well-known essential function, is of tremendous practical importance in ruminant animals. Prior to discovery of the need for Co in the 1930's in Australia, ruminants could not be successfully produced on large acreages in many parts of the world owing to severe Co deficiencies. The deficiencies were widespread in grazing ruminants on a wide range of soils and climates in areas from the tropics to cool temperate regions (Underwood, 1966).

Acute Co deficiency had many names in different locations including coast disease (South Australia), bush sickness (New Zealand), wasting disease (Western Australia), pining (Great Britain), Grand Traverse disease (Michigan), salt sick, and others (Underwood, 1966, 1971). All of these in some way describe the conditions appropriately termed "enzootic maramus" (meaning a disease indigenous to a certain area characterized by extreme emaciation or loss of flesh) (Underwood, 1966). Borderline deficiencies occur far more frequently than severe deficiencies and are of vastly greater economic importance.

The only clearly defined essential function of Co is as a component of vitamin B_{12} which can be synthesized by rumen microbes. Accordingly, cattle must consume the Co because injected Co is not effective. In this respect Co is unique among the essential mineral elements; all others are required in the body tissues rather than just in the digestive tract.

Cattle cannot store significant amounts of Co in a usable form. Even though vitamin B_{12} synthesis declines quickly when dietary Co is deficient, cattle can store considerable vitamin B_{12}, especially in the liver. This reserve supply of vitamin B_{12} is slowly depleted during a Co deficiency. Thus, when Co intake is inadequate, deficiency symptoms develop slowly, often over a period of several months. Since the essential function of Co is for vitamin B_{12} synthesis, a Co deficiency is really a deficiency of that vitamin. Without adequate Co, the ruminant animal virtually starves itself.

Symptoms of a Co deficiency include decreased appetite and feed consumption, listlessness, decreased growth or loss of weight, and lowered milk production. With an extreme and/or extended deficiency, symptoms can include extreme emaciation or wasting of the muscles, pale skin and mucous membranes, incoordination of the muscles, rough hair coat, a stumbling gait, and a high death rate especially among calves (Fig. 5.30) (Neal and Ahmann, 1937; NRC, 1978' Underwood, 1977). The starvation resulting from the cobalt–vitamin B_{12} deficiency is caused, at least partially, by the inability of the affected animals to

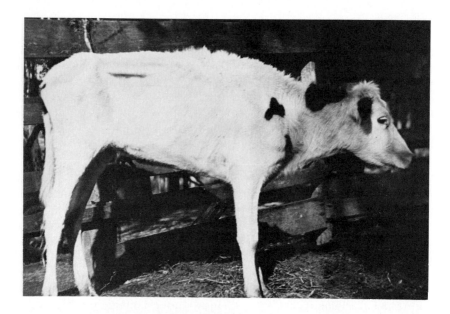

Fig. 5.30 Top photo shows a 12-month-old Holstein heifer with typical cobalt-deficiency symptoms including low blood hemoglobin (6.6 gm per 100 ml). Note the severe emaciation. In bottom photo, the same heifer fully recovered on the same pasture after being given supplemental cobalt had blood hemoglobin of 12.4 gm/100 ml. (See Becker *et al.*, 1965.) (Courtesy of R. B. Becker, Univ. of Florida.)

metabolize propionate (Underwood, 1977). This is associated with inadequate methylmalonyl-Coenzyme-A isomerase, a vitamin B_{12}-dependent enzyme. Propionate is an intermediate product of rumen fermentation (see Chapters 1 and 2). The great dependency of ruminants on volatile fatty acids, including propionate, explains why appetite is more severely affected by a cobalt–vitamin B_{12} deficiency in ruminants than in nonruminants.

The minimum Co requirement of dairy cattle is about 0.10 ppm in dry ration (Ammerman, 1970; NRC, 1978; Underwood, 1977). Since many forages and some concentrates contain less than this level, supplemental Co is needed in many areas. The symptoms of a borderline Co deficiency are largely nonspecific. A positive diagnosis of a deficiency can be attained by measuring the response in appetite, weight gains, and temperaments when supplemental Co is fed. If the unthriftiness was caused by Co deficiency, there will be considerable recovery within a few weeks.

Relatively effective biochemical measures for Co status have been developed (Miller and Stake, 1974a,b; Underwood, 1966, 1977). Liver vitamins B_{12} provides a reliable biochemical assessment of the Co status (Table 5.26). Likewise, determining the Co content of the feed is useful. Values substantially below 0.1 ppm in dry matter indicate Co is inadequate to maintain adequate vitamin B_{12} synthesis. When the intake is above 0.1 ppm, a deficiency should not occur. Since Co deficiency is an area problem (Fig. 5.31), a knowledge of soil (cobalt, pH, etc.) also is useful (Netherlands Committee on Mineral Nutrition, 1973). Extractable soil Co appears to be a very useful measure.

Supplemental Co has been supplied to cattle in a number of ways including feeding in concentrate, trace mineralized salt, and as a Co bullet. Treatment of soil with Co salts has resulted in variable success. Uptake of Co by plants from alkaline soils may be quite low.

Cobalt toxicity is not known to be a practical problem except where excessive amounts of supplemental Co are carelessly or accidentally fed. In one experiment, Co equivalent to approximately 30 ppm in the diet was not detrimental to dairy calves (Keener *et al.,* 1949). This is 300 times the minimum requirement.

TABLE 5.26
Vitamin B_{12} Content of the Liver as Biochemical Indication of the Cobalt Status of Ruminants[a]

Vitamin B_{12} in fresh liver (ppm)	Cobalt status of animal
<0.07	Severe Co deficiency
0.07–0.10	Moderate Co deficiency
0.11–0.19	Mild deficiency
<0.19	Sufficiency

[a] Adapted from Miller and Stake (1974b) and Underwood (1966).

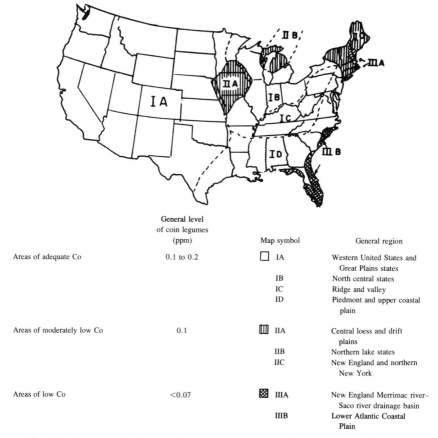

Areas of adequate Co	General level of coin legumes (ppm)	Map symbol	General region
Areas of adequate Co	0.1 to 0.2	☐ IA	Western United States and Great Plains states
		IB	North central states
		IC	Ridge and valley
		ID	Piedmont and upper coastal plain
Areas of moderately low Co	0.1	▥ IIA	Central loess and drift plains
		IIB	Northern lake states
		IIC	New England and northern New York
Areas of low Co	<0.07	▨ IIIA	New England Merrimac river-Saco river drainage basin
		IIIB	Lower Atlantic Coastal Plain

Fig. 5.31 Regional distribution of low, moderately low, and adequate areas of cobalt in the United States. (Courtesy of J. Kubota, SCS, United States Plant, Soil and Nutrition Laboratory, Ithaca, N.Y.) (Kubota and Allaway, 1972.)

The 10 ppm accepted as a safe level of cobalt (NRC, 1978) provides a considerable safety margin. Occasionally toxic amounts of Co have been fed. Symptoms include reduced appetite, slower growth, decreased water intake, lack of muscular coordination, rough hair coat, increased hemoglobin, higher packed red cell volume, and increased liver cobalt (Dickson and Bond, 1974; Ely *et al.*, 1948; Keener *et al.*, 1949; NRC, 1978).

Although the role as a component of vitamin B_{12} is the only essential function established for Co, benefits have been shown in at least one other special situation (Underwood, 1977). Added cobalt protects ruminants from the toxic condition, ''phalaris staggers,'' or Ronpha Staggers, which sometimes occurs when the perennial grass *Pharalis tuberosa* or one of several other related species is eaten (Underwood, 1977). The toxic principal may be an alkaloid. This detoxification role of Co is unrelated to its role in vitamin B_{12}.

5.26. IODINE

Iodine, like cobalt, is required for only one clearly established essential biochemical function, and has great practical importance in cattle feeding. This element is an essential component of the thyroid hormone, thyroxine, which regulates the rate of energy metabolism in all cells of the body. An iodine (I) deficiency, however, affects almost every major aspect of life from reproduction and lactation to the ability to resist stress.

The classical symptom of I deficiency, goiter, occurs in many species of animals. Goiter, an enlargement of the thyroid, often is seen in newborn calves when there is no symptom of inadequate I in the dam (Hemken *et al.*, 1971). Although many calves are born with obvious enlargements of the thyroid, often thyroids four or five times normal size could not be diagnosed definitely until the calf was slaughtered (Hemken, 1971; Hemken *et al.*, 1971). Iodine-deficient calves may be born blind, hairless, weak, or dead with the exact symptoms depending on the severity of the deficiency (Fig. 5.32) (Hemken, 1970; Underwood, 1966). Likewise, an I deficiency may reduce reproduction at any stage causing irregular or suppressed estrus, abortions, stillbirths, and/or increased incidence of retained placentas. Male fertility may also be affected (Underwood, 1966). Dairy cattle fed insufficient I are less able to resist stress and may even have a higher incidence of ketosis (Hemken, 1970).

Results of the University of Maryland experiment are of special interest for several reasons (Hemken, 1971; Hemken *et al.*, 1971). Although appearing normal, cows fed corn silage as the only forage and soybean meal as the protein supplement produced calves with enlarged thyroid glands (Fig. 5.32) and gave 8 pounds per day less milk than controls given supplemental iodine. The unsupplemented cows had more retained placentas, a lower percentage of fat in milk, appeared to be more subject to stress, and were less able to "milk off" the body fat in early lactation (Hemken, 1971).

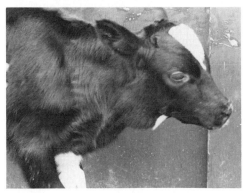

Fig. 5.32 Iodine-deficient calf showing an enlarged thyroid, typical of goiter. (Courtesy, R. W. Hemken, Univ. of Kentucky.)

If supplemental I is not fed, at least a marginal deficiency may occur in many areas. The available I in the soil has a major effect on the I content of plants; thus I deficiency is often an area problem. Regions which are a substantial distance from the ocean, such as the Great Lakes, Great Plains and Western Mountain regions of North America, usually have low soil I. Likewise, many areas close to the ocean, including parts of Maryland, New Zealand, and England, also have low soil I. Generally soil I is low in areas subjected to a glaciation that was recent in geological time, or low rainfall. In addition to the influence soil I has on the I content of plants, the species and variety or strain of plants (Table 5.6), and the fertilization, have a major effect (Underwood, 1971; Hemken, 1970, 1971). Higher nitrogen fertilization reduces I in forages (Hemken, 1970). Normally, forages and roughages contain more I than grains or oil seed meals (Underwood, 1977). Urea contains little or no I.

Perhaps the presence of goitrogenic substances is of equal or greater importance than low feed I as a contributing factor toward I deficiencies. *Brassica* species of plants such as rape, kale, and turnips contain high levels of goitrogens. Goitrogenic substances are much more prevalent in feeds than is generally recognized. For instance, soybean meal is somewhat goitrogenic (Hemken, 1970). The actions of goitrogenic substances vary. The soybean-meal goitrogenic factor(s) interferes with reabsorption of organic I thereby increasing fecal loss. In contrast, some goitrogens decrease the uptake of I by the thyroid.

Essentially all inorganic I in the diet is absorbed by cattle (Hemken, 1970). A substantial portion, 30% when intake just meets minimum requirements, of this will be trapped by the thyroid with a lower percentage on high-I diets (J. K. Miller *et al.*, 1975; NRC, 1978). Although urine is the major excretion route, appreciable I may appear in the feces (Hemken, 1970). Usually I excretion is delayed by considerable recycling through the abomasum (J. K. Miller *et al.*, 1975). Because of the recycling, on a very low-I diet, as much as 65% may be bound by the thyroid (Lengemann and Swanson, 1957).

A number of approaches are useful in determining whether I is deficient. Certainly, if calves are born with enlarged thyroids, an I deficiency is indicated, but a substantial enlargement of the thyroid is needed before this can be detected in the live calf. Observing the response to supplemental I is often considered to be one of the most useful procedures (Netherlands Committee on Mineral Nutrition, 1973). Estimates of the thyroid secretion rate may also be useful in determining iodine status (Miller and Stake, 1974a,b). Analyses of I content of the milk appears to be a relatively easy and quite helpful way of determining whether I intake is adequate (Netherlands Committee on Mineral Nutrition, 1973). Although I content of milk from high-producing cows is lower when dietary intake is sufficient, I in herd milk below 0.020 to 0.025 ppm suggests inadequate dietary I (Hemken, 1970; Iwarsson, 1973; NRC, 1978).

A substantial portion, perhaps 70–80%, of the total I in the animal body is in the thyroid. When the diet is adequate, there will be substantial reserves, so that an extended period on a low-I diet may occur before major deficiency effects appear (Swanson, 1972).

Among the essential mineral elements, I is unique in that I content of the milk increases in almost direct proportion to the amount consumed (see Section 5.7.d and Table 5.12). Approximately 10% of dietary I goes into milk; however the precentage may increase with higher milk production (J. K. Miller *et al.*, 1975). There are indications that the "normal" I content of commercial milk has increased from 0.05 to 0.40 ppm during the last 20 or 30 years (Hemken, 1975b). This increase is apparently due to the use of I for nonnutritional purposes and should be of substantial benefit in alleviating marginal iodine deficiencies in people.

It is estimated that 0.25 ppm iodine in the feed dry matter will meet the minimum needs of growing or nonlactating dairy cattle when no goitrogenic substances are eaten (see Appendix, Table 3). Because of the great amount of I in milk, lactating cows require about 0.5 ppm (see Appendix, Table 3). Likewise, due to the critical importance of I to the fetus, 0.5 ppm is recommended for the last two months of gestation. The I requirement of dairy cattle is dependent on the amount of goitrogenic substances in the diet. If the diet contains as much as 25% strongly goitrogenic feed on dry basis, the I in the feed should be increased at least double (see Appendix, Table 3).

Some inorganic I compounds are unstable and/or volatile, whereas I in some organic compounds is not available or usable by cattle (see Section 5.6.d). Thus, special care should be given to the way supplemental I is given. Iodine compounds such as potassium iodide are strong catalysts and may cause destruction of some vitamins, drugs, and antibiotics (Miller and Stake, 1972).

Potassium iodide, sodium iodide, and calcium iodate are readily available to cattle but will leach or evaporate from salt blocks. Pentacalcium orthoperiodate (PCOP) is equally available to cattle but is not as rapidly lost from salt blocks (J. K. Miller *et al.*, 1968). Supplying I in liquid feed supplements presents special stability, volatilization and solubility problems. EDDI (ethylenediamine dihydriodide) is often used in the liquid feed supplements and apparently meets these needs (Miller and Stake, 1972). Some of the iodine used in teat dipping (or applied to skin) is absorbed and available to meet nutritional needs, etc. (Hemken, 1975b; Underwood, 1971).

The maximum safe dietary level of I for dairy cattle is estimated to be 50 ppm (see Appendix, Table 3) (NRC, 1978). Although the maximum safe levels approach 100 times minimum requirements, I toxicity is of some practical approach with dairy cattle. The principal reason for this is the use of I for pharmaceutical purposes such as the prevention of foot rot. Even though I toxicity in cattle has been reported by veterinarians, its extent is not known.

Iodine-toxicity symptoms include excessive tears from the eyes, more saliva than normal, a watery nasal discharge, tracheal congestion resulting in coughing, decreased feed intake, lowered growth rate, and reduced milk production (NRC, 1978). Care should be taken to be sure that I intake, including that used for nonnutritional purposes, does not exceed the maximum safe level for any sustained period of time. In most practical situations the I content of the feed and the extent of some goitrogenic substance present are not known. Thus, adequate supplemental I generally should be provided to meet the minimum recommended NRC requirements. This can be done for lactating cows with 1% iodized salt containing 0.01% I if concentrates and forages are fed in equal amounts.

5.27. SELENIUM

Although effects of Se toxicity were observed in Chinese farm animals as early as 1295 by Marco Polo, Se was not known to be the causative agent until 1934 (Ammerman and Miller, 1975; Krehl, 1970). The essentiality of Se was not established until 1957 (Schwarz and Foltz, 1957). Selenium deficiency in farm animals is much more widespread and economically important than the toxicity (Ammerman and Miller, 1975; Muth *et al.*, 1967; Underwood, 1971). Because of the long and well-known history of toxicity, providing supplemental Se to meet dietary needs is subject to special and unusual regulatory problems. Although fairly well established (Underwood, 1977), the beneficial effects of adequate Se in protecting animals against cancer are not well known by the general population.

Both deficiencies and toxicities of Se do occur in dairy cattle but the extent, especially of borderline deficiencies, has not been established. Recent research in Ohio has shown that the incidence of retained placenta was greatly reduced by an injection of Se and vitamin E during the dry period (Table 5.27) (Conrad *et al.*, 1976; Julien *et al.*, 1976). Probably a deficiency of Se is not as important a factor in the practical feeding of dairy cattle as with swine and poultry. Although modified by other factors, Se deficiencies and toxicities in farm animals are closely related to the Se content in the feed. The toxicity occurs in limited areas in many countries. In the United States these are predominantly areas of arid and alkaline soils located throughout parts of the western plains and rocky mountain states (Fig. 5.33). Apparently the high Se levels resulted from volcanic eruptions in the geological formation of the soils (Krehl, 1970). Low-Se areas cover much of the Northeastern, Southeastern and Northwestern USA (Fig. 5.33) (Kubota *et al.*, 1967). In the United States feeds grown west of the Mississippi (except the Northwest) usually have sufficient Se to fully meet the needs of dairy cattle.

The amount of Se available for uptake by plants is a major factor in determining the Se content in feeds (Table 5.5). Higher humidity and more acid soils decrease Se content of plants (Allaway, 1974; National Research Council, 1971b).

TABLE 5.27
Effect of Selenium and Vitamin E Treatments on Incidence of Retained Placenta in Parturient Cows[a]

Se in ration (ppm)	Treated[b]			Control		
	No. cows	Retained placenta		No. cows	Retained placenta	
		No.	%		No.	%
0.04	53	6	11.3	39	16	41
0.02	37	4	10.8	23	12	52
0.035	14	0	0	9	7	77.7
—	9	0	0	9	6	66.6
	113	10	8.8[c]	80	41	51.2[c]

[a] Adapted from Julien *et al.* (1976).
[b] Given intramuscular injection of sodium selenite and α-tocopherol acetate (vitamin E) 20, or 20 and 40 days before calving.
[c] Percentage of total.

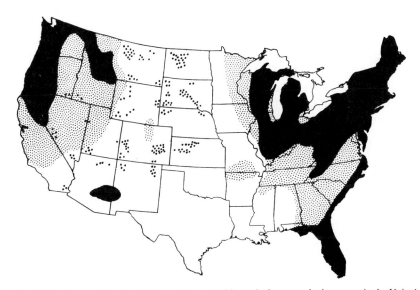

Fig. 5.33 Geographical distribution of low, variable, and adequate selenium areas in the United States. ■, Low—approximately 80% of all forage and grain contain <0.05 ppm of selenium; ▣, variable—approximately 50% contains >0.1 ppm; ☐, adequate—80% of all forages and grain contain >0.1 ppm of selenium; ●, local areas where selenium accumulator plants contain >50 ppm. (Courtesy, J. Kubota, United States Plant, Soil and Nutrition Laboratory, Ithaca, N.Y.) (Kubota and Allaway, 1972.) ("Micronutrients in Agriculture," p. 542, 1972; by permission of the Soil Science Society of America.)

Selenium in plants also is influenced by the plant species, the chemical form of Se in soils, and other factors (Ammerman and Miller, 1975). Special "accumulator" plants may contain extremely high amounts of Se, often exceeding 1000 ppm Se (NRC, 1971b; Underwood, 1966). Except for those near starvation, native dairy cattle rarely eat such accumulator plants. Since dairy cattle seldom experience extreme hunger, Se toxicity is less prevalent than in beef cattle.

Contrary to popular belief, the range between the minimum Se requirements and the maximum safe level is relatively wide; namely, at least 30- to 50-fold. Dairy cattle require about 0.1 ppm Se in the diet (Fig. 5.34) (NRC, 1978). The requirement is appreciably influenced by the chemical form of Se and the levels of interacting factors in the diet including vitamin E, sulfur, lipids, proteins, amino acids, and several microelements (Ammerman and Miller, 1975). The maximum safe level is about 3–5 ppm, depending upon the protein, sulfur, and arsenic contents of the diet (higher amounts of each decreases the toxicity) as well as the chemical form of Se (NRC, 1978). Naturally occurring organic Se of plants is much more toxic than inorganic Se forms used to supplement feeds.

Selenium is essential for such body functions as growth, reproduction, prevention of various diseases, and protecting the integrity of muscles. This element is an essential component of the enzyme, glutathione peroxidase (Rotruck *et al.,* 1973). Among other functions, this enzyme aids in protecting cellular and subcellular membranes from oxidative damage (Hoekstra, 1975). Selenium probably is involved in other important enzyme systems, other biochemical functions, and many metabolic processes (Allaway, 1974; Hoekstra, 1974).

The best known syndrome of severe Se deficiency in cattle is "white muscle

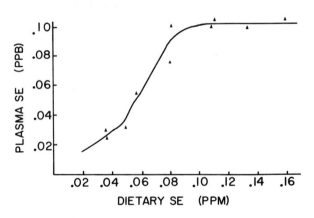

Fig. 5.34 Plasma selenium of dairy cows as affected by changes in dietary selenium. Note the plateau in plasma selenium when cows are fed more than about 0.1 ppm needed to meet the requirement. (Adapted from Conrad *et al.,* 1976.)

Fig. 5.35 Heart muscle from calf with white muscle disease showing extreme calcification which frequently accompanies selenium deficiencies. (Courtesy of O. H. Muth and J. E. Oldfield, Oregon State Univ.)

disease," nutritional muscular dystrophy, in young calves (Ammerman and Miller, 1975; Andrews *et al.,* 1968; Hartley and Grant, 1961; NRC, 1971b). White muscle disease has been reported in many countries including Australia, Bulgaria, Canada, Finland, Italy, Japan, New Zealand, Norway, Russia, Scotland, South Africa, Sweden, and the United States (NRC, 1971b). In this "disease," heart and skeletal muscles develop chalky white striations, degeneration and necrosis (Fig. 5.35) (NRC, 1978). The heart muscle is most frequently affected in calves (Ammerman and Miller, 1975). Likewise, heart failure, paralysis usually of the hind legs, and a dystrophic tongue may be evident. Selenium-responsive unthriftiness ranging from reduced growth to progressive loss of weight, is often associated with diarrhea and may occur in cattle of all ages (Andrews *et al.,* 1968). Likewise, in subclinical Se deficiency, performance may be reduced with slower gains and lower reproductive efficiency involving an increased number of services needed per conception and the birth of premature, weak, or dead calves (Andrews *et al.,* 1968; Underwood, 1977). Also the incidence of retained placenta may be elevated with a borderline deficiency (Conrad *et al.,* 1976).

A number of approaches are useful in diagnosing the possibility of a Se deficiency in cattle. The presence of white muscle disease is evidence of a severe deficiency. Selenium content of certain tissues, of which kidney and liver Se are the best indicators, declines with a deficiency (Ammerman and Miller, 1975). Although completely definitive standards are not available for cattle, calves with white muscle disease had selenium concentrations in the liver and kidney cortex of 0.035 and 0.36 ppm respectively (Andrews *et al.,* 1968; Hartley, 1967). On a comparable basis, Se values above 0.1 and 1.0 ppm for liver and kidney cortex are normal in sheep (Underwood, 1971).

The enzymes serum glutamic oxaloacetic transaminase (SGOT) and lactic dehydrogenase (LDH) increase to abnormally high levels with a severe Se defi-

TABLE 5.28
SGOT (Serum Glutamic Oxaloacetate Transaminase) Values in
Normal Calves and Those with White Muscle Disease[a]

	Normal	White muscle disease
No. calves	69	4
SGOT activity[b]		
Average (units/ml)	57	1313
Range	19–99	295–2360

[a] Adapted from Blincoe and Dye (1958).
[b] An increase in SGOT activity indicated tissue damage.

ciency (Table 5.28) (Underwood, 1971, 1977). These are general indicators of tissue damage. Thus, they are not sensitive measures of a borderline deficiency nor are they specific for Se deficiency. For instance, in human heart attacks resulting in severe heart damage comparable enzyme changes occur.

The most sensitive and specific biochemical measure yet discovered for Se deficiency is tissue glutahione peroxidase that is an enzyme containing Se (Ammerman and Miller, 1975). When sufficient detailed research is completed concerning the most suitable tissues to sample, other factors affecting the enzyme level, and suitable standards, glutathione peroxidase promises to be an excellent way to diagnose Se status of dairy cattle (Hoekstra, 1975). Other useful indicators of Se deficiency are the response of animals to supplemental Se and certain electrocardiographic changes observed in a subclinical deficiency (Ammerman and Miller, 1975).

Selenium toxicity, generally, is classified as either acute, often called blind staggers, or chronic, which frequently is termed alkali diaease (NRC, 1971b; Oldfield et al., 1974; Underwood, 1977). When cattle consume plants containing extremely high amounts (400–800 ppm) of Se, acute Se toxicity may occur (NRC, 1971b). Symptoms include slight ataxia; a characteristic posture with the head lowered and ears drooped; elevated temperature; rapid, weak pulse; labored breathing; bloody froth in the mouth and nose; a dark, watery diarrhea and blindness (NRC, 1971b). The animals are very lethargic and become prostrate prior to death, usually from respiratory failure. Accidental administration of excessive amounts of Se can produce a similar effect. The LD_{50} for acute Se toxicity is about 23 mg/100 lb liveweight (Shortridge et al., 1971).

Symptoms of chronic Se toxicity that develop when cattle eat plants containing moderately high-Se feeds (often 5–20 ppm) for prolonged periods include loss of vitality, elongated hooves, loss of hair from the switch, sore feet, progressive weight loss, stiffness and lameness, severe pain, starvation, atrophy and cirrhosis of the liver, chronic nephritis, or kidney degeneration, and rapid degeneration and necrosis of the myocardium (heart muscle). Reproduction may be reduced at Se levels below those needed to produce severe clinical symptoms (Olson, 1969; Underwood, 1977). With chronic Se toxicity, the hooves and hair as well as such tissues as the kidney contain high concentrations of Se.

The chemical form of Se has a major effect on the development of Se toxicity as well as on the metabolism of selenium (NRC, 1971b; Underwood, 1977). The Se which causes the chronic toxicity generally observed under practical conditions is in the organic form. Some attempts to produce Se toxicity with the inorganic sodium selenite and sodium selinate failed (NRC, 1971b). However, the symptoms were produced when heifers were fed 15 ppm Se as sodium selenite for 231 days (Olsen and Embry, 1973). Apparently, the relative toxicity in decreasing order appears to be selenium as wheat, corn, barley, selenite and selenate. In this discussion it seems important to reemphasize that Se toxicity *does not appear to be a major practical problem with dairy cattle.*

Apparently most of the net absorption of Se takes place from the lower part of the small intestine. Selenium, as is true with many trace elements, becomes firmly bound to organic compounds, especially protein. Selenium may also replace sulfur in the amino acids cystine or methionine and form selenocystine or selenomethionine. Selenium is excreted through the feces, urine, and respiration. Usually a substantial part of the dietary Se is not absorbed and thus is eliminated via feces. Apparently, most of the endogenous excretion of Se is through urine in ruminants (Lopez *et al.*, 1969). The amounts eliminated in the breath are very small except at high intakes (Lopez *et al.*, 1969). Much of the Se excreted in urine appears to be in the form of trimethylselenium which is not very available either to animals or, if in the soil, to plants.

The metabolism of Se is exceedingly complex (Krehl, 1970) partially due to its many different valence states (4 in all) (Fig. 5.36). (Several other essential trace elements, including manganese and iron, have more than one valence

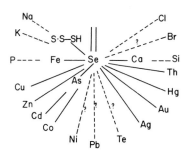

Fig. 5.36 This chart illustrates some of the complexity of selenium in animal nutrition and metabolism. (The interactions between selenium and organic compounds are omitted from this adaptation.) Note the very large number of elements with which selenium interacts and the uncertainties as shown by the broken lines and question marks. (Courtesy of D. V. Frost as adapted from Frost, 1976.)

state.) The multiple valence states make more different types of reactions possible as well as interactions with a larger number of nutrients and substances. Highest concentrations of Se occur in the kidney followed by the pancreas, pituitary and liver (Underwood, 1977). Whereas, muscle, bone and blood are relatively low in Se; fatty tissue has even less.

Selenium content of milk varies appreciably, with values from 0.003 to 0.067 ppm observed (Ammerman and Miller, 1975; Underwood, 1971; Waite *et al.,* 1975). The lower concentrations are associated with low-Se diets. In one experiment 1.5% of absorbed radioactive Se was recovered in milk (Waite *et al.,* 1975). In another study, 0.7% of dietary selenium was secreted into milk (Hogue, 1970). A considerable portion of the selenium in milk, probably 60–75%, occurs in the protein fraction (Hogue, 1970; Underwood, 1977).

One of the most important and best known interactions in nutrition is that between Se and vitamin E. Although neither can replace the other completely, higher levels of vitamin E reduce the requirement for Se and vice versa. Cattle receiving lush, young forage containing a high vitamin E (tocopherol) content probably have a comparatively low Se requirement (Allaway, 1974).

Other nutrients interacting with Se include sulfur, sulfur amino acids, proteins, and several trace elements including arsenic, mercury, and cadmium (Fig. 5.1) (NRC, 1978). As an illustration of the powerful effect of Se in certain situations, very small amounts of Se are quite effective in protecting animals against cadmium toxicity (Parizek *et al.,* 1974). While some of the interactions are quite complex, generally higher levels of sulfur, proteins, or arsenic will decrease the toxicity of Se (NRC, 1971b).

The practical nutrition of Se presents certain unusual features because of public opinion. Largely because of early research reports (later often severely criticized for key scientific weaknesses) suggesting a possible causitive relation of selenium to cancer in laboratory animals, the use of Se for nutritional purposes in the United States is subject to Food and Drug Administration regulations. More recent research indicates that adequate dietary selenium probably reduces the incidence of cancer (Allaway, 1974; Underwood, 1977). Thus, prohibiting the use of supplemental selenium in feeds appears to have an effect opposite to that intended. Even with unreasonably high concentrations of Se in the feed of dairy cattle, it appears that only the kidney and liver might accumulate more than 3 or 4 ppm (NRC, 1971b). Thus, even feeding toxic levels to dairy cattle does not appear to offer a potential hazard to humans.

5.28. FLUORINE

Research with laboratory animals suggests that fluorine (F) is an essential element for all farm animals (Messer *et al.,* 1974; Schwarz, 1974; Underwood,

1977). However, F deficiency has never been produced or observed in dairy cattle. Likewise, there is no reason to suspect inadequate F would ever be a practical problem for cattle. This is in contrast to the beneficial effects of providing supplemental F, usually added to water, to humans in many large areas of the world. Fluorine increases the resistance of teeth to decay in children and decreases osteoporosis of bones in adults (Messer et al., 1974; Underwood, 1977). The desirable influence of F to humans probably are due to pharmacological effects rather than a correction of a strict deficiency (Messer et al., 1974).

The practical importance of F in dairy cattle feeding is due to its toxicity (National Research Council, 1974a; Stoddard et al., 1973; Suttie et al., 1961). Among all domestic animals, dairy cattle are the most sensitive to high F intakes (NRC, 1974a). Although most feeds used for dairy cattle do not contain hazardous amounts of F, there are three main sources of excess F which may cause fluorosis (F toxicity) (NRC, 1974a; Underwood, 1971, 1977). (1) Nondefluorinated mineral supplements, especially phosphates often contain high amounts of F (Table 5.29). Rock phosphate usually has 2–5% F (NRC, 1974a). Thus, most rock phosphate used for feeding to dairy cattle is defluorinated (see Table 5.16). To be classified as defluorinated phosphate, the F content must be no more than 1% of the phosphorus (NRC, 1974a). (2) In limited areas around certain types of mines and industrial plants, such as reduction plants or steel processing operations, the forage, water, and soil may be contaminated from F dust and/or fumes

TABLE 5.29
Typical Fluoride Content of Certain High Fluoride Phosphate Compounds[a]

Compound	Phosphorus content (%)	Calcium content (%)	Fluoride content (%)	Fluoride contribution from compound at 0.25% phosphorus[b] (ppm)
Soft rock phosphate	9.0	17.0	1.2	334
Ground rock phosphate	13.0	35.0	3.7	710
Ground low-fluorine rock phosphate	14.0	36.0	0.45	81
Triple superphosphate	21.0	16.0	2.0	238
Diammonium phosphate (fertilizer grade)	20.0	0.5	2.0	250
Wet-process phosphoric acid (undefluorinated)	23.7	0.2	2.5	264

[a] Adapted from NRC (1974a).

[b] Phosphate compounds at dietary levels to furnish 0.25% P would contribute fluorine levels as indicated.

(Underwood, 1971). Most of the F in these areas comes from surface contamination but forages may take up excessive amounts from the soil. (3) In certain areas of the world, including parts of Australia, India, Coastal Plains of North Africa, South Africa, and North America, water from deep wells has excessive amounts of F (NRC, 1974a; Underwood, 1966). Often in these areas, forages and grain crops irrigated with surface water do not contain high F levels. Rather, the water from deep wells derives the F from underground rock formations.

In dairy cattle acute F toxicity is comparatively rare (NRC, 1974a). Most fluorosis of dairy cattle develops slowly over a long period of time, usually years, and is referred to as chronic. Dairy cattle have two major physiological mechanisms which provide substantial protection against F toxicity. These are the deposition of substantial amounts in bones and teeth without adverse effects and excretion of a considerable portion of the excess intake via urine.

The absorption of dietary F is closely related to solubility of the source (NRC, 1974a). When excess F is absorbed, a substantial proportion accumulates in the bones and teeth. Typically about 99% of the total F in the body is in close association with calcium in the skeleton and teeth where it exists as the inorganic mineral fluoroapatite (NRC, 1974a). The mineral structure of bones and teeth becomes more stable as F content increases. This explains the beneficial effects of added F in humans.

The primary pathway of F excretion after absorption is via urine. In fact, urinary excretion increases with intake and remains elevated as long as content in the bone is excessive, thus providing a mechanism of reducing body F with time. Usually older animals excrete a higher proportion of the absorbed F in urine (NRC, 1974a). There is an upper limit in the ability of the animal to excrete F in urine (Underwood, 1966). Likewise, the ability to accumulate F in the bone without adverse effects also is limited. Generally no toxicity has been observed when the F in the compact bone on a fat free dry basis was below 4500 ppm (Underwood, 1977). Although values above 5500 ppm are associated with toxicity, the development of F toxicosis is insidious with no easily defined point of toxicity.

The effects of F toxicity depend somewhat on the age of the animals. Developing teeth, especially the incisors, before eruption are the most sensitive tissue to excess F. Sufficiently high levels, even for short periods, can result in permanent damage to the teeth. In cattle, the highly sensitive ages are from 6 months to 3 years (NRC, 1974a). This is the reason for the lower maximum safe levels for cattle of these ages (see Appendix, Table 3).

Gross F toxicity lesions in teeth include mottling (white, chalky patches or striations in the enamel), chalkiness (dull white), and defective calcification (Figs. 5.37, 5.38) (NRC, 1974a). Likewise, the size, shape, orientation, and structure of the teeth are affected, and excessive erosion and wear of enamel occurs (Underwood, 1966). With varying F toxicity, there is considerable range

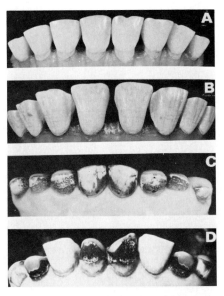

Fig. 5.37 Effects of excess fluoride on incisor teeth of cattle. (A) Normal, 7-yr-old; (B) moderate dental fluorosis in 8-yr-old animal; (C) severe dental fluorosis in 4-yr-old due to constant excessive fluorine intake; (D) severe dental fluorosis due to intermittent periods of excessive fluoride intake. (Courtesy of J. L. Shupe and A. E. Olson, Utah State Univ.)

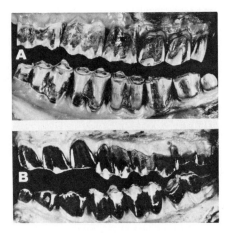

Fig. 5.38 Effects of excess fluoride on premolar and molar teeth. (A) Normal 8-yr-old animal; (B) moderate dental fluorosis in 4-yr-old. (Courtesy of J. L. Shupe and A. E. Olson, Utah State Univ.)

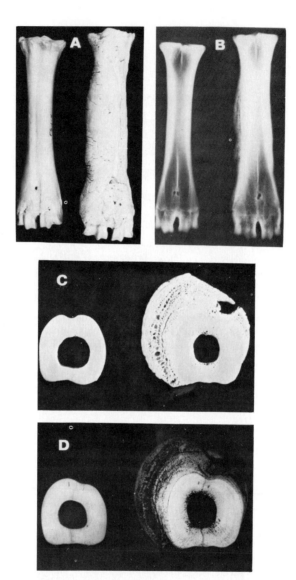

Fig. 5.39   Effects of excessive fluoride intake on cattle bones. (A) and (C) metatarsal bones; (left) normal, (right) severe osteofluorosis; (B) and (D) radiograph of same bones as in (A) and (C). (Courtesy of J. L. Shupe and A. E. Olson, Utah State Univ.)

in the extent of tooth defects. Changes in teeth are often used to measure the degree of fluorosis. Fluorosis also causes major lesions in the bones, including enlarged areas known as exostoses and a chalky, white appearance (Fig. 5.39; NRC, 1974a; Underwood, 1966, 1977).

Fig. 5.40   Severely lame 7-yr-old Holstein cow due to excessive fluorine intake. The cow also had severe bone lesions. (Courtesy of J. L. Shupe and A. E. Olson, Utah State Univ.) (Also see NRC, 1974a.)

The effects of F toxicity on teeth and bones can cause intermittent stiffness and lameness (Fig. 5.40) reduced feed intake leading to decreased general health, lower milk production, or rate of growth. Neither the reproduction nor the health of the calf are directly affected by F toxicity (NRC, 1974a). Of course when general health is sufficiently impaired, reproduction can be indirectly affected.

The maximum safe level of F in dairy cattle rations is influenced by several variables. Approximately 30 ppm F in the dry matter of the total diet can be fed

TABLE 5.30
Relation of Dietary Fluorine Content to Development of Toxicity Symptoms of Cattle[a]

Symptom	Total fluorine in diet (ppm)			
	20–30	30–40	40–50	50
Discernible dental mottling[c]	Yes[b]	Yes	Yes	Yes
Enamel hypoplasia (score number)[c]	No	No	Yes	Yes
Slight gross periosteal hyperostosis	No	Yes	Yes	Yes
Moderate gross periosteal hyperostosis	No	No	Yes	Yes
Significant incidence of lameness	No	No	No	Yes
Decreased milk production	No	No	No	Yes
Skeletal F equivalent to 5000 ppm at 5 yr[d]	No	No	No	Yes
Urine F of 25 ppm[e]	No	No	Yes	Yes

[a] Adapted from NRC (1974a).
[b] Yes or No indicates if the symptom would be reproducibly seen at this level.
[c] Only if fluoride is present during formative period of the tooth.
[d] Metacarpal or metatarsal bone, dry, fat-free basis.
[e] Values taken after 2–3 yr of exposure.

to dairy cattle without decreasing performance, even over long periods (NRC, 1974a, 1978). If the elevated F is first fed after cattle are 3–4 years old, 40 ppm will cause little or no decrease in performance for two or three lactations. The desirable maximum safe level for continuous feeding to heifers and breeding bulls (6 months to 3 years of age) is 20 ppm. At F levels of 20–30 ppm histologic changes may develop in bone and tooth structure. Histological changes occur with less F than required to affect performance (Table 5.30). Because of the short duration of the average feeding period, finishing cattle can be safely fed 100 ppm F. All these recommendations for maximum safe levels are based on highly soluble sources of F such as sodium fluoride or other soluble fluorides. Because F from less soluble sources is absorbed in smaller amounts, cattle can tolerate a higher dietary level. The maximum safe level is increased somewhat when the F comes from most phosphates (NRC, 1974a).

When excessive F is given for only intermittent periods, a higher maximum level may be fed than if the intake is uniformly high. Thus, except for finishing cattle, the maximum safe levels suggested here are intended to be those which could be fed continuously for the entire life of the animal without adverse effect on performance or health (Table 5.30). To a limited degree the toxic effects of F can be reduced by dietary aluminum salts, high calcium levels, green forages, or liberal grain feeding (NRC, 1974a).

Diagnosis of F toxicity can be made with a combination of measures (Table 5.30) (NRC, 1974a; Underwood, 1977). Among the most useful are analyses of F in the diet, urine, bones, and teeth along with the gross effects on teeth such as chalkiness or mottling, erosion, and wear, and changes in color and shape (NRC, 1974a; Underwood, 1977). Likewise, evidences of systemic effects such as reduced appetite, loss of condition, and bone changes are helpful. Also, increased bone alkaline phosphatase, an enzyme, occurs with F toxicity. These diagnostic aids must be used with an understanding of their meaning and of F metabolism. For example, the effects on teeth generally indicate toxicity during tooth formation rather than current intake. Since urine F will remain high as long as the bones contain large amounts, it is not a measure of current F intake.

Although milk F content increases with intake, the level remains relatively low (NRC, 1974a). For example, cattle fed 3–5 ppm F had about 0.1 ppm in milk, whereas increasing the intake to 50 ppm resulted in a milk content of 0.4 ppm (NRC, 1974a; Suttie et al., 1957). Fluorine does not accumulate in the muscle or other soft tissues of cattle. Thus, F toxicity does not directly lower the value of the milk and meat for human consumption.

5.29. MANGANESE

In the practical feeding of dairy cattle, Mn requires less attention than such trace elements as iodine, cobalt, and copper. However, a marginal deficiency of

Mn can occur in some practical situations. In addition, several aspects of Mn nutrition and metabolism aid considerably in understanding the whole subject of mineral nutrition. Most soils contain comparatively high levels of Mn (see Section 5.7.d and Table 5.1). In contrast, the animal body has a very low average Mn concentration. The average Mn content of only about 0.3 ppm in a dairy animal is very unevenly distributed with liver, bones, pituitary, pancreas, and kidney having relatively high levels and muscles having very little. The Mn concentration of most tissues is quite characteristic and not very responsive to changes in Mn intake. In the animal, Mn usually is combined with other compounds especially with proteins in soft tissues.

Milk has about 0.03 ppm Mn, which is less than 1% of zinc. The level can be increased 2- to 4-fold by feeding large amounts of Mn (Underwood, 1971). Although extremely variable, most dairy cattle feeds have a comparatively high Mn content (Adams, 1975; Underwood, 1977). Corn grain is very low, typically around 5 ppm, and barley moderately low at perhaps three times that of corn.

The metabolism and homeostasis of Mn in many ways contrasts sharply with that of zinc. Some of the aspects illustrate how animals have adapted to the environment during evolution. The relatively high Mn content in feed and the low average amount in body tissues of dairy cattle is achieved by a low absorption, with 3–4% being typical, and very rapid endogenous excretion of any excess absorbed. Most of the endogenous excretion of Mn is through the feces. Tissue Mn concentration is quite closely regulated. Manganese is relatively unique, in that a rapid endogenous fecal excretion is a major homeostatic control route for regulation of body Mn. Changes in absorption also are an important factor. Manganese is an extremely dynamic element: its metabolism is a "high speed" chain of events. Absorption, tissue turnover, and reexcretion of Mn proceeds at a very rapid pace.

Manganese is essential for the functioning of several enzyme systems. Usually these are metal enzyme complexes in which the association between the enzyme and the Mn is quite loose, in contrast to the behavior of some other trace elements.

Generally symptoms of a severe Mn deficiency include reduced growth, abnormalities of the bones and skeleton, depressed and disturbed reproduction, and the birth of abnormal calves (Fig. 5.41) (NRC, 1978; Underwood, 1977). Current information indicates that in cattle the Mn requirement for reproduction and birth of normal calves is substantially higher than for normal growth (Anke and Groppel, 1970; Anke et al., 1973; Bentley and Phillips, 1951; NRC, 1978; Rojas et al., 1965). At Washington State University, all calves born from cows given 16–17 ppm Mn for 12 months were deformed at birth (Rojas et al., 1965). The abnormalities included characteristically enlarged joints, weak twisted legs and pasterns, stiffness, more easily broken bones, and general weakness. At the dietary level of 16–17 ppm where these deficiency effects occurred, usually growth rate and probably milk production would not be impaired seriously. Cows

Fig. 5.41 This deformed calf was born from a cow fed a low-manganese diet. Such calves have enlarged joints, stiffness, twisted legs, and a general physical weakness. (Courtesy of I. A. Dyer, Washington State Univ.)

and heifers are slower to exhibit estrus and more likely to have silent heats with low Mn rations. Conception rate also is reduced. An increased ratio of male to female calves and goats has been reported from Mn-deficient dams (Table 5.31) (Anke *et al.*, 1973). When cows were given only 7–10 ppm Mn for long periods of time, abscessed livers, and an abnormally low amount of bile were observed (Bentley and Phillips, 1951). Field observations, unconfirmed by critical experimentation, suggest that certain hoof and lameness problems of dairy cows fed substantial amounts of corn silage and grain have responded favorably to supplemental Mn (J. E. Osborn, personal communication, 1976). Such herds often had breeding problems which also appeared to be corrected by the Mn.

No biochemical measures are available which permit the diagnosis of a borderline deficiency of Mn in dairy cattle (Miller and Stake, 1974b). Since the highest requirement for Mn appears to be for normal reproduction and the birth of healthy calves, with current methodology, experiments to critically determine the minimum Mn requirement would have to depend on measuring these effects. Because of the variability in reproduction and birth deformaties, determining minimum Mn needs would require large numbers of cattle under a variety of situations. Such research appears to be prohibitively expensive. Accordingly, the minimum requirement for Mn must be estimated.

The minimum dietary Mn requirement for dairy cattle is estimated to be 40 ppm of dry matter in feed (see Appendix, Table 3) (NRC, 1978). This includes a reasonable safety margin to cover possible genetic variability of cattle, and some interacting substances in the feed. The estimated amounts are believed to be fully adequate for normal reproduction under almost any practical farm conditions. Adverse effects may not consistently occur when somewhat lower levels are fed, especially if given for limited times. Apparently a relatively long period is

TABLE 5.31
Effect of Mn Deficiency on Reproduction and Mortality in Goats and Cattle[a]

Animals	Control diet	Low-Mn diet
Goats		
No.	27	28
Pregnant (%)	100	93
Insemination per pregnant goat	1.07	1.42
Abortions (%)	0	23
Ratio of female to male kids	1:1.5	1:2.3
Birth weight of kids (lb)	7.2	5.8
Mortality of dams (%)	18	43
Cattle (498 total calves)		
Ratio of female to male calves	1:1	1:1.3

[a] Adapted from Anke and Groppel (1970). Low-Mn goats fed 20 ppm Mn first year and 5 the second year compared to 100 ppm for controls.

needed for the effects of a marginal deficiency to be exhibited. As with most minerals, the effects of the genetic makeup of dairy cattle on Mn requiremens and metabolism have not been investigated. High concentrations of calcium and phosphorus in the diet appear to increase the Mn requirement (Vagg and Payne, 1971; Hawkins et al., 1955). However, the extent of this interaction has not been determined with cattle.

Manganese along with iron and zinc, are among the least toxic of the essential trace elements. Dairy cattle probably can tolerate 1000 ppm dietary Mn without adverse effects (NRC, 1978). The first observed effects of excessive dietary Mn are reduced feed consumption and slower growth, presumably caused by less feed intake (Underwood, 1971). In one study, performance was not affected by 820 ppm but was decreased with 2460 ppm supplemental Mn (G. N. Cunningham et al., 1966).

In the practical feeding of dairy cattle, it appears that usually only limited attention need to be given Mn. However, if corn grain is a major portion of the ration for extended periods or if other low Mn diets are used, sufficient Mn should be added to have at least 40 ppm Mn in the total ration dry matter.

5.30. CHROMIUM, SILICON, VANADIUM, NICKEL, AND TIN

Research with small laboratory animals indicates that chromium and silicon are essential (Carlisle, 1974; Miller and Neathery, 1977). Similar work suggests

Fig. 5.42 These two calves were the same age but the smaller one had been fed 1000 ppm of nickel for 8 weeks. Even though calves fed the high-nickel ration consumed less feed and lost weight, they were not emaciated but appeared younger than the controls. (Courtesy of G. D. O'Dell, Clemson Univ., Clemson, S.C.) (See O'Dell *et al.*, 1970b.)

that vanadium, tin, and nickel, probably are required (Hopkins and Mohr, 1974; Miller, 1974; Miller and Neathery, 1977; Nielsen and Ollerich, 1974; Schwarz, 1974). Since all animals usually require trace mineral elements for the same biochemical functions, generally an element needed for laboratory animals is required by dairy cattle. Accordingly, chromium, silicon, vanadium, tin, and nickel probably are essential for cattle. However, a deficiency has never been produced for any of these in ruminants. Likewise, there is no information to suggest that a deficiency will ever be a practical problem for dairy cattle (Miller and Neathery, 1977).

As discussed later (see Section 12.6.d), high levels of silica (silicon) in certain forages depresses digestibility. This appears to be a physical effect making other nutrients less available rather than toxicity of silica per se. With usual practical feeding conditions there is no reason to suspect that either chromium, vanadium, tin, or nickel levels will be high enough in feeds to adversely affect performance of dairy cattle. For example, more than 62 ppm nickel was needed to adversely affect growth in dairy calves (O'Dell *et al.*, 1970b). Higher levels of nickel progressively decreased feed intake and weight gains (Fig. 5.42). Performance of lactating cows was not depressed by 100 ppm in the total dry matter ration (O'Dell *et al.*, 1970a). These amounts of nickel are far higher than normally found in feeds.

5.31. TOXIC MINERALS

As discussed earlier (see Section 5.10), all nutrients and all elements when consumed in large enough amounts are toxic. In the sections above some of the essential mineral elements especially molybdenum, selenium, fluorine, and copper, which need special consideration due to their toxicity in practical feeding situations, were discussed. A few elements not shown to be essential, including cadmium, lead, and mercury, are of practical interest in dairy cattle feeding because of their toxicity. Thus, the concentration found in milk and meat also may be important.

In addition to those discussed in this book, some limited practical possibilities exist for the toxicity of other mineral elements including arsenic, vanadium, rubidium, strontium, and bromine.

5.32. CADMIUM

On the basis of present evidence, cadmium (Cd) is not an essential element for dairy cattle (NRC, 1978). Although it is highly toxic to dairy cattle, with most feeding, management, and environmental conditions, cattle do not obtain enough Cd to make toxicity a practical problem (NRC, 1978). Thus, for efficient dairy production, Cd requires little attention. Cadmium is a highly useful industrial metal which is being utilized in increasing amounts. Likewise, Cd has been implicated circumstantially as a possible important factor in human health, especially relative to heart and blood vessel disease problems which are the No. 1 killers of men 45 to 74 years of age (Miller, 1971a; Neathery and Miller, 1975, 1976c).

Because of the public concern with Cd pollution, it is important to understand the metabolism of Cd by dairy cattle and to know the amounts that may be found in milk and meat (Miller, 1971a). Even though there is no reason to suspect even a marginal Cd toxicity under usual practical conditions, a borderline toxicity appears possible where animals are fed certain types of recycled waste such as sewerage sludge in which Cd is concentrated (Neathery and Miller, 1975, 1976a). Substantial amounts of Cd may also be obtained by animals grazing around zinc mines and smelters (Neathery and Miller, 1976a).

Chemically and metabolically, Cd is closely related to zinc; both are in the same group in the periodic chart of elements. Industrially, Cd is obtained as a by-product of zinc mining. In the metabolism of dairy cattle, Cd and zinc are antagonists. Some of the symptoms of Cd toxicity are similar to zinc deficiency (Powell et al., 1964). Likewise, one of the effects of a marginal Cd toxicity is an increased requirement for zinc as well as for iron and copper.

Whereas most dairy cattle feeds contain less than 1 ppm Cd, calves given 40 ppm exhibited no toxicity symptoms (Powell et al., 1964). When 160 ppm was fed, growth, feed intake, and water consumption were reduced. Likewise, milk production was drastically decreased in lactating cows given 3 gm of Cd per day (Miller et al., 1967).

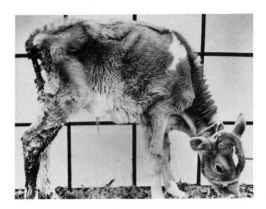

Fig. 5.43 This 10-week-old calf had been fed 2650 ppm cadmium for 14 days. (Courtesy, W. J. Miller, Univ. of Georgia.) (See Powell *et al.,* 1964.)

Symptoms of Cd toxicity include decreased feed intake, reduced growth, increased mortality, retarded testicular development or degeneration, enlarged joints, liver and kidney damage, scaly skin, and anemia (Figs. 5.43, 5.44) (Powell *et al.,* 1964). In other species some of the toxicity effects have been diminished or prevented by cobalt, selenium, thio compounds, and zinc (Flick *et al.,* 1971; Gunn and Gould, 1967; NRC, 1978). The mechanisms of these interactions are not well understood.

Cattle have much more effective mechanisms for keeping Cd out of the body tissues than for excreting it once it is absorbed (Table 5.32). Because of the very low percentage, Cd absorption must be measured by direct analyses of the

Fig. 5.44 These two 18-week-old calves were fed the same diet for 8 weeks and weighed the same at that time. From 9 through 18 weeks of age the smaller calf was fed a ration containing 640 ppm cadmium. (Courtesy, W. J. Miller, Univ. of Georgia.) (See Powell *et al.,* 1964.)

TABLE 5.32
Kidney, Liver, and Muscle ^{109}Cd, 14 Days after Oral or Intravenous Dosing in Goats[a]

Tissue	% of dose/kg fresh tissue	
	Oral	Intravenous
Liver	0.6	160
Kidney	1.1	49
Muscle	0.03	0.6

[a] Adapted from Neathery and Miller (1975).

amounts retained by tissues (Miller, 1975a). Using this approach, in one study, lactating cows retained only 0.75% of radioactive Cd given orally (Table 5.33) (Neathery *et al.*, 1974). In a comparable study, young goats absorbed and retained only 0.34% (Table 5.33) (Miller *et al.*, 1969). Absorption of inhaled Cd is far higher. Based on data from other species perhaps 10–40% of inhaled Cd is absorbed.

Once absorbed, animals retain Cd for a very long time: the endogenous excretion rate, mainly in feces, is very slow (Miller, 1973). Much of the retained Cd is in the liver and kidney, which together contain as high as 75% of the total body content. Highest concentrations are in the kidneys, but due to the far larger size, a greater total amount may be found in the liver. Muscles contain very little Cd.

Unlike the essential elements, dairy cattle do not have homeostatic control mechanisms for controlling tissue levels of Cd as intake varies (Miller, 1973). Thus, the amount of Cd in tissues varies directly with the amount consumed and/or inhaled. However, in addition to the very low percentage absorbed, cattle

TABLE 5.33
^{109}Cadmium in Tissues and Total Retention in Goats and Lactating Cows 2 Weeks after a Single Oral Dose

Body components	% of dose/whole organ or section	
	Goats[a]	Lactating cows[b]
Liver	0.17	0.24
Kidney	0.08	0.08
GI tract tissue and contents	0.06	0.26
Muscle	—	0.07
Other tissues	0.03	0.12
Total body	0.34	0.75

[a] Adapted from Miller *et al.* (1969).
[b] Adapted from Neathery *et al.* (1974).

have a mechanism(s) for limiting the toxicity of Cd. Relatively soon after absorption, Cd combines with protein(s), perhaps metallothionein, which greatly reduces its toxicity (Neathery and Miller, 1975). The amount of the metalloprotein available to inactivate the Cd increases with the amount of Cd absorbed. This inactivation of Cd has important practical effects. When a toxic level of Cd is removed from the diet, after being fed for a period of time, the toxicity symptoms decline rapidly before there is any substantial decrease in the total amount in the body.

Dairy cattle are very effective in keeping Cd out of the milk. In one study less than 0.0008%, which was the lowest detectible limit, of a dose of radioactive Cd appeared in the milk (Neathery et al., 1974). By comparison with a typical ration, 8–12% of the dietary zinc may be obtained in the milk (Miller et al., 1965a; Neathery et al., 1973c). Milk samples taken from various United States cities average from 0.017 to 0.030 ppm Cd (Murthy and Rhea, 1968). Other evidence suggests that a large proportion of the very low level of Cd in commercial milk may come from equipment through contamination (Miller et al., 1967).

The practical aspects of Cd may be summarized as follows. Although highly toxic, Cd is not a problem in normal feeding of dairy cattle. Due to low absorption and an effective screening mechanism, dairy cattle are very effective in keeping Cd levels in milk and muscle meat very low. Thus, dairy cattle protect the consumer from the possible adverse effect of Cd.

5.33. LEAD (Pb)

Lead is one of the most frequently encountered causes of poisoning in cattle. Among farm animals ruminants are most often affected. Their natural curiosity and licking habits are an important factor in the comparatively high incidence, relative to other farm animals, of lead poisoning in cattle. In cattle, most lead poisoning is acute toxicity rather than a chronic problem. Lead-based paint, either peeling or flaking, and discarded paint cans, are the most common sources of lead toxicity for cattle (Allcroft, 1950). Other major sources are used motor oil, discarded oil filters, storage batteries, certain types of grease, putty, and linoleum (Allcroft, 1950; National Research Council, 1972). Although there is some lead contamination in feeds including limited amounts taken up from contaminated soils, these sources are of minor practical importance.

Calves were killed by 90–180 mg per pound of body weight in 4–6 days (Allcroft, 1951). Older cattle can tolerate about twice as much lead per pound of body weight as calves (Buck, 1970; Neathery and Miller, 1975, 1976c). As little as 2.7 mg per pound of body weight per day, or about 300 ppm in the total diet, for 60 days was fatal to cattle (Hammond and Aronson, 1964).

Acute lead poisoning symptoms include dullness, loss of appetite, and abdominal pain with constipation possibly followed by diarrhea. In advanced stages, sometimes two to three days after a lethal dose, cattle may bellow, stagger, have snapping eyelids, twitching muscles, frothing mouth, and convulsive seizures (Allcroft and Blaxter, 1950; Merck Veterinary Manual, 1973). Lead poisoning causes derangement of the digestive tract and central nervous system as well as interfering with red blood cell formation.

Lead poisoning in cattle can be diagnosed best by considering the clinical symptoms, the presence of a lead source, the lead content of blood and feces if the animal survives or if dead, lead content of kidney cortex and liver (Allcroft and Blaxter, 1950; NRC, 1978). High fecal and blood lead may only establish recent exposure which may or may not be responsible for death. For many weeks following a single oral dose whole blood lead may remain above a normal level of 0.13 ppm (Allcroft, 1950). Lead values of 22 ppm in fresh kidney cortex and 9 ppm in liver indicate death from lead poisoning (Allcroft and Blaxter, 1950).

Although data with cattle are limited, it is evident that lead absorption is very low. Values of 1.3% absorption were obtained with sheep (Blaxter, 1950a). At least in some species, lead absorption is decreased with higher calcium or phosphorus intake (Neathery and Miller, 1975, 1976C). After absorption, a small portion of the lead may go through the lymphatic system into the general circulation (Cantarow and Trumper, 1944; Neathery and Miller, 1976c). Most of the lead, apparently, goes to the liver via the portal circulation; and then through the bile into the small intestine for excretion in feces. This route of lead metabolism, called the "liver bypass system," lowers lead concentrations in the systemic circulation and thereby gives some protection against toxicity.

Feces is the predominant route of endogenous lead excretion from the body after absorption from the digestive tract. However, urinary excretion is more important when lead enters through the respiratory tract. The distribution of lead among tissues is affected by the amount involved, the route it enters the animal and the chemical form of the lead. Generally, most of the lead absorbed from the digestive tract goes to the skeleton. As higher amounts are taken in, apparently a larger proportion is distributed throughout the body, especially in the kidney and liver (Cantarow and Trumper, 1944). Very little lead is found in muscles (Blaxter, 1950b; Neathery and Miller, 1976c).

As long as most of the lead is confined to the skeleton there is little toxicity. In contrast, severe symptoms occur when substantial amounts enter the kidney, liver, and central nervous system. In this respect and in its mobilization to and from bone, lead resembles fluorine (see Section 5.28).

Generally, the lead content of cow's milk is quite low. Murthy et al. (1967) observed an average of 0.049 ppm with a range of 0.023–0.070 ppm in market milk samples throughout the United States. When substantial amounts of lead are fed, only a very small percentage is secreted into the milk (Lynch et al., 1974). Only 0.16% of the dietary lead was found in milk over a two week period (Lynch *et al.*, 1974). As with cadmium, dairy cattle provide an effective screen against excess lead going into either the milk or the muscle meat. Thus, the consumer is given considerable protection.

5.34. MERCURY

Present information indicates that mercury is not an essential element. Although very toxic, generally, mercury poisoning is uncommon in dairy cattle because of relatively low exposure (NRC, 1978). As with cadmium, mercury toxicity is of some concern in human health. Thus the main consideration with dairy cattle is to make sure that the meat does not contain hazardous amounts of mercury and is not falsely accused. In dairy cattle, mercury toxicity only occurs sporadically, usually from accidental overdosing with medicines containing mercury, absorption from too liberally applied mercury skin ointments, or from seed grains treated with organic mercury fungicide(s) (Blood and Henderson, 1968; NRC, 1978).

The toxicity, tissue distribution, and metabolism are almost totally different for inorganic mercury and organic mercury compounds (Table 5.34). Generally, organic mercury, especially in the methylmercury form, is more toxic for dairy cattle than inorganic mercury (Neathery and Miller, 1975, 1976b). Likewise, the toxic effects are different. Methylmercury primarily affects the nervous system producing symptoms similar to those of polioencephalomacia in calves, except

TABLE 5.34

Mercury in Calves Dosed Orally and Intravenously with Methylmercury Chloride ($CH_3{}^{203}HgCl$) and Mercuric Chloride ($^{203}HgCl_2$) and in Cows Given Methylmercury Chloride[a]

	% of dose/kg fresh tissue[b]						
	Calves						Cows
	Oral			Intravenous			Oral
Tissues	Methylmercury chloride	Mercuric chloride	Ratio	Methylmercury chloride	Mercuric chloride	Ratio	Methylmercury chloride
Kidney	4.4	0.55	8	10.3	82.1	0.12	1.75
Liver	2.2	0.15	15	2.8	9.9	0.3	0.87
Brain	0.6	0.0028	214	0.6	0.15	4	0.16
Muscle	1.9	0.0032	594	1.8	0.30	6	0.50

[a] Adapted from Neathery and Miller (1976b).
[b] Radioactive mercury.

TABLE 5.35
Metabolism of Radioactive Organic (Methyl) Mercury (^{203}Hg) in Lactating Cows and Goats from a Single Oral Dose[a]

	% of dose		
		Cows	
^{203}Hg	Goats[b]	Sell, *et al.*[b]	Neathery *et al.*[c]
Net absorption	69	65	59
Urinary	1.4	1.3	1.1
Milk	0.28	0.0	0.17

[a] Adapted from Neathery and Miller (1976b), Neathery *et al.* (1974), and Sell and Davison (1973).
[b] Goat and cow balance studies for 13 and 7 days.
[c] 14–day balance study.

thiamin does not alleviate the symptoms (Davies *et al.*, 1965, Herigstad *et al.*, 1972; NRC, 1978). Symptoms of methylmercury toxicity may include incoordination, tetanic-like spasms when excited, twitching of eyelids, excessive salivation, recumbency (prone lying position), and inability to eat or drink (NRC, 1978). Convulsions and death may follow.

Inorganic mercury is very caustic to the mouth and parts of the digestive tract. Among the symptoms of chronic poisoning with inorganic mercury in cattle are a depressed appearance, decreased appetite, weight loss, a stiff stilted gait, possibly some paralysis, loss of hair and scabby lesions around the anus, tender gums, loss of teeth, and possibly diarrhea (Blood and Henderson, 1968; NRC, 1978).

The greater toxicity of methylmercury, relative to inorganic mercury, is related to the far higher absorption (Table 5.34), slower turnover rate in tissues, and the much longer retention time in the body after absorption (Ansari *et al.*, 1973; Neathery *et al.*, 1974; Friberg and Vostal, 1972). Absorption of inorganic mercury appears to be less than 2% (Neathery and Miller, 1976b; Potter *et al.*, 1972). In contrast, methylmercury absorption by lactating cows was 59% (Neathery *et al.*, 1974) and 65% (Sell and Davison, 1973) (Table 5.35).

Most of the very small amounts of inorganic mercury absorbed and retained by cattle is in the kidney and liver (Ansari *et al.*, 1973). With methylmercury, highest concentrations are in the kidney and liver, but appreciable amounts are found in muscle and brain (Neathery *et al.*, 1974). In one study, 72% of the total methylmercury in the lactating cows was in the muscles (Neathery *et al.*, 1974). Accordingly, muscle meat from cattle recently given considerable methylmercury, may be somewhat hazardous to humans.

Very little mercury, from any chemical form, is secreted into the milk. For example, only 0.17% of an oral dose of radioactive methylmercury was recovered in the milk in 14 days (Table 5.34) (Neathery *et al.*, 1974). In another study with inorganic mercury, the milk contained only 0.01% of that given orally (Potter *et al.*, 1972).

It is well known that certain microorganisms in stream sediments can convert inorganic mercury to organic mercury. Fortunately, the microbes in the digestive tract (rumen) of cattle do not convert inorganic mercury to methylmercury in measurable amounts (Ansari *et al.*, 1973; Neathery and Miller, 1976b).

REFERENCES

Adams, R. S. (1975). *J. Dairy Sci.* **58,** 1538–1548.
Agricultural Research Council (ARC) (1965). "The Nutrient Requirements of Farm Livestock," No. 2. Ruminants. ARC, London.
Aines, P. D. and S. E. Smith (1957). *J. Dairy Sci.* **40,** 682–688.
Albert W. W., U. S. Garrigus, R. M. Forbes, and H. W. Norton (1956). *J. Anim. Sci.* **15,** 559–569.
Allaway, W. H. (1969). *Proc. Cornell Nutr. Conf. Feed Manuf.* pp. 12–16.
Allaway, W. H. (1974). *Proc. Arkansas Nutr. Conf.* pp. 78–84.
Allcroft, R. (1950). *J. Comp. Pathol. Ther.* **60,** 190–208.
Allcroft, R. (1951). *Vet. Rec.* **63,** 583–590.
Allcroft, R., and K. L. Blaxter (1950). *J. Comp. Pathol. Ther.* **60,** 209–218.
Allcroft, R., and G. Lewis (1957). *J. Sci. Food Agric.* **8,** Suppl. Issue, S96–S103.
Allen, R. D. (1977). *Feedstuffs* **49** (No. 30), 31.
Ammerman, C. B. (1970). *J. Dairy Sci.* **53,** 1097–1107.
Ammerman, C. B., and S. M. Miller (1975). *J. Dairy Sci.* **58,** 1561–1577.
Ammerman, C. B., J. M. Wing, B. G. Dunavant, W. K. Robertson, J. P. Feaster, and L. R. Arrington (1967). *J. Anim. Sci.* **26,** 404–410.
Andresen, E., T. Flagstad, A. Basse, and E. Brummerstedt (1970). *Nord. Veterinaer med.* **22,** 473–485.
Andrews, E. D., W. J. Hartley, and A. B. Grant (1968). *N.Z. Vet. J.* **16,** 3–17.
Anke, M., and B. Groppel (1970). *Trace Elem. Metab. Anim., Proc. WAAP/IBP Int. Symp., 1969* pp. 133–136.
Anke, M., B. Groppel, W. Reissig, H. Lüdke, M. Grün, and G. Dittrich (1973). *Arch. Anim. Nutr.* **23,** 197–211.
Ansari, M. S., W. J. Miller, R. P. Gentry, M. W. Neathery, and P. E. Stake (1973). *J. Anim. Sci.* **36,** 415–419.
Arnold, P. T. Dix, and R. B. Becker (1936). *J. Dairy Sci.* **19,** 257–266.
Babcock, S. M. (1905). *Wis., Agric. Exp. Stn., Annu. Rept.* **22,** 129–156.
Becker, R. B., W. M. Neal, and A. L. Shealy (1933a). *Fla., Agric. Exp. Stn., Tech. Bull.* **262.**
Becker, R. B., W. M. Neal, and A. L. Shealy (1933b). *Fla., Agric. Exp. Stn., Bull.* **264.**
Becker, R. B., P. T. Dix Arnold, W. G. Kirk, G. K. Davis, and R. W. Kidder (1953). *Fla., Agric. Exp. Stn., Bull.* **513.**
Becker, R. B., J. R. Henderson, and R. B. Leighty (1965). *Fla., Agric. Exp. Stn., Tech. Bull.* **699.**
Beeson, W. M., D. W. Bolin, C. W. Hickman, and R. F. Johnson (1941). *Idaho, Agric. Exp. Stn., Bull.* **240.**
Belyea, R. L., C. E. Coppock, and G. B. Lake (1976). *J. Dairy Sci.* **59,** 1068–1077.

Bentley, O. G., and P. H. Phillips (1951). *J. Dairy Sci.* **34,** 396–403.

Blaxter, K. K. (1950a). *J. Comp. Pathol. Ther.* **60,** 140–159.

Blaxter, K. L. (1950b). *J. Comp. Pathol. Ther.* **60,** 177–189.

Blaxter, K. L., and R. F. McGill (1956). *Vet. Rev.* **2,** 35–55.

Blaxter, K. L., and J. A. F. Rook (1954). *J. Comp. Pathol.* **64,** 176–186.

Blaxter, K. L., J. A. F. Rook, and A. M. Macdonald (1954). *J. Comp. Pathol.* **64,** 157–174.

Blaxter, K. L., G. A. M. Sharman, and A. M. MacDonald (1957). *Br. J. Nutr.* **11,** 234–246.

Blincoe, C., and W. B. Dye (1958). *J. Anim. Sci.* **17,** 224–226.

Block, R. J., and J. A. Stekol (1950). *Proc. Soc. Exp. Biol. Med.* **73,** 391–394.

Blood, D. C., and J. A. Henderson (1968). "Veterinary Medicine," 3rd ed., pp. 771–772. Williams & Wilkins, Baltimore, Maryland.

Bouchard, R., and H. R. Conrad (1973a). *J. Dairy Sci.* **56,** 1276–1282.

Bouchard, R., and H. R. Conrad (1973b). *J. Dairy Sci.* **56,** 1429–1434.

Bouchard, R., and H. R. Conrad (1973c). *J. Dairy Sci.* **56,** 1435–1438.

Bouchard, R., and H. R. Conrad (1974). *Can. J. Anim. Sci.* **54,** 587–593.

Bremner, I., and A. C. Dalgarno (1973a). *Br. J. Nutr.* **29,** 229–243.

Bremner, I., and A. C. Dalgarno (1973b). *Br. J. Nutr.* **30,** 61–76.

Buck, W. B. (1970). *J. Am. Vet. Med. Assoc.* **156,** 1468–1472.

Bull, L. S., and J. H. Vandersall (1973). *J. Dairy Sci.* **56,** 106–112.

Cantarow, A., and M. Trumper (1944). "Lead Poisoning." Williams & Wilkins, Baltimore, Maryland.

Carlisle, E. M. (1974). *Fed. Proc.* **33,** *Fed. Am. Soc. Exp. Biol.* 1758–1766.

Chalupa, W., R. R. Oltjen, L. L. Slyter, and D. A. Dinus (1971). *J. Anim. Sci.* **33:** 278 (abstr.).

Chapman, H. L., Jr., S. L. Nelson, R. W. Kidder, W. L. Sippel, and C. W. Kidder (1962). *J. Anim. Sci.* **21,** 960–962.

Chesters, J. K., and M. Will (1978). *Trace Elem. Metab. Man Anim., 3rd, 1977* pp. 211–214.

Conrad, H. R., A. L. Moxon, and W. E. Julien (1976). *Proc. Distill. Feed Res. Counc.* **31,** 49–51.

Coppock, C. E., R. W. Everett, and R. L. Belyea (1976). *J. Dairy Sci.* **59,** 571–580.

Cunningham, G. N., M. B. Wise, and E. R. Barrick (1966). *J. Anim. Sci.* **25,** 532–538.

Cunningham, I. J. (1950). *Johns Hopkins Univ., McCollum-Pratt Inst., Contrib.* A symposium on copper **5,** 246–273.

Cunningham, I. J. (1960). *N. Z. Dep. Agric., Bull.* **378.**

Cunningham, I. J., K. G. Hogan, and B. M. Lawson (1959). *N. Z. J. Agric. Res.* **2,** 145–152.

Davies, E. T., A. H. Pill, D. F. Collings, J. A. J. Venn, and G. D. Bridges (1965). *Vet. Rec.* **77,** 290.

Davis, G. K., R. Jorden, J. Kubota, H. A. Laitinen, G. Matrone, P. M. Newberne, B. L. O'Dell, and J. S. Webb (1974). *In* "Geochemistry and the Environment," Vol. I. Natl. Acad. Sci., Washington, D.C.

Dennis, R. J., R. W. Hemken, and D. R. Jacobson (1976). *J. Dairy Sci.* **59,** 324–328.

Dickson, J., and M. P. Bond (1974). *Aust. Vet. J.* **50,** 236.

Dunkley, W. L., A. A. Franke, J. Robb, and M. Ronning (1968). *J. Dairy Sci.* **51,** 863–866.

Eckles, C. H., R. B. Becker and L. S. Palmer (1926). *Minn., Agric. Exp. Stn., Bull.* **229.**

Eckles, C. H., T. W. Gullickson, and L. S. Palmer (1932). *Minn., Agric. Exp. Stn., Tech. Bull.* **91.**

Elam, C. J. (1975). *Feedstuffs* **47**(35), 23–25 and 48.

Ellenberger, H. B., J. A. Newlander, and C. H. Jones (1950). *Vt., Agric. Exp. Stn., Bull.* **558.**

Ely, R. E., K. M. Dunn, and C. F. Huffman (1948). *J. Anim. Sci.* **7,** 239–246.

Flick, D. F., H. F. Kraybill, and J. M. Dimitroff (1971). *Environ. Res.* **4,** 71–85.

Fontenot, J. P. (1972). *In* "Magnesium in the Environment -Soils, Crops, Animals and Man" (J. B. Jones, M. C. Blount, and S. R. Wilkinson, eds), Chapter 5, pp. 131–151. Taylor County Printing Co., Reynolds, Georgia.

Friberg, L., and J. Vostal (1972). "Mercury in the Environment." CRC Press, Cleveland, Ohio.

Frost, D. V. (1976). *Feedstuffs* **48**(52), 19–31.

Ganther, H. E., and C. A. Baumann (1962). *J. Nutr.* **77**, 408–414.

Garrigus, U. S. (1970). *In* "Symposium: Sulfur in Nutrition" (O. H. Muth and J. E. Oldfield, eds.), pp. 126–152. Avi Publ. Co., Westport, Connecticut.

Grieve, D. G., C. E. Coppock, W. G. Merrill and H. F. Tyrrell (1973). *J. Dairy Sci.* **56**, 218–223.

Gunn, S. A., and T. C. Gould (1967). *In* "Symposium: Selenium in Biomedicine" (O. H. Muth, J. E. Oldfield, and P. H. Weswig, eds.), pp. 395–413. Avi Publ. Co., Westport, Connecticut.

Haase, G., and W. L. Dunkley (1970). *Milchwissenschaft* **25**(11), 656–661.

Hammond, P. B., and A. L. Aronson (1964). *Ann. N. Y. Acad. Sci.* **111**, 595–611.

Hansard, S. L., C. L. Comar, and M. P. Plumlee (1954). *J. Anim. Sci.* **13**, 25–36.

Hansard, S. L., H. M. Crowder, and W. A. Lyke (1957). *J. Anim. Sci.* **16**, 437–443.

Hartley, W. J. (1967). *In* "Symposium: Selenium in Biomedicine" (O. H. Muth, J. E. Oldfield and P. H. Weswig, eds.), pp. 77–96. Avi Publ. Co., Westport, Connecticut.

Hartley, W. J., and A. B. Grant (1961). *Fed. Proc., Fed. Am. Soc. Exp. Biol.* **20**, 679–688.

Hartmans, J. (1974). *Trace Elem. Metab. Anim., 2nd, 1973* pp. 261–273.

Hawkins, G. E., Jr., G. H. Wise, G. Matrone, and R. K. Waugh (1955). *J. Dairy Sci.* **38**, 536–547.

Hemken, R. W. (1970). *J. Dairy Sci.* **53**, 1138–1143.

Hemken, R. W. (1971). *Hoard's Dairyman* **116**, 11.

Hemken, R. W. (1975a). *Feed Manage.* **26**(9), 7–8.

Hemken, R. W. (1975b). *Feed Manage.* **26**(12), 17–18.

Hemken, R. W., J. H. Vandersall, B. A. Sass, and J. W. Hibbs (1971). *J. Dairy Sci.* **54**, 85–88.

Herigstad, R. R., C. K. Whitehair, N. Beyer, O. Mickelsen, and M. J. Zabik (1972). *J. Am. Vet. Med. Assoc.* **160**, 173–182.

Hoekstra, W. G. (1974). *Trace Elem. Metab. Anim., 2nd, 1973* pp. 61–77.

Hoekstra, W. G. (1975). *Fed. Proc., Fed. Am. Soc. Exp. Biol.* **34**, 2083–2089.

Hogan, A. G., and J. L. Nierman (1927). *Mo., Agric. Exp. Stn., Res. Bull.* **107.**

Hogue, D. E. (1970). *J. Dairy Sci.* **53**, 1135–1137.

Hopkins, L. L., Jr., and H. E. Mohr (1974). *Fed. Proc., Fed. Am. Soc. Exp. Biol.* **33**, 1773–1775.

Iwarsson, K. (1973). Dissertation, Royal Veterinary College, Stockholm, Sweden.

Jacobson, D. R., J. W. Barnett, S. B. Carr, and R. H. Hatton (1967). *J. Dairy Sci.* **50**, 1248–1254.

Johnson, J. M., and G. W. Butler (1957). *Physiol. Plant.* **10**, 100–111.

Johnson, W. H., R. D. Goodrich, and J. C. Meiske (1971). *J. Anim. Sci.* **32**, 778–783.

Jorgensen, N. A. (1974). *J. Dairy Sci.* **57**, 933–944.

Julien, W. E., H. R. Conrad, and A. L. Moxon (1976). *J. Dairy Sci.* **59**, 1960–1962.

Keener, H. A., G. P. Percival, K. S. Morrow, and G. H. Ellis (1949). *J. Dairy Sci.* **32**, 527–533.

Kemp, A. (1960). *Neth. J. Agric. Sci.* **8**, 281–304.

Kemp, A. (1963). *Tijdschr. Diergeneeskd.* **88**, 1154–1172.

Kemp, A. (1964). *Neth. J. Agric. Sci.* **12**, 263–280.

Kemp, A., W. B. Keijs, O. J. Hemkes, and A. J. H. Van Es (1961). *Neth. J. Agric. Sci.* **9**, 134–149.

Kincaid, R. L., W. J. Miller, P. R. Fowler, R. P. Gentry, D. L. Hampton, and M. W. Neathery (1976). *J. Dairy Sci.* **59**, 1580–1584.

King, R. L., and W. L. Dunkley (1959). *J. Dairy Sci.* **42**, 420–427.

Kirchgessner, M. (1959). *Z. Tierphysiol., Tierernaehr. Futtermittelkd.* **14**, 270–278 (From Underwood, 1966).

Kirchgessner, M., and W. A. Schwarz (1975). *Zentralbl. Veterinaer med., Reihe A* **22**, 572–582.

Kirchgessner, M., and W. A. Schwarz (1976). *Arch. Tierernaehr.* **26**, 3–16.

Kirchgessner, M., W. A. Schwarz, and H. P. Roth (1978). *Trace Elem. Metab. Man Anim., 3rd, 1977* pp. 116–121.

Krehl, W. A. (1966). *Nutr. Today* **1**, December, pp. 16–24.

Krehl, W. A. (1967). *Nutr. Today* **2,** September, pp. 16–20.

Krehl, W. A. (1970). *Nutr. Today* **5,** Winter, pp. 26–31.

Kroneman, J., G. J. W. von der Mey, and A. Helder (1975). *Zentralbl. Veterinaer med., Reihe A* **22,** 201–208.

Krook, L., L. Lutwak, and K. McEntee (1969). *Am. J. Clin. Nutr.* **22,** 115–118.

Krook, L., L. Lutwek, K. McEntee, P. A. Henrikson, K. Braun, and S. Roberts (1971). *Cornell Vet.* **61,** 625–639.

Kubota, J., and W. H. Allaway (1972). *In* "Micronutrients in Agriculture" (J. J. Mortvedt, P. M. Giordano, and W. L. Lindsey, eds.), pp. 525–554. Soil Sci. Soc. Am., Madison, Wisconsin.

Kubota, J., W. H. Allaway, D. L. Carter, E. E. Cary, and V. A. Lazar (1967). *Agric. Food Chem.* **15,** 448–453.

Lengemann, F. W., and E. W. Swanson (1957). *J. Dairy Sci.* **40,** 215–224.

Lopez, P. S., R. L. Preston, and W. H. Pfander (1969). *J. Nutr.* **97,** 123–132.

Lynch, G. P., D. G. Cornell, and D. F. Smith (1974). *Trace Elem. Metab. Animals, 2nd, 1973* pp. 470–472.

McCullough, M. E. (1973). "Optimum Feeding of Dairy Animals for Meat and Milk." University of Georgia Press, Athens.

MacDougall, D. B., I. Bremner, and A. C. Dalgarno (1973). *J. Sci. Food Agric.* **24,** 1255–1263.

Manston, R. (1967). *J. Agric. Sci.* **68,** 263–268.

Matrone, G., C. Conley, G. H. Wise, and R. K. Waugh (1957). *J. Dairy Sci.* **40,** 1437–1447.

Merck Veterinary Manual (1973). "Lead Poisoning" (O. H. Siegmund, ed.), 4th ed., pp. 935–937. Merck and Co., Inc., Rahway, New Jersey.

Messer, H. H., W. D. Armstrong, and L. Singer (1974). *Trace Elem. Metab. Animals, 2nd, 1973* pp. 425–437.

Meyer, J. H., W. C. Weir, N. R. Ittner, and J. D. Smith (1955). *J. Anim. Sci.* **14,** 412–418.

Miller, J. K. (1967). *J. Nutr.* **93,** 386–392.

Miller, J. K., and W. J. Miller (1962). *J. Nutr.* **76,** 467–474.

Miller, J. K., B. R. Moss, E. W. Swanson, P. W. Aschbacher, and R. G. Cragle (1968). *J. Dairy Sci.* **51,** 1831–1835.

Miller, J. K., E. W. Swanson, and G. E. Spalding (1975). *J. Dairy Sci.* **58,** 1578–1593.

Miller, W. J. (1970a). *Ga., Nutr. Conf. Feed Ind.* pp. 32–42.

Miller, W. J. (1970b). *J. Dairy Sci.* **53,** 1123–1135.

Miller, W. J. (1971a). *Feedstuffs* **43**(29), 24–26.

Miller, W. J. (1971b). *In* "Mineral Studies with Isotopes in Domestic Animals," pp. 23–41. IAEA, Vienna.

Miller, W. J. (1973). *Fed. Proc., Fed. Am. Soc. Exp. Biol.* **32,** 1915–1920.

Miller, W. J. (1974). *Fed. Proc., Fed. Am. Soc. Exp. Biol.* **33,** 1747.

Miller, W. J. (1975a). *J. Dairy Sci.* **58,** 1549–1560.

Miller, W. J. (1975b). *Feedstuffs* **47**(8), 40–41.

Miller, W. J., and R. L. Kincaid (1975). *Prof. Nutr.* **7,** 2–4.

Miller, W. J., and M. W. Neathery (1977). *BioScience* **27,** 674–679.

Miller, W. J., and P. E. Stake (1972). *Proc. AFMA, Liquid Feed Symp. 1st, 1971,* pp. 54–64.

Miller, W. J., and P. E. Stake (1974a). *Feedstuffs* **46**(28), 24, 25, and 35.

Miller, W. J., and P. E. Stake (1974b). *Proc. Ga. Nutr. Conf. Feed Ind.* pp. 25–43.

Miller, W. J., C. M. Clifton, and N. W. Cameron (1963). *J. Dairy Sci.* **46,** 715–719.

Miller, W. J., C. M. Clifton, P. R. Fowler, and H. F. Perkins (1965a). *J. Dairy Sci.* **48,** 450–453.

Miller, W. J., W. J. Pitts, C. M. Clifton, and J. D. Morton (1965b). *J. Dairy Sci.* **48,** 1329–1334.

Miller, W. J., G. W. Powell, and J. M. Hiers, Jr. (1966). *J. Dairy Sci.* **49,** 1012–1013.

Miller, W. J., B. Lampp, G. W. Powell, C. A. Salotti, and D. M. Blackmon (1967). *J. Dairy Sci.* **50,** 1404–1408.

Miller, W. J., Y. G. Martin, R. P. Gentry, and D. M. Blackmon (1968). *J. Nutr.* **94,** 391–401.

Miller, W. J., D. M. Blackmon, R. P. Gentry, and F. M. Pate (1969). *J. Dairy Sci.* **52,** 2029–2035.

Miller, W. J., D. M. Blackmon, R. P. Gentry, and F. M. Pate (1970). *J. Nutr.* **100,** 893–902.

Miller, W. J., W. M. Britton, and M. S. Ansari (1972a). *In* "Magnesium in the Environment—Soils, Crops, Animals, and Man" (J. B. Jones, Jr., M. C. Blount and S. R. Wilkinson, eds.), Chapter 4, pp. 109–130. Taylor Printing Co., Reynolds, Georgia.

Miller, W. J., J. W. Lassiter, and J. B. Jones, Jr. (1972b). *Proc. Ga. Nutr. Conf. Feed Ind.* pp. 94–106.

Miller, W. J., R. L. Kincaid, M. W. Neathery, R. P. Gentry, M. S. Ansari, and J. W. Lassiter (1978). *Trace Elem. Metab. Man Animals, 3rd, 1977.* pp. 175–178.

Moir, R. J. (1970). *In* "Symposium: Sulfur in Nutrition" (O. H. Muth and J. E. Oldfield, eds.), pp. 165–181. Avi Publ. Co., Westport, Connecticut.

Moir, R. J., M. Somers, and A. C. Bray (1968). *Sulphur Inst. J.* **3,** 15–18.

Möllerberg, L. (1974). "Studies in Normal and Iron-Deficiency Anemic Calves." Depts. of Medicine II and Clinical Biochemistry, Royal Veterinary College, Stockholm, Sweden.

Moore, L. A., E. T. Hallman, and L. B. Sholl (1938). *Arch. Pathol.* **26,** 820–838.

Mortimer, C. E. (1967). "Chemistry, A Conceptual Approach." Van Nostrand-Reinhold, Princeton, New Jersey.

Murthy, G. K., and U. Rhea (1968). *J. Dairy Sci.* **51,** 610–612.

Murthy, G. K., U. Rhea, and J. T. Peeler (1967). *J. Dairy Sci.* **50,** 651–654.

Muth, O. H., and J. E. Oldfield, eds. (1970). "Symposium: Sulfur in Nutrition." Avi Publ. Co., Westport, Connecticut.

Muth, O. H., J. E. Oldfield, and P. H. Weswig, eds. (1967). "Symposium: Selenium in Biomedicine." Avi Publ. Co., Westport, Connecticut.

National Research Council (NRC) (1969). "U.S.-Canadian Tables of Feed Composition," 2nd rev. ed., Publ. No. 1684. Natl. Acad. Sci., Washington, D.C.

National Research Council (NRC) (1970). "Nutrient Requirements of Beef Cattle," 4th rev. ed. Natl. Acad. Sci., Washington, D.C.

National Research Council (NRC) (1971a). "Nutrient Requirements of Dairy Cattle," 4th rev. ed. Natl. Acad. Sci., Washington, D.C.

National Research Council (NRC) (1971b). "Selenium in Nutrition." Natl. Acad. Sci., Washington, D.C.

National Research Council (NRC) (1972). "Lead. Airborne Lead in Perspective." Natl. Acad. Sci., Washington, D.C.

National Research Council (NRC) (1974a). "Effects of Fluorides in Animals." Natl. Acad. Sci., Washington, D.C.

National Research Council (NRC) (1974b). "Feed Phosphorus Shortage," pp. 12–17. Natl. Acad. Sci., Washington, D.C.

National Research Council (NRC) (1976). "Nutrient Requirements of Beef Cattle," 5th rev. ed. Natl. Acad. Sci., Washington, D.C.

National Research Council (NRC) (1978). "Nutrient Requirements of Dairy Cattle," 5th rev. ed. Natl. Acad. Sci., Washington, D.C.

Neal, W. M., and C. F. Ahmann (1937). *J. Dairy Sci.* **20,** 741–753.

Neathery, M. W., and W. J. Miller, (1975). *J. Dairy Sci.* **58,** 1766–1781.

Neathery, M. W., and W. J. Miller (1976a). *Feedstuffs* **48**(3), 30–32.

Neathery, M. W., and W. J. Miller (1976b). *Feedstuffs* **48**(5), 21–22.

Neathery, M. W., and W. J. Miller (1976c). *Feedstuffs* **48**(7), 36–41.

Neathery, M. W., and W. J. Miller (1977). *Feedstuffs* **49**(36), 18, 19, and 34.

Neathery, M. W., W. J. Miller, D. M. Blackmon, F. M. Pate, and R. P. Gentry (1973a). *J. Dairy Sci.* **56,** 98–105.

Neathery, M. W., W. J. Miller, D. M. Blackmon, R. P. Gentry, and J. B. Jones (1973b). *J. Anim. Sci.* **37,** 848–852.

Neathery, M. W., W. J. Miller, D. M. Blackmon, and R. P. Gentry (1973c). *J. Dairy Sci.* **56,** 212–217.

Neathery, M. W., W. J. Miller, R. P. Gentry, P. E. Stake, and D. M. Blackmon (1974). *J. Dairy Sci.* **57,** 1177–1183.

Netherlands Committee on Mineral Nutrition (1973). ''Tracing and Treating Mineral Disorders in Dairy Cattle.'' Centre for Agricultural Publishing and Documentation, Wageningen, The Netherlands.

Niedermeier, R. P., N. N. Allen, R. D. Lance, E. H. Rupnow, and R. W. Bray (1959). *J. Anim. Sci.* **18,** 726–731.

Nielsen, F. H., and D. A. Ollerich (1974). *Fed. Proc., Fed. Am. Soc. Exp. Biol.* **33,** 1767–1772.

Newton, G. L., J. P. Fontenot, R. E. Tucker, and C. E. Polan (1972). *J. Anim. Sci.* **35,** 440–445.

O'Dell, G. D., W. J. Miller, W. A. King, J. C. Ellers, and H. Jurecek (1970a). *J. Dairy Sci.* **53,** 1545–1548.

O'Dell, G. D., W. J. Miller, W. A. King, S. L. Moore, and D. M. Blackmon (1970b). *J. Nutr.* **100,** 1447–1454.

Oldfield, J. E., W. H. Allaway, H. A. Laitinen, A. W. Lakin, and O. H. Muth (1974). *In* ''Geochemistry and the Environment,'' Vol. I, pp. 57–63. Natl. Acad. Sci., Washington, D.C.

Olson, O. E. (1969). *Proc. Ga. Nutr. Conf. Feed Manuf.* pp. 68–78.

Olson, O. E., and L. B. Embry (1973). *Proc. S.D. Acad. Sci.* **52,** 50–58.

Oltjen, R. R. (1975). *Proc. Ga. Nutr. Conf. Feed Ind.* pp. 31–40.

Ott, E. A., W. H. Smith, R. B. Harrington, and W. M. Beeson (1966). *J. Anim. Sci.* **25,** 419–423.

Parizek, J., J. Kalouskova, A. Babicky, J. Benes, and L. Pavlik (1974). *Trace Elem. Metab. Anim., 2nd, 1973* pp. 119–131.

Parkash, S., and R. Jenness (1967). *J. Dairy Sci.* **50,** 127–134.

Peeler, H. T. (1972). *J. Anim. Sci.* **35,** 695–712.

Perry, T. W., W. M. Beeson, W. H. Smith, and M. T. Mohler (1968). *J. Anim. Sci.* **27,** 1674–1677.

Pope, A. L. (1971). *Feed Manage.* **22** (12), 29 and 32.

Potter, G. D., D. R. McIntyre, and G. M. Vattuone (1972). *Health Phys.* **22,** 103–106.

Powell, G. W., W. J. Miller, J. D. Morton, and C. M. Clifton (1964). *J. Nutr.* **84,** 205–214.

Pradhan, K., and R. W. Hemken (1968). *J. Dairy Sci.* **51,** 1377–1381.

Ricketts, R. E., J. R. Campbell, D. E. Weinman, and M. E. Tumbleson (1970). *J. Dairy Sci.* **53,** 898–903.

Rojas, M. A., I. A. Dyer, and W. A. Cassatt (1965). *J. Anim. Sci.* **24,** 664–667.

Rook, J. A. F., and J. E. Storry (1962). *Nutr. Abstr. Rev.* **32,** 1058.

Rosas, H., and M. C. Bell (1969). *Radiat. Res.* **39,** 164–176.

Rotruck, J. T., A. L. Pope, H. E. Ganther, A. B. Swanson, D. G. Hafeman, and W. G. Hoekstra (1973). *Science* **179,** 588–590.

Schellner, G., M. Anke, H. Lüdke, and A. Henning (1971). *Arch. Exp. Veterinaer med.* **25,** 823–827.

Schroeder, H. A. (1965). *J. Chronic Dis.* **18,** 217–228.

Schwarz, K. (1974). *Fed. Proc., Fed. Am. Soc. Exp. Biol.* **33,** 1748–1757.

Schwarz, K., and C. M. Foltz (1957). *J. Am. Chem. Soc.* **79,** 3292–3293.

Schwarz, W. A., and M. Kirchgessner (1975). *Arch. Tierernaehr.* **25,** 597–608.

Sell, J. L., and K. L. Davison (1973). *J. Dairy Sci.* **56,** 671 (abstr.).

Shortridge, E. H., P. J. O'Hara, and P. M. Marshall (1971). *N. Z. Vet. J.* **19,** 47–50.

Smith, A. M., G. L. Holck, and H. B. Spafford (1966). *J. Dairy Sci.* **49,** 239–243.

Smith, S. E., and P. D. Aines (1959). *N.Y., Agric. Exp. Stn., Bull.* **938,** Ithaca.

Smith, S. E., and G. J. St. Laurent (1970). *Proc. Cornell Nutr. Conf. Feed Mfg.* pp. 77–84.

Smith, S. E., F. W. Lengemann, and J. T. Reid (1953). *J. Dairy Sci.* **36,** 762–765.

Somers, M., and E. J. Underwood (1969). *Aust. J. Agric. Res.* **20,** 899–903.

Stake, P. E. (1974). *Proc. Ga. Nutr. Conf. Feed. Ind.* pp. 63–68.

Standish, J. F., C. B. Ammerman, F. C. Neal, A. Z. Palmer, and C. F. Simpson (1968). *J. Anim. Sci.* **27,** 1177 (abstr.).

Standish, J. F., C. B. Ammerman, C. F. Simpson, F. C. Neal, and A. Z. Palmer (1969). *J. Anim. Sci.* **29,** 496–503.

Stoddard, G. E., G. Q. Bateman, L. E. Harris, J. L. Shupe, and D. A. Greenwood (1963). *J. Dairy Sci.* **46,** 720–726.

Suttie, J. W., R. F. Miller, and P. H. Phillips (1957). *J. Nutr.* **63,** 211–224.

Suttie, J. W., R. Gesteland, and P. H. Phillips (1961). *J. Dairy Sci.* **44,** 2250–2258.

Suttle, N. F., and K. W. Angus (1976). *J. Comp. Pathol.* **86,** 595–608.

Swanson, E. W. (1972). *J. Dairy Sci.* **55,** 1763–1767.

Taylor, A. N. (1973). *Proc. Ga. Nutr. Conf. Feed Ind.* pp. 77–87.

Thomas, W. E., J. K. Loosli, H. H. Williams, and L. A. Maynard (1951). *J. Nutr.* **43,** 515–523.

Thornton, I., G. F. Kershaw, and M. K. Davies (1972). *J. Agric. Sci.* **78,** 165–171.

Todd, J. R. (1970). *Trace Elem. Metab. Anim., Proc. WAAP/IBP Int. Symp., 1969* pp. 448–451.

Todd, J. R., and R. H. Thompson (1965). *Brit. Vet. J.* **121,** 90–97.

Tomas, F. M., and B. J. Potter (1976). *Br. J. Nutr.* **36,** 37–45.

Ullrey, D. E. (1974). *Trace Elem. Metab. Anim., 2nd, 1973* pp. 275–293.

Underwood, E. J. (1966). "The Mineral Nutrition of Livestock." Central Press, Ltd., Aberdeen.

Underwood, E. J. (1970). *Trace Elem. Metab. Anim., Proc. WAAP/IBP Int. Symp., 1969* pp. 5–21.

Underwood, E. J. (1971). "Trace Elements in Human and Animal Nutrition," 3rd ed. Academic Press, New York.

Underwood, E. J. (1977). "Trace Elements in Human and Animal Nutrition," 4th ed. Academic Press, New York.

Vagg, M. J., and J. M. Payne (1971). *In* "Mineral Studies with Isotopes in Domestic Animals," pp. 121–123.

Vanderveen, J. E., and H. A. Keener (1964). *J. Dairy Sci.* **47,** 1224–1230.

Van't Klooster, A. Th. (1976). Z. Tierphysiol. Tierernaehr. FutterMittekd **37,** 169–182.

Van Ulsen, F. W. (1973). *Tijdschr. Diergeneeskd.* **98,** 543–546.

Voelker, H. H., N. A. Jorgensen, G. P. Mohanty, and M. J. Owens (1969). *J. Dairy Sci.* **52,** 119 (abstr.).

Wacker, W. E. C., and A. F. Parisi (1968). *N. Engl. J. Med.* **278,** 658–663, 712–717, and 772–776.

Waite, R., H. R. Conrad, and A. L. Moxon (1975). *J. Dairy Sci.* **58,** 749–750.

Ward, G. M. (1966). *J. Dairy Sci.* **49,** 268–276.

Weeth, H. J., and L. H. Haverland (1961). *J. Anim. Sci.* **20,** 518–521.

Whanger, P. D., and G. Matrone (1970). *In* "Symposium: Sulfur in Nutrition" (O. H. Muth and J. E. Oldfield, eds.), pp. 153–164. Avi Publ. Co., Westport, Connecticut.

Wise, M. B., S. E. Smith, and L. L. Barnes (1958). *J. Anim. Sci.* **17,** 89–99.

Wise, M. B., A. L. Ordoveza, and E. R. Barrick (1963). *J. Nutr.* **79,** 79–84.

Zurcher, T. (1970). Ph.D. Dissertation, Purdue University, Lafayette, Indiana.

6

Vitamin Requirements of Dairy Cattle

6.0. INTRODUCTION

Dairy cattle require the same vitamins as nonruminants such as poultry, swine, and rats. However, because of synthesis in the rumen and tissues, most vitamins are not needed in the diet of dairy cattle which have a functioning rumen (National Research Council, 1978).

Rumen microbes synthesize enough B vitamins and vitamin K to meet the usual needs of dairy cattle except for young calves. Before the rumen begins functioning, young dairy calves must have a dietary source of these vitamins. Vitamin C is synthesized in the tissues of cattle and, thus, is not needed in the diet. Dairy cattle of all ages must have a source of vitamins A and E in the diet. Likewise, vitamin D must either be in the diet or synthesized in the skin under the influence of ultraviolet irradiation in sunlight. Some research suggests rumen synthesis of choline may not be adequate to meet the needs of older cattle in certain unusual circumstances (Rumsey, 1975) (see Section 6.7). Additional vitamin K also may be beneficial when excessive amounts of dicumarol are consumed (see Section 6.5).

In practical feeding of dairy cattle, vitamin deficiencies should not be a major problem. With many typical diets, no supplemental vitamins are needed for either baby calves or older animals. However, added vitamins are essential for optimum health and performance with certain types of diets. Because the amounts needed are extremely small, the cost, generally, is minor for adding needed supplemental vitamins. Fortunately, the safety margins between minimum needs and maximum safe levels are very wide. Since cattle are able to store sufficient reserves of the fat-soluble vitamins (A, D, E, and K) to meet their needs for a period of time, continuous supplementation is not mandatory.

6.1. THE INDIVIDUAL VITAMINS

Because they are essential for the optimum performance and well-being of dairy cattle, each of the vitamins is discussed individually. The amount of information presented varies substantially, reflecting the huge differences in the practical importance of the vitamins and in the information known. Vitamins are classified as fat-soluble vitamins which include A, D, E, K, and water-soluble ones which are the B vitamins and vitamin C.

6.2. VITAMIN A

6.2.a. Functions and Deficiency Effects

Vitamin A is essential for maintenance of normal epithelial tissues in numerous parts of the body (Harris, 1975; NRC, 1978). Thus, a deficiency is characterized by stratified keratinization (development of horny layer), degeneration, and drying out of the epithelial or mucosal tissue of the intestinal tract, mouth, salivary glands, eyes, tear glands, urethra, kidneys, vagina, gonads, and respiratory tract (NRC, 1978). When affected by lack of vitamin A, these tissues are highly susceptible to infections of all types including colds and pneumonia. Because of the nature of the deficiency symptoms, often vitamin A is known as the antiinfection vitamin. Other symptoms associated with vitamin A deficiency in dairy cattle include diarrhea, reduced appetite, loss of weight or lower gains, rough hair coat, and unthrifty appearance.

In a severe vitamin A deficiency, characteristic changes occur in the eye including excessive watering, keratitis, softening and cloudiness of the cornea, and development of xerophthalmia that is characterized by drying of the conjunctiva. Blindness may follow the eye infections caused by vitamin A deficiency.

Since the visual process requires vitamin A, a deficiency reduces the ability of the animal to adapt to dim light. This condition, known as night blindness, is one of the earliest symptoms of a deficiency. Animals affected with night blindness can be readily detected if driven among obstacles in dim light (NRC, 1978).

In an advanced vitamin A deficiency the cerebrospinal fluid pressure is elevated and may result in a staggering gait and convulsive seizures probably caused by the increased cerebrospinal fluid pressure (NRC, 1978). Vitamin A deficiency lowers reproductive efficiency in both males and females. Key indications of the deficiency are shortened pregnancies either as abortions or reduced gestation length; a high incidence of retained placenta; and the birth of dead, weak, incoordinated, or permanently blind calves caused by bone abnormalities in the optic foramen which constricts the optic nerve (NRC, 1978).

One of the most sensitive ways to detect a vitamin A deficiency in growing calves is through the elevation of cerebrospinal fluid pressure (NRC, 1978). Results of early studies suggested that values higher than 120 mm of saline indicated a deficiency (Moore *et al.*, 1948). Subsequent research demonstrated that the point at which the cerebrospinal fluid pressure begins to increase is a more accurate indicator (Eaton *et al.*, 1964, 1972). This change in cerebrospinal fluid pressure can be used as a measure of vitamin A adequacy.

Both liver vitamin A and plasma vitamin A are relatively good indicators of the status of this vitamin in cattle. With Holstein calves, a plasma vitamin A level less than 20 μg/100 ml suggests a deficiency (Eaton *et al.*, 1970). A plasma level of 10 μg/100 ml indicates an advanced deficiency. Likewise, liver vitamin A values below 1.0 μg/gm is indicative of a critical deficiency (NRC, 1978).

6.2.b. Sources and Requirements of Vitamin A and Precursors (Carotene)

Most of the vitamin A obtained from usual feedstuffs is in the form of precursors or provitamins (Harris, 1975). Several compounds in plants including carotenes can be converted to vitamin A in the small intestines of cattle. Of these provitamin A compounds, β-carotene is the most active, the most important, and thus the one generally considered in nutrition. Often β-carotene is referred to as just carotene.

Vitamin A does not exist in plants, but green forages have a very high content of carotene, and yellow corn grain contains appreciable cryptoxanthin which is a provitamin A. Most of the vitamin A activity in milk and animal products is vitamin A. Likewise, most of the supplemental vitamin A used in animal feeds is vitamin A. "Vitamin A activity" is used to denote the total effective amounts of vitamin A and provitamin A in diets or other materials.

Historically, the vitamin A requirements of dairy cattle have been expressed in terms of carotene because it was the primary source of vitamin A activity in cattle diets. With the increased usage of supplemental, synthetic vitamin A, it is important to express the requirements as vitamin A also. Unfortunately, some problems are associated in converting carotene requirements to vitamin A units. For rat growth, 1.0 mg β-carotene is equivalent to 1,667 IU (international units) of vitamin A. β-Carotene is converted to vitamin A by cattle with much less efficiency. The conversion ratio is influenced by the level of carotene consumed, the sources of the carotene, the vitamin A status of the animal and by the genetics of the animal (NRC, 1978). For cattle, 1.0 mg of β-carotene has been assumed to be equivalent to 400 IU of vitamin A which is only 24% of the rat value. As the level of carotene increases, the number of units of vitamin A equivalent decreases below this level.

The estimated minimum maintenance requirement of vitamin A is 1900 IU/100 lb of body weight each day (see Appendix, Tables 1, and 2) (NRC,

1978). Using 400 IU per mg of β-carotene, this would be 4.8 mg β-carotene. The 4.8 mg of β-carotene per 100 lb of body weight was based on data obtained with calves by Eaton *et al.* (1964) indicating this was the minimum needed for normal cerebrospinal fluid pressure (NRC, 1978). Results of a more recent study (Eaton *et al.*, 1972) suggest that 4400 IU of vitamin A per 100 lb of body weight daily is necessary to prevent any elevation of cerebrospinal fluid pressure in growing Holstein bull calves (NRC, 1978). Because no evidence of field problems had appeared, the suggested minimum requirement levels were not increased in the most recent revision of the Nutrient Requirements of Dairy Cattle (NRC, 1978). Thus, the values selected represent minimums (NRC, 1978). Vitamin requirements are higher under stressful conditions such as abnormal temperatures or exposure to infective bacteria (NRC, 1978). Likewise, many other factors may possibly affect the metabolism and increase the requirements of vitamin A. These include free nitrates in feeds, inadequate protein, a zinc deficiency, and low dietary phosphorus (Harris, 1975).

Additional vitamin A (or carotene) is needed during the last 2 to 3 months of gestation (NRC, 1978). Although a vitamin A deficiency reduces fertility in bulls, there is no indication that the requirement is higher than for maintenance.

The amount of vitamin A activity in the milk is greatly affected by the vitamin A and β-carotene content of the diet. Since there is no indication that higher amounts are needed for normal health and performance of cows (Bratton *et al.*, 1948), the listed requirements for lactating cows is the same as for maintenance of nonlactating cows (NRC, 1978).

6.2.c. Destruction of Vitamin A and Carotene

The amounts of carotene in fresh green forages is very high relative to the dietary requirements. Thus, a small quantity of fresh pasture forage will fully supply the needs of dairy cattle. In contrast to the general high stability of energy, protein, and minerals, both vitamin A and carotene are very easily oxidized to compounds having no vitamin A activity (Fig. 6.1). Heat, moisture, and especially light are potent agents in the oxidation and destruction of carotene. Thus, hay usually has only a small proportion of the carotene content of fresh grass. Even so, well preserved hay can be a fairly good source with the amount of green color giving some indication of the potency. Carotene content in hay declines continuously in storage with very rapid destruction when exposed to light. Generally, the acid medium of silage is a reasonably good preservative for carotene.

The rate at which supplemental vitamin A is oxidized and destroyed depends on the nature of the feed, storage conditions, and the degree to which the vitamin was stabilized. Most commercially sold vitamin A is stabilized to resist oxida-

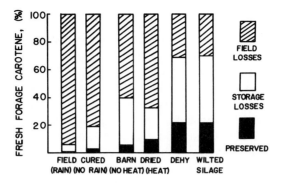

Fig. 6.1 Carotene losses during harvesting and storage of alfalfa as hay or silage. Note the very small percentage of the original carotene which was preserved for feeding in rain damaged field cured hay. Even with the best preservation methods, most of the carotene of fresh forage is lost. (Adapted from Shepherd *et al.*, 1954.)

tion. Conditions which speed oxidation are high temperatures, moisture, light, and the presence of oxidative catalysts such as copper and certain other minerals and compounds. An acid medium as in many liquid supplements, substantially offsets the effects of the moisture which otherwise would result in rapid oxidation. Even under the best of conditions, it is prudent to avoid storage of feeds for substantial periods after mixing.

6.2.d. Body Reserves of Vitamin A and Carotene

During periods of surplus intake, dairy cattle can store substantial amounts of vitamin A, especially in the liver. A dairy animal with high reserves, can use these to meet all its needs for several weeks. When vitamin A reserves are high, the use and decline rates are far more rapid than when they are low.

At birth, the calf usually does not have sufficient reserves to provide for its needs for any substantial time. Accordingly, it is important that the calf receive colostrum, which generally is rich in vitamin A, or another source of the vitamin within a few days after birth. If the cow has received a diet low in vitamin A activity, the newborn calf is likely to be susceptible to a vitamin A deficiency because the body reserves are low and the colostrum will have a subnormal content.

Although most of the vitamin A reserves are in the liver, when carotene intake is high, some is stored in fat. Yellow body fat is associated with the carotene. The amount of yellow fat is strongly influenced by the breed of cattle with Guernsey and Jerseys having more than Holsteins.

6.2.e. Vitamin A Toxicity

The efficiency of carotene conversion to vitamin A declines progressively with increasing intakes (see Section 6.2b). This appears to be a natural "homeostatic control mechanism" (see Section 2.6 for definition of homeostasis), which protects cattle from any harmful effects due to the great abundance of carotene present in high quality, fresh forages when they are the major feed for long periods. Likewise, the comparatively rapid disposal of very high levels of stored vitamin A is a protective mechanism. On a practical basis, toxicity is more easily caused by vitamin A than by carotene. Even so, vitamin A toxicity is not a practical problem, except when unreasonably large amounts are given accidentally or from lack of planning.

Extended vitamin A toxicity causes decreased growth; lower feed consumption; enlarged liver, heart, and kidney; elevated heart rate; bone changes, including enlargement of the frontal sinus; possibly increased vitamin E requirements; and perhaps reduced cerebrospinal fluid pressure (Hazzard *et al.*, 1964). The dietary level at which damage occurs varies among affected tissues with some changes observed in bones when as little as 60,000 IU of vitamin A per 100 lb of body weight per day is given (Hazzard *et al.*, 1964). This level is only about 30 times the requirement. Since 1.0 IU of vitamin A alcohol is 0.3 μg, the 60,000 IU would be only 0.018 gm, and could easily be added to feed accidently. In contrast, weight gains were depressed above 400,000 IU of vitamin A per 100 lb of body weight. When fed about 800,000 IU of vitamin A per 100 lb of body weight per day, all calves developed characteristic vitamin A toxicity symptoms (Hazzard *et al.*, 1964).

6.2.f. Other Practical Considerations

Partially because of the very small quantities, the cost of providing supplemental vitamin A is very low—only a few cents per million units. Accordingly, the decision to add vitamin A to the diet should be based mainly on whether or not a deficiency could be a practical problem. As with most other nutrients, a borderline deficiency is much more likely than a severe deficiency. Likewise, a marginal deficiency adversely affecting performance by a few percentage points is not easily detected.

When cattle receive a modest amount of fresh, green pasture forage, there is little likelihood of a deficiency. Likewise, with a substantial amount of good silage made from green forage, or with liberal feeding of fresh hay with a good green color, a deficiency would not occur.

Conditions where supplemental vitamin A may be needed include (1) feeding of poor quality forage or forage with little or no green color, (2) diets composed primarily of concentrates and no green pasture, (3) feeding mainly corn silage and a concentrate mixture low in vitamin A activity (Jordan *et al.*, 1963), (4) in young calves fed milk from cows on a low intake of vitamin A or carotene, and (5) when calves are fed relatively little whole milk or colostrum (NRC, 1978).

6.3. VITAMIN D

6.3.a. Functions and Deficiency Effects

Vitamin D is essential for normal calcium and phosphorus metabolism and influences the metabolism of certain other minerals (NRC, 1978). This vitamin has a key role in the synthesis of a special calcium-binding protein that is involved with calcium absorption (Dobson and Ward, 1974; Taylor, 1973). Vitamin D also facilitates deposition of calcium and phosphorus in bone.

In young dairy cattle, a vitamin D deficiency results in rickets which is characterized by retarded calcification of bones, accumulation of osteoid tissue, and easily broken bones (Bechtel *et al.*, 1936; NRC, 1978). Other clinical signs include swollen or enlarged joints such as knees and hocks, arched back, straight pasterns, and beading at the end of the ribs (NRC, 1978).

Clinical signs observed in advanced stages of rickets include a stiff gait, dragging of the hind feet, accumulation of fluid in the joints, irritability, tetany, fast and labored breathing, greatly decreased appetite, weakness, and slower growth (NRC, 1978).

In older dairy cattle, a vitamin D deficiency causes osteomalacia characterized by reabsorption of the mineral from the already formed bone. If the deficiency is continued for an extended time, the bones become soft, weak, and easily broken (Dobson and Ward, 1974; NRC, 1978). Likewise, milk production may be decreased and estrus inhibited by inadequate vitamin D. If vitamin D is marginally deficient, the dietary requirement for calcium and phosphorus are increased (Dobson and Ward, 1974; NRC, 1978).

Among the first indications of rickets caused by a vitamin D deficiency are lower amounts of calcium and inorganic phosphorus in the blood plasma and elevated phosphatase levels (NRC, 1978). With a prolonged deficiency, retention of calcium, phosphorus, and nitrogen decrease whereas metabolic rate increases (Colovos *et al.*, 1951; NRC, 1978). Rickets and ostemalacia can also be caused by a deficiency of calcium or phosphorus, or by an unbalanced ratio of calcium and phosphorus (see Chapter 5).

6.3.b. Sources and Metabolism of Vitamin D

Several different forms of vitamin D are known but only vitamin D_2 and D_3 are important in the normal feed supply of dairy cattle. These two are equally effective for dairy cattle. Vitamin D_2 (ergocalciferol or calciferol), often called the plant form, is found mainly in the sun-dried forages and dead leaves of growing plants. The vitamin D_2 is derived from the precursor or provitamin ergosterol under the influence of ultraviolet light.

Vitamin D_3, cholecalciferol, originally was obtained largely from fish-liver oil, and thus is sometimes known as the animal form. In the animal, the provitamin D_3, 7-dehydrocholesterol, is converted to vitamin D_3 by the ultraviolet rays of sunlight. If dairy cattle obtain sufficient sunlight with an adequate amount of the effective ultraviolet rays with wavelengths of 280–297 nm, all the requirement for vitamin D will be met with none added to the diet. The effective rays are filtered out by glass, clouds, or dust, Also, the amount reaching the earth, when the sun is low on the horizon as in the early morning, late evening, and in the winter in far northern or far southern climates, is greatly reduced. These irradiations are more effective in cattle with light-colored skin.

Sun-cured hay is a good source of vitamin D, but in silages the content is dependent on the amount of sun-drying plus that in dead leaves. Synthetic vitamin D is readily available commercially and exceedingly low in price.

Recent research indicates that the dietary forms of vitamin D, D_2 and D_3, are not the forms used in the tissues (DeLuca, 1973, 1974). These studies demonstrate that the liver converts vitamin D_3 to 25-hydroxy-vitamin D_3, a metabolite of vitamin D_3. This product is about 4 times as active for rats as vitamin D_3 (NRC, 1978). In sequence the kidney converts the 25-hydroxy-vitamin D_3 to 1,25-dihydroxy-vitamin D_3 which is approximately 20 times as active as the vitamin D_3 (DeLuca, 1973; Taylor, 1973). The formation of the 1,25-dihydroxy-vitamin D_3 is regulated by feedback mechanisms. Low levels of plasma calcium or phosphorus and high levels of parathyroid hormone stimulate synthesis of more 1,25-dihydroxy-vitamin D_3 (NRC, 1978).

6.3.c. Vitamin D Requirements

The recommended vitamin D requirements of calves is 300 IU per 100 lb of body weight (Bechdel et al., 1938; NRC, 1978). With adequate effective sunlight, no supplementation is needed. Although vitamin D is clearly essential for maintenance, reproduction, and lactation, the amount needed is not well defined. Wallis (1944) found that 5000 to 6000 IU per cow per day prevented deficiency symptoms.

6.3.d. Body Reserves of Vitamin D

Cattle can store considerable vitamin D; thus, a daily supply is not essential. Even so, the ability of the body to store vitamin D is much less than for vitamin A. The liver is the major storage area for vitamin D with some in the lungs, kidneys, and other tissues. If the cow has adequate vitamin D, the calf will have sufficient reserves of this vitamin to meet its needs for a short period after birth.

6.3.e. Vitamin D Toxicity

Very high levels of vitamin D cause high blood plasma calcium, deposition of calcium in many soft tissues including the heart and arteries, and other pathological changes which can become sufficiently severe to cause death. Unfortunately, the maximum safe amount of supplemental vitamin D which can be fed to dairy cattle has not been accurately established. There is no indication that harmful amounts of vitamin D are synthesized in the animal under the influence of sunlight.

6.3.f. Other Practical Considerations

Feeding massive amounts of vitamin D for short periods has been used successfully in the prevention of milk fever (see Section 18.1). Twenty million IU of vitamin D fed daily, beginning 3–5 days before expected calving and continuing until one day after calving with a maximum of 7 days, reduces the incidence of milk fever (Hibbs and Conrad, 1966). The 7–days maximum is used to avoid the toxicity effects of the vitamin. The role of massive doses of vitamin D in reducing the incidence of milk fever apparently is not a nutritional effect but a pharmacological effect (Miller, 1970).

Research is in progress to determine if one of the more active vitamin D metabolites, 25-hydroxy-vitamin D_3, at much lower levels than vitamin D, might be effective in preventing milk fever without comparable toxicity effects (NRC, 1978).

6.4. VITAMIN E

Several tocopherols have vitamin E activity with α-tocopherol being the most active. These "fat-soluble" substances are widely distributed in cattle feeds especially in green forages but also in the germ of grains (Gullickson et al., 1949). In animal products the amount of vitamin E is greatly influenced by the dietary intake.

6.4.a. Vitamin E Deficiency Effects, Functions, and Relation to Oxidized Milk

A condition known as white muscle disease or nutritional muscular dystrophy has been observed under field conditions and produced experimentally in calves receiving inadequate vitamin E (see Section 5.27 for other details) (NRC, 1978). The amount of vitamin E required to prevent the white muscle disease is closely related to the amount of selenium in the diet. With higher amounts of selenium in the diet, a lower level of vitamin E is required and vice versa.

White muscle disease is characterized by generalized weakness, stiffness, and deterioration of muscles with the leg muscles affected first. Calves may walk with a typical crossing of the hind legs, relaxation of the pastern, and splaying of the toes (NRC, 1978). When the tongue muscles are afflicted, the calf may not be able to suckle; in advanced stages, the calf may be unable to stand or hold up its head. The heart muscle is characteristically affected. Often death occurs suddenly from heart failure with postmortem examination revealing severe damage to the heart muscle.

Functionally, vitamin E and selenium are closely related: both have antitoxidant properties. Rancid fats or high levels of polyunsaturated fats increase the requirement for vitamin E and selenium and accentuate any tendency toward a deficiency (Adams et al., 1959; NRC, 1978). The use of cod liver oil or other fish oil increases the vitamin E requirement. More degenerative muscle lesions, apparently due to vitamin E deficiency, have been observed in calves fed fish protein concentrate containing only 1.43% fat than in those given milk replacers with dried skim milk, even though all received the 2/mg/lb of vitamin E (Michel et al., 1972).

In dairy cattle, other than young calves, severe vitamin E deficiency is uncommon and the effects have not been fully described. Attempts to establish a practical role for vitamin E in reproduction of both males and females have been largely unsuccessful suggesting that most practical diets may have adequate vitamin E (NRC, 1978). The relationships to reproduction is of special interest since early rat research demonstrated that reproductive failure was a key feature of vitamin E deficiency. Four generations of female and male dairy cattle were fed a diet containing too little vitamin E to support reproduction in the rat (Gullickson et al., 1949). Although growth, reproduction, and milk production were normal, several cattle died suddenly of apparent heart failure between 21 months and 5 years of age.

High levels of vitamin E (400 to 1000 mg per cow per day) may be beneficial in reducing the incidence of oxidized flavor in milk (King, 1968; NRC, 1978). Less than 2% of the dietary vitamin E is transferred from feed to the milk (King, 1968; Tikriti et al., 1968). Generally the amount of vitamin E in natural feedstuffs is reduced during storage. For example, in one study 80% of the vitamin E was lost in hay making (King et al., 1967). Ensiling or rapid dehydration retains most of the vitamin.

Cows fed only stored feed for long periods may produce milk that is more susceptible to oxidized flavor. Even so, the milk of cows fed stored feed for 15 months in one study did not have a significantly lower vitamin E content or increased oxidized flavor (NRC, 1978).

6.4.b. Vitamin E Requirements

A minimum vitamin E level of 300 ppm in calf milk replacer is recommended (see Appendix Table 3) (NRC, 1978). One mg of *dl*-α-tocopherol acetate is 1 mg of vitamin E and is equal to 1 IU. A minimum requirement of vitamin E has not been suggested for other classes of dairy cattle because of inadequate research information. Difficulties in establishing the amount required include the major influences of dietary selenium and polyunsaturated fat (see Section 6.4.a).

6.4.c. Stability, Body Reserves, and Other Practical Considerations

Vitamin E is relatively stable in most feeds, very resistant to heat, but is readily oxidized and thus, easily destroyed by rancid fats. The body of cattle can store appreciable vitamin E with the largest amount in the liver, but also some in other tissues and organs. A short period of low dietary vitamin E will not cause a deficiency in animals which have just received adequate vitamin E. Placental transfer of vitamin E is very inefficient, so the young calf generally has low tissue levels.

Unlike A and D, vitamin E is relatively expensive. Thus, combined with the low transfer rate to milk, providing large amounts of supplemental vitamin E to reduce oxidized flavor in milk, or for other purposes, can be an important cost item. Due to the high cost and relatively low absorption, vitamin E is sometimes injected rather than fed to calves.

6.5. VITAMIN K

Vitamin K is essential for normal blood clotting, primarily through its role in the synthesis of prothrombin in the liver. Under most situations, a deficiency of this vitamin is not a practical problem in feeding dairy cattle because of ample synthesis by the rumen microbes. Likewise, vitamin K is widely distributed in forages. There is no indication that added vitamin K is beneficial in most practical diets (NRC, 1978).

A conditioned vitamin K deficiency called "sweet clover disease" sometimes occurs in dairy cattle fed moldy sweet clover. This condition is caused by dicumarols in the moldy sweet clover which reduces the prothrombin necessary for normal blood clotting. The symptoms are generalized hemorrhaging and slow clotting from even minor injuries. Supplemental vitamin K will prevent or alleviate the sweet clover disease condition.

6.6. B VITAMINS

Dairy cattle tissues must have B vitamins as cofactors with various enzyme systems for many important biochemical reactions in the metabolism of various nutrients and in other crucial functions. Even though all the B vitamins are synthesized by the rumen microbes in the young calf before rumen function is active, these vitamins are needed in the diet. Deficiencies of thiamin, riboflavin, pyridoxine, pantothenic acid, biotin, nicotinic acid, and vitamin B_{12} have been produced experimentally in young ruminants. In isolated situations, supplemental thiamin has corrected metabolic disturbances caused by very unusual conditions. Except for a possible growth stimulation with choline, in certain special conditions, there is no indication of a need for dietary B vitamins by dairy cattle over about 6 weeks of age with normal functioning rumen microbes.

Many ordinary feeds including milk are good sources of the B vitamins. Thus, a need for supplemental B vitamins has not been established in conventionally fed calves. However, in milk replacers composed of plant proteins, there is a possibility for deficiencies of some of these vitamins (Benevenga and Ronning, 1965; NRC, 1978).

6.6.a. Thiamin

In the young calf, symptoms of thiamin deficiency include reduced appetite, loss of weight and/or condition, muscular weakness, poor coordination of the legs especially the forelimbs, inability to rise and stand, possibly irregular heart beat, a progressively impaired functioning of the nervous system known as polyneuritis, severe diarrhea, followed by dehydration and death (Johnson et al., 1948; National Research Council, 1971, 1978). Upon lying down after exertion, the head may be retracted along the shoulder, giving a star-gazing appearance. In a thiamin deficiency there is a large increase in blood pyruvate and lactate, and a drastic drop in urinary thaimin.

If the thiamin deficiency is not too advanced, feeding or injecting thiamin will rapidly and dramatically alleviate all the deficiency signs (NRC, 1978). Feeding 3 mg of thiamin hydrochloride per 100 lb of live weight per day is adequate to prevent the clinical and biochemical signs of a deficiency (Johnson et al., 1948; Benevenga et al., 1966).

Under typical feeding conditions, dairy cattle with a functioning rumen do not require dietary thiamin. However, in several practical situations a polioencephalomalacia, prevented or corrected by thiamin therapy, appears to be produced by a thiamin inhibitor (Blackmon *et al.*, 1970). One such chemical, amprolium, has been studied by several researchers (Pill *et al.*, 1966; Blackmon *et al.*, 1970). The condition is observed in feedlot cattle and in calves a few months of age.

6.6.b. Riboflavin

In the young calf, an experimentally produced riboflavin deficiency is characterized by excessive tear and saliva formation, soreness in the corners of the mouth and edge of the lips, blood congestion in the mucosa of the mouth, diarrhea, loss of hair, loss of appetite, decreased growth, and often death (NRC, 1978; Roy, 1969; Wiese *et al.*, 1947).

The riboflavin requirement is not well defined. From different studies values ranging from 3.0–20 mg per 100 lb of body weight per day have been indicated as the "minimum" requirement (NRC, 1978; Roy, 1969). The whey fraction of milk and green forages are rich sources of riboflavin. This vitamin is stable to heat in acid or neutral solutions but is destroyed by alkali and ultraviolet light. There is no indication that a riboflavin deficiency is a practical problem in dairy cattle feeding.

6.6.c. Pyridoxine

A deficiency of pyridoxine has been studied in young calves fed a synthetic diet. Gross signs which developed over a period of 3.5–12 weeks include loss of appetitie, a decrease or cessation of growth, and diarrhea (NRC, 1978). After a prolonged period, some calves had convulsive seizures including wild thrashing of legs and head, grinding of teeth, and even death. Pathological studies indicated some loss of myelin around the peripheral nerves and hemorrhages in the epicardium. Calves fed 3 mg of pyridoxine per 100 lb of liveweight per day were normal (Johnson *et al.*, 1950; NRC, 1978). There is no reason to suspect a pyridoxine-deficiency problem in the practical feeding of dairy cattle.

6.6.d. Biotin

An experimentally produced biotin deficiency in young calves was characterized by paralysis of the hindquarters (Wiese *et al.*, 1946; NRC, 1978). No symptoms were observed in calves fed 0.045 mg of biotin per 100 lb of liveweight per day. Other researchers failed to produce the deficiency even though they fed raw egg white containing the biotin antagonist, avidin (Kon and Porter, 1951). A deficiency of biotin is unlikely to be a practical problem in feeding dairy cattle.

6.6.e. Pantothenic Acid

The symptoms of experimentally produced pantothenic acid deficiency include scaly inflammation of the skin around the eyes and muzzle (NRC, 1978). After 11–20 weeks on the deficient diet, loss of appetite, diarrhea, weakness, inability to stand, and possibly convulsions developed. Likewise, the calves are susceptible to infection of the mucosal cells especially in the respiratory tract. When 6 mg of calcium pantothenate was fed per 100 lb of body weight per day, no symptoms developed (NRC, 1978; Sheppard and Johnson, 1957).

6.6.f. Nicotinic Acid (Niacin)

Symptoms of nicotinic acid deficiency include a sudden loss of appetite, severe diarrhea, dehydration, and death (NRC, 1978). Supplementation with 1.2 mg of nicotinic acid per pound of synthetic milk, free of nicotinic acid and tryptophan, prevented the deficiency (Hopper and Johnson, 1955). The amino acid tryptophan can be converted to nicotinic acid by the animal; thus, a deficiency will not occur when the diet contains adequate tryptophan (NRC, 1978).

6.6.g. Vitamin B_{12}

With a diet containing no animal protein, a vitamin B_{12} deficiency was developed in calves by 6 weeks of age (Lassiter *et al.*, 1953). Calves had a poor appetite, slow growth, muscular weakness, and poor general condition (Fig. 6.2). The suggested vitamin B_{12} requirement is between 0.015 and 0.030 mg per 100 lb of liveweight per day (NRC, 1978). Cobalt is an essential component of vitamin B_{12} which is synthesized by the microbes of cattle with a functioning rumen (see Section 5.25).

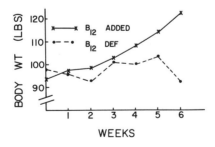

Fig. 6.2 Growth of calves maintained on vitamin B_{12}-free, synthetic milk with and without supplemental vitamin B_{12}. (Adapted from Lassiter *et al.*, 1953.)

6.6.h. Other B Vitamins

Based on work with laboratory animals, it is believed that the tissues of dairy cattle require folic acid, lipoic acid, inositol, and p-aminobenzoic acid. Deficiencies of these have not been characterized in cattle and presently there is no reason to suspect that any of them will become a problem in the practical feeding of dairy cattle.

6.7. CHOLINE

Although choline may not meet all the criteria for a true vitamin, it is usually listed with the vitamins for convenience. This compound is required for normal growth and metabolism of dairy cattle under certain conditions. The amino acid methionine and certain other compounds can at least partially replace choline in the diet.

In young calves, a deficiency of choline on certain purified diets causes fatty livers, reduced growth rate, extreme weakness, and labored breathing (Johnson et al., 1951; NRC, 1978). Most practical diets appear to contain sufficient choline to meet the requirement of young calves. Choline is a component of lecithins found in most natural fats. Likewise, grains and green forages are good sources. In addition, the rumen microbes synthesize choline.

Whether supplemental choline may improve the performance of feedlot cattle in certain situations is not clear. Some of the work at Washington State University (Swingle and Dyer, 1970) indicates a growth response. Other research suggests no benefit from choline supplementation.

6.8. VITAMIN C

Cattle synthesize vitamin C (ascorbic acid) in their tissues. Thus, they do not require a dietary source and a deficiency is unknown. Only primates, guinea pigs, some birds, and certain fish (Wilson and Poe, 1973) require dietary vitamin C; although under certain conditions vitamin C supplementation has been beneficial with the pig.

REFERENCES

Adams, R. S., J. H. Sautter, T. W. Gullickson, and J. E. Gander (1959). *J. Dairy Sci.* **42,** 1580–1591.

Bechdel, S. I., N. W. Hilston, N. B. Guerrant, and R. A. Dutcher (1938). *Pa., Agric. Exp. Stn., Bull.* **364.**

Bechtel, H. E., E. T. Hallman, C. F. Huffman, and C. W. Duncan (1936). *Mich., Agric. Exp. Stn., Tech. Bull.* **150.**

Benevenga, N. J., and M. Ronning (1965). *Hilgardia* **36,** 333–346.

Benevenga, N. J., R. L. Baldwin, and M. Ronning (1966). *J. Nutr.* **90,** 131–140.

Blackmon, D. M., G. Pope, and W. J. Miller (1970). *Proc. Ga. Nutr. Conf. Feed Ind.* pp. 83–84.

Bratton, R. W., G. W. Salisbury, T. Tanabe, C. Branton, E. Mercier, and J. K. Loosli (1948). *J. Dairy Sci.* **31,** 779–791.

Colovos, N. F., H. A. Keener, A. E. Teeri, and H. A. Davis (1951). *J. Dairy Sci.* **34,** 735–742.

DeLuca, H. F. (1973). *Proc. Ga. Nutr. Conf. Feed Ind.* pp. 88–108.

DeLuca, H. F. (1974). *Fed. Proc., Fed. Am. Soc. Exp. Biol.* **33,** 2211–2219.

Dobson, R. C., and G. Ward (1974). *J. Dairy Sci.* **57,** 985–991.

Eaton, H. D., J. E. Rousseau, Jr., C. G. Woelfel, M. C. Calhoun, S. W. Nielsen, and J. J. Lucas (1964). *Conn., Storrs Agric. Exp. Stn., Bull.* **383.**

Eaton, H. D., J. J. Lucas, S. W. Nielsen, and C. F. Helmboldt (1970). *J. Dairy Sci.* **53,** 1775–1779.

Eaton, H. D., J. E. Rousseau, Jr., R. C. Hall, Jr., H. I. Frier, and J. J. Lucas (1972). *J. Dairy Sci.* **55,** 232–237.

Gullickson, T. W., L. S. Palmer, W. L. Boyd, J. W. Nelson, F. C. Olson, C. E. Calverley, and P. D. Boyer (1949). *J. Dairy Sci.* **32,** 495–508.

Harris, B., Jr. (1975). *Feedstuffs* **47**(48), 42–43.

Hazzard, D. G., C. G. Woelfel, M. C. Calhoun, J. E. Rousseau, Jr., H. D. Eaton, S. W. Nielsen, R. M. Grey, and J. J. Lucas (1964). *J. Dairy Sci.* **47,** 391–401.

Hibbs, J. W., and H. R. Conrad (1966). *J. Dairy Sci.* **49,** 243–246.

Hopper, J. H., and B. C. Johnson (1955). *J. Nutr.* **56,** 303–310.

Johnson, B. C., T. S. Hamilton, W. B. Nevens, and L. E. Boley (1948). *J. Nutr.* **35,** 137–145.

Johnson, B. C., J. A. Pinkos, and K. A. Burke (1950). *J. Nutr.* **40,** 309–322.

Johnson, B. C., H. H. Mitchell, and J. A. Pinkos (1951). *J. Nutr.* **43,** 37–48.

Jordan, H. A., G. S. Smith, A. L. Neumann, J. E. Zimmerman, and G. W. Breniman (1963). *J. Anim. Sci.* **22,** 738–745.

King, R. L. (1968). *J. Dairy Sci.* **51,** 1705–1707.

King, R. L., F. A. Burrows, R. W. Hemken, and D. L. Bashore (1967). *J. Dairy Sci.* **50,** 943–944.

Kon, S. K., and J. W. G. Porter (1951). *Rep. Natl. Inst. Res. Dairy* p. 83 (cited by Roy, 1969).

Lassiter, C. A., G. M. Ward, C. F. Huffman, C. W. Duncan, and H. D. Webster (1953). *J. Dairy Sci.* **36,** 997–1005.

Michel, R. L., D. D. Makdani, J. T. Huber, and A. E. Sculthorpe (1972). *J. Dairy Sci.* **55,** 498–506.

Miller, W. J. (1970). *Proc. Ga. Nutr. Conf. Feed Ind.* pp. 32–42.

Moore, L. A., J. F. Sykes, W. C. Jacobson, and H. G. Wiseman (1948). *J. Dairy Sci.* **31,** 533–538.

National Research Council (NRC) (1971). "Nutrient Requirements of Dairy Cattle," 4th rev. ed. Natl. Acad. Sci., Washington, D.C.

National Research Council (NRC) (1978). "Nutrient Requirements of Dairy Cattle," 5th rev. ed. Natl. Acad. Sci., Washington, D.C.

Pill, A. H., E. T. Davies, D. F. Collings, and J. A. J. Venn (1966). *Vet. Rec.* **78,** 737–738.

Roy, J. H. B. (1969). *In* "Nutrition of Animals of Agricultural Importance. Part 2" (D. Cuthbertson, ed.), pp. 645–716. Pergamon, Oxford.

Rumsey, T. S. (1975). *Feedstuffs* **47**(7).

Shepherd, J. B., H. G. Wiseman, R. E. Ely, C. G. Melin, W. J. Sweetman, C. H. Gordon, L. G. Schoenleber, R. E. Wagner, L. E. Campbell, G. D. Roane, and W. H. Hosterman (1954), *U.S., Dep. Agric., Tech. Bull.* **1079.**

Sheppard, A. J., and B. C. Johnson (1957). *J. Nutr.* **61,** 195–205.

Swingle, R. S., and I. A. Dyer (1970). *J. Anim. Sci.* **31,** 404–408.
Taylor, A. N. (1973). *Proc. Ga. Nutr. Conf. Feed Ind.* pp. 77–87.
Tikriti, H. H., F. A. Burrows, A. Weisshaar, and R. L. King (1968). *J. Dairy Sci.* **51,** 979 (abstr.).
Wallis, G. C. (1944). *S.D., Agric. Exp. Stn., Bull.* **372.**
Wiese, A. C., B. C. Johnson, and W. B. Nevens (1946). *Proc. Soc. Exp. Biol. Med.* **63,** 521–522.
Wiese, A. C., B. C. Johnson, H. H. Mitchell, and W. B. Nevens (1947). *J. Nutr.* **33,** 263–270.
Wilson, R. P., and W. E. Poe (1973). *J. Nutr.* **103,** 1359–1364.

7

Fat (Lipids) and Water Requirements and Utilization by Dairy Cattle

7.0. FATS AND LIPIDS

Lipids are a major class of compounds and nutrients in feeds which are soluble in various organic solvents. Of the lipids, fats make up the greatest quantity, but other lipids are of some importance in dairy cattle nutrition. The fat-soluble vitamins A, D, E, and K, and their precursors are included in this broad classification (see Chapter 6). Likewise, choline and phospholipids, cholesterol and other steroids, waxes, and the very abundant chlorophylls in green forages are lipids.

The "true fats" have an energy value about 2.25 times that of carbohydrates and proteins. Thus, utilizable energy in feeds increases rapidly with the fat content. However, except in young calves, dairy cattle are much less able to efficiently utilize substantial amounts of fats than most nonruminants (see Section 7.3).

Generally the lipid content of feeds is determined as "ether extract" (see Section 12.2.c) and may be listed either as "fat" or as "ether extract." When the feed ether extract is largely from grains and oil meals, it is mostly true fats. Many forages, however, contain substantial amounts of "ether extractable" lipid material which is not true fat but compounds such as chlorophyll and other plant pigments. The usable energy value of these substances is much less than for true fats, owing both to lower digestibility and less energy in that digested.

7.1. METABOLISM OF FATS AND FATTY ACIDS BY DAIRY CATTLE

A molecule of true fat consists of three fatty acids attached to a glycerol molecule. The three fatty acids may be any combination of a substantial number

204

of different fatty acids, including many known as "saturated" and others as "unsaturated fatty acids." These terms refer to whether or not the fatty acids contain all the hydrogen atoms of which they are capable. An unsaturated fat can be hydrogenated by the rumen microbes and thus become saturated. The amount of unsaturation is measured by the number of "double bonds." For each double bond, two hydrogen atoms are required to "saturate" the fatty acid molecule. Fatty acids having two or more double bonds are called polyunsaturated fatty acids (PUFA's). The unsaturated fatty acids lower the melting point of the fat and chemically are more reactive. Thus, fats with fatty acids having a high degree of unsaturation are highly reactive, quite unstable, and oxidize and become rancid easily.

Fatty acids are classified and characterized in several ways. One of these is by the length of the carbon chain which varies in length from 2 atoms to more than 20 carbons. The volatile fatty acids, acetic, propionic, and butyric, contain 2, 3, and 4 carbon atoms (see Section 1.1.c). These short chained fatty acids are metabolized very differently from the longer chained ones. In fact, the differences are so great that often the VFA's are not considered in the same context with other fatty acids. Except for certain silages, generally, the amounts of the volatile fatty acids in dietary fats are very small.

7.1.a. Digestion and Hydrogenation of Unsaturated Fatty Acids in the Rumen

In most situations, before they can be absorbed, fats must be separated (hydrolyzed) into fatty acids and glycerol (see Section 1.2). A substantial portion of the fats are hydrolyzed to glycerol and fatty acids in the rumen (Lough, 1970), where much of the glycerol is fermented to other products such as volatile fatty acids, especially propionic acid (Church, 1970). Apparently most of the fatty acids move on down the digestive tract for eventual absorption in the small intestine (see Section 1.3).

In the fats of many feeds including forages (Church, 1970; Dawson and Kemp, 1970; Oksanen and Thafvelin, 1965), and such seeds as corn, soybeans, safflower, and peanuts, a substantial portion of the fatty acids are unsaturated. The degree of unsaturation varies considerably in different feeds. Although the content of highly unsaturated fatty acids, including linolenic with 3 double bonds, is very high in timothy forage at harvest, the value may decrease considerably in curing especially if there are rains (Oksanen and Thafvelin, 1965).

The rumen microbes hydrogenate many of the unsaturated fatty acids. Thus the fatty acids subsequently absorbed are much more saturated than the fatty acids in the feed (Tove and Mochrie, 1963; Dawson and Kemp, 1970). This hydrogenation has a major influence on the composition and physical properties of milk and meat (see Sections 7.4 and 7.5). When very high amounts of highly unsaturated fats and oils are fed, some of the fatty acids escape hydrogenation (Cameron and Hogue, 1968) (see Section 7.6).

7.1.b. Digestion and Absorption of Fats in the Small Intestine

Any fat which remains intact when the digesta reaches the small intestine is hydrolyzed to glycerol and fatty acids under the influences of lipase(s) (see Section 1.2). A substantial portion of the fatty acids hydrolyzed in the small intestine plus those from hydrolysis in the rumen are absorbed from the small intestine (Lough, 1970) (see Section 1.3). In absorption, the fatty acids and the glycerol move into the mucosal cells of the small intestine where they recombine to form true fats. These fats, unlike most other nutrients, move through the lymphatic system rather than the blood. Bile plays an important role in the absorption of fatty acids through its emulsification action.

7.2. EFFECTS OF DIETARY FATS ON BODY AND MILK COMPOSITION

In nonruminants, the fatty acid composition of the body fat and the milk fat partially reflect the composition of the dietary fatty acids. For instance, when pigs are fed a substantial amount of peanuts containing appreciable oil with a high proportion of unsaturated fatty acids, the pork fat has a low melting point and is known as soft pork. In the young calf, before the rumen microbes become very active, the fatty acids available for absorption reflect the feed fatty acids. Considerable fat can be synthesized by the rumen microbes (see Section 7.6). Thus, the fatty acids mixture at the absorption site does not necessarily reflect the characteristics of the feed fat. Accordingly, the fatty acids of cattle, other than young calves, usually are highly saturated, regardless of the fatty acid composition in the feed. Likewise, the fatty acids of milk fat are substantially different from those in the feed. Characteristically, milk fat has relatively high amounts of fatty acids having 4 to 10 carbon atoms and those with odd-length carbon chains (Church, 1970). Apparently, most of the unusual fatty acids found in milk originate from rumen microbial synthesis.

7.3. RUMEN BYPASS OF FATS AND LIPIDS

In recent years, there has been a widespread popular belief that fats having a high percentage of polyunsaturated fatty acids are more desirable than others for human health. (Although detailed discussion of this subject is beyond the scope of this book, it seems to this author and numerous other scientists that this belief is based on unsatisfactory scientific evidence. Likewise, there are substantial reasons to think it is a false notion. In fact, there is appreciable reason to think that a large amount of PUFA's may be quite damaging.) Because of the "supposedly beneficial effects" of the polyunsaturated fatty acids and the very low amount in the fat of normal beef and cow's milk, methods have been devised to materially change the composition of the fats (Astrup *et al.*, 1976; Goering *et al.*, 1976, 1977; Macleod *et al.*, 1977; Palmquist, 1976; Scott *et al.*, 1970; Wrenn *et al.*, 1976).

In Section 3.10, the advantages, technology, etc., of having a substantial amount of protein bypass the rumen were discussed. Similar procedures have been developed to permit special fats to bypass the actions of the rumen microbes (Scott *et al.*, 1970). When these procedures are used with fats having a high amount of PUFA's, the fatty acids are not hydrogenated in the rumen and thus reach the absorption site in the small intestine. The nature of the fat in the milk and the meat can be materially changed in this way. This not only alters the fatty acid composition of these fats but the keeping, flavor, and physical characteristics of the products are adversely modified. When these adverse effects on "quality of the products" from the handling standpoint are considered along with the extra cost, and the very debatable nutritional benefits, the modifications do not appear to offer a great deal of practical merit.

7.4. SYNTHESIS OF FATTY ACIDS AND FATS FROM CARBOHYDRATES AND OTHER NUTRIENTS IN THE RUMEN AND THE ANIMAL BODY

Since they are able to synthesize the fatty acids, fats, and other lipids needed for the cell protoplasm, rumen microbes do not require a source of dietary fat. Likewise, the fats of the rumen microbes can supply the needs of the cow. Both ruminants and nonruminants can synthesize fatty acids and fats from carbohydrates and other nutrients in their body tissues. This is illustrated by the well-known fact that when consumed in sufficient amounts, carbohydrates are fattening.

7.5. ARE FATS OR FATTY ACIDS ESSENTIAL IN THE DIET OF DAIRY CATTLE?

Although fatty acids and fats are synthesized in the tissues of nonruminants, research has shown that these animals require a dietary source of certain fatty acids. These essential fatty acids (EFA) are linoleic, linolenic, and arachidonic.

However, linoleic acid is the key one as it can fully meet the needs of animals for the essential fatty acids. Linolenic and arachidonic acids can partially substitute for the linoleic acid.

As with other nonruminants, the baby calf requires the essential fatty acids in the diet (Lambert *et al.*, 1954). Calves fed a fat-free ration developed deficiency effects including retarded growth; scaly dandruff; long, dry hair; excessive loss of hair; and diarrhea (Lambert *et al.*, 1954). The amount of essential fatty acids required does not appear to be clearly established. As the rumen develops and the rumen microbes begin to synthesize them, the dietary need of the calf for essential fatty acids disappears.

Although dairy cattle with a functioning rumen do not require fat or fatty acids in the diet, there are several advantages to including some fat. As mentioned earlier (see Section 7.0), fats have a higher energy content per pound than any other type nutrient. A moderate amount of fat in concentrates greatly improves the handling properties of the feed, as one which has little or no fat will be extremely dusty. Likewise, a moderate amount of fat reduces wear on feed handling machinery and equipment. Some fat is beneficial in feed as a carrier for the fat-soluble vitamins.

In practical feeding, it is quite rare for a dairy ration to be devoid or extremely low in fat, as most feed ingredients contain a small percentage of fat. Thus, only with very unusual ingredients or when the fat is extracted, would the diet be fat free. In earlier years, considerable attention was given to the percentage of fats needed in dairy feeds. Although often milk production increased with the added fat, apparently most of the benefits were due to higher energy consumption.

7.6. EFFECTS OF HIGH LEVELS OF DIETARY FAT

The young calf is able to effectively utilize a very high level of dietary fat. On a dry matter basis, whole milk often exceeds 30% fat. Thus, more than half of the digestible energy content in whole milk may come from fat. When mixed with the dry feed, usually, even the young calf does not make effective use of an extremely high amount of fat, indicating the importance of the physical form of the fat. Proper emulsification is essential if substantial amounts are to be used effectively (Hopkins *et al.*, 1959). Likewise, it appears that large amounts should be fed in liquid form in such a way as to achieve rumen bypass (see Section 3.10 for discussion of esophageal groove closure). The fat content in milk replacers usually is much lower than in whole milk. In the dry milk replacer about 10% fat is sufficient to supply the needed essential fatty acid, carry fat soluble vitamins, and furnish extra energy (National Research Council, 1978). A higher fat content will increase growth rate of veal calves.

Typical dairy cattle rations will contain around 2–4% fat. The addition of 2–5%, depending somewhat on the amount already present, may be utilized quite efficiently. However, more than this may have adverse effects, including interferences with rumen microbial action such as reducing fiber digestion and nitrogen metabolism. Because of the reduced protein metabolism, adding both fat and urea to dairy feeds may be undesirable. Likewise, with high amounts of fat, the character of the body fat and the milk fat may be changed. For example, when 15% corn oil, an unsaturated fat, was fed to lambs, unsaturated fatty acids in the body fat was materially increased (see Section 7.1.a) (Cameron and Hogue, 1968). Excessive amounts of fats increase the quantity of fatty acids which combine with calcium to form soaps, resulting in a lower digestibility and wastage of fatty acids, calcium, and fat-soluble vitamins (Davison and Woods, 1961).

The effects of added fat depend a great deal on the type of fat. For example, certain fats, including those that are relatively saturated, may increase the fat percentage in milk of lactating cows. In contrast, feeding highly unsaturated fats such as fish oils can cause a substantial reduction in the milk fat percentage. As little as 250 ml of fish oils daily can depress fat percentage (NRC, 1978). Since highly unsaturated fats become rancid very easily, their inclusion in the diet increases the vitamin E requirement (see Section 6.4).

Silages contain substantial amounts of shorter carbon chain length fatty acids such as acetic, propionic, butyric, and lactic which are utilized quite efficiently as energy sources by dairy cattle. On a practical basis, usually these are not considered to be fats (see Section 7.1).

7.7. WATER REQUIREMENTS OF DAIRY CATTLE

Dairy cattle require several times more water than all other nutrients combined. Likewise, the performance of dairy cattle will be depressed sooner from inadequate water intake than by a deficiency of any other nutrient. Because of its very low cost, often water is given little or no attention in feeding dairy cattle.

Water serves many key functions in dairy cattle nutrition and metabolism and is required in most of the biochemical reactions and physiological functions of the animal. A high proportion of the blood, which transports other nutrients, is water. Likewise, many of the other tissues of the body contain appreciable water. For instance, 75–80% of fat-free muscle is water (Cullison, 1975). Thus, water is a key factor giving the animal body form. In the regulation of body temperature, water has a key role, especially with high environmental temperatures. A large amount of water is needed for the excretion of waste products in urine, feces, perspiration, and respiration.

7.7.a. Sources of Water for Dairy Cattle

Dairy cattle obtain water (1) by drinking it, (2) as a part of their feed, and (3) in the chemical metabolism of the feed (Agricultural Research Council, 1965). The last of these is largely of academic importance, except that it helps explain the higher water requirement when certain types of feed are used. For each pound of fat utilized by dairy cattle for energy, 1.19 lb of water are obtained metabolically. Similarly, for each pound of carbohydrate and protein so used, about 0.56 and 0.45 lb of water respectively are formed. In measuring the water needs of dairy cattle, only that in feeds and the drinking water are included.

Many feeds contain a substantial amount of water. For instance, very lush, young pasture frequently consists of as much as 85% water. Silages generally will have 60–75% water, whereas, most dry forages and concentrates usually contain only 8–15% water. By far the most important source of water is that which cattle drink. In determining the amount of water required by cattle, it is convenient to add that in the feed to that which is drunk.

7.7.b. Water Requirements of Dairy Cattle

The estimated water needs of dairy cattle are based on measurements of the amounts actually consumed when it was readily available. The degree to which slight restrictions on intake would affect performance, is not clearly established.

Many factors affect the water consumption of dairy cattle including dry matter consumption, environmental temperature, humidity, level of milk production, pregnancy, activity of the animal, breed, amount of salt in the diet, and protein content of the feed (ARC, 1965; NRC, 1978). A key factor is the amount of feed eaten. Because of the large number of functions associated with feed intake, the water requirements are proportional to the amount of dry matter consumed in the feed. When dry matter consumption is considered, the size of the animal usually can be ignored. Generally, if the environmental temperature is between 10° and 40°F, about 3–4 lb of water are consumed for each pound of dry matter eaten (ARC, 1965; NRC, 1978). These values are based on nonpregnant, nonlactating cows fed indoors. As the air temperature increases above about 50°F, water consumption goes up fairly rapidly and almost linearly. For example, at 85°F approximately 90% more water is consumed than at 40°F (ARC, 1965). With temperatures above 80°F, the rate of increase in water consumption accelerates (NRC, 1978).

Although definitive data are largely inadequate, it has been suggested that the water requirement from feed and drinking for grazing cattle should be calculated as 50% higher than that for fed animals (ARC, 1965). Thus, with temperatures below 50°F about 4.5–6 lb of water would be needed for each pound of dry matter eaten.

To the estimated water needs based on dry matter consumption, including the grazing factor if applicable, an additional amount is required by lactating cows. For each pound of milk produced, about 0.87 pounds more water is estimated to be needed (Fig. 7.1) (ARC, 1965). Since the lactating cow also eats much more dry matter, the increase in water intake is much greater than the water going into the milk. During the last 3–4 months of pregnancy, the needs apparently increase substantially (ARC, 1965; NRC, 1978).

Young calves are believed to need additional water; about 6 lb of water per pound of feed dry matter consumed (ARC, 1965). This higher need of young calves is based partially on research showing reduced performance when the dry matter content of the milk replacers is higher than 15% (Pettyjohn *et al.*, 1963).

When certain types of feeds are used, larger amounts of water are consumed. The two most notable examples are high-protein feeds and those with a high salt content (ARC, 1965). Cattle are able to tolerate a comparatively high amount of salt in the diet but the content which can be safely included in water is far more limited (about 1.0%) because they cannot adapt to the extra salt by drinking more water (see Section 5.19).

7.7.c. Other Practical Considerations

Because water is cheap, it is important that this should not be a a limiting factor in dairy cattle production. If dairy animals, especially lactating cows, do

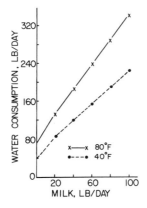

Fig. 7.1   Estimated minimum water consumption of a 1300-lb cow that was neither gaining nor losing weight as affected by the level of (3.5% fat) milk production. Calculations based on 3.1 lb of water at 40°F (5.2 lb at 80°F) for each pound of dry matter consumed, plus 0.87 lb of water for each pound of milk produced (ARC, 1965; NRC, 1978). See text for other factors affecting water consumption.

not consume sufficient water, performance will be quickly reduced. When the water is excessively cold, too warm, distasteful, or inconvenient to the cow, consumption may be reduced and thereby milk production decreased. To illustrate the effects of water restrictions on performance, one only need recall that one of the most effective ways to turn a cow dry is to severely restrict water consumption. This is much more effective than reducing feed intake. Although dairy cows do not have well-developed sweat glands, they lose considerable water by insensible perspiration. At 50°F this can be about 1 lb/hr.

In supplying water for dairy cattle, it is important that the supply should not be contaminated with disease organisms, hazardous chemicals, or other pollutants that are either adverse to the health of the cow or would be translocated to the milk or meat. Results of several studies indicate that neither water consumption, nor performance of cows are affected by hard versus soft water (Allen *et al.*, 1958; NRC, 1978).

REFERENCES

Agricultural Research Council (ARC) (1965). "The Nutrient Requirements of Farm Livestock," No. 2. Ruminants. ARC, London.
Allen, N. N., D. Ausman, W. N. Patterson, and O. E. Hays (1958). *J. Dairy Sci.* **41,** 688–691.
Astrup, H. N., L. Vik-Mo, A. Ekern, and F. Bakke (1976). *J. Dairy Sci.* **59,** 426–430.
Cameron, C. W., and D. E. Hogue (1968). *J. Anim. Sci.* **27,** 553–556.
Church, D. C. (1970). "Digestive Physiology and Nutrition of Ruminants," Vol. 1. D. C. Church Publ., distributed by Oregon State University Book Store, Corvallis.
Cullison, A. E. (1975). "Feeds and Feeding." Reston Publ. Co., Reston, Virginia.
Davison, K. L., and W. Woods (1961). *J. Anim. Sci.* **20,** 532–536.
Dawson, R. M. C., and P. Kemp (1970). *In* "Physiology of Digestion and Metabolism in the Ruminant" (A. T. Phillipson, ed.), pp. 504–518. Oriel Press, London.
Goering, H. K., C. H. Gordon, T. R. Wrenn, J. Bitman, R. L. King, and F. W. Douglas, Jr. (1976). *J. Dairy Sci.* **59,** 416–425.
Goering, H. K., T. R. Wrenn, L. F. Edmondson, J. R. Weyant, D. L. Wood, and J. Bitman (1977). *J. Dairy Sci.* **60,** 739–748.
Hopkins, D. T., R. G. Warner, and J. K. Loosli (1959). *J. Dairy Sci.* **42,** 1815–1820.
Lambert, M. R., N. L. Jacobson, R. S. Allen, and J. H. Zaletel (1954). *J. Nutr.* **52,** 259–272.
Lough, A. K. (1970). *In* "Physiology of Digestion and Metabolism in the Ruminant" (A. T. Phillipson, ed.), pp. 519–528. Oriel Press, London.
Macleod, G. K., Y. Yu, and L. R. Schaeffer (1977). *J. Dairy Sci.* **60,** 726–738.
National Research Council (NRC) (1978). "Nutrient Requirements of Dairy Cattle," 5th rev. ed. Natl. Acad. Sci., Washington, D.C.
Oksanen, H. E., and B. Thafvelin (1965). *J. Dairy Sci.* **48,** 1305–1309.
Palmquist, D. L. (1976). *J. Dairy Sci.* **59,** 355–363.
Pettyjohn, J. D., J. P. Everett, Jr., and R. D. Mochrie (1963). *J. Dairy Sci.* **46,** 710–714.
Scott, T. W., L. J. Cook, K. A. Ferguson, I. W. McDonald, R. A. Buchanan, and G. L. Hills (1970). *Aust. J. Sci.* **32,** 291–293.
Tove, S. B., and R. D. Mochrie (1963). *J. Dairy Sci.* **46,** 686–689.
Wrenn, T. R., J. R. Weyant, D. L. Wood, J. Bitman, R. M. Rawlings, and K. E. Lyon (1976). *J. Dairy Sci.* **59,** 627–635.

8

Fiber Utilization and Requirements
of Dairy Cattle

8.0. INTRODUCTION

Because a higher fiber content of feedstuffs generally decreases their economic value, often fiber has been regarded as a necessary evil. However, an adequate amount of digestible fiber is needed in the diet of dairy cattle for efficient performance and health. Unlike many nonruminants, ruminants make considerable use of fiber as an energy source. Although a full discussion is beyond the scope of this book, an adequate amount of fiber is an important factor in human nutrition and health (Anonymous, 1975a,b; Kritchevsky, 1974).

8.1. FIBER COMPONENTS, TYPES OF FIBER

In contrast to many of the nutrients discussed earlier, fiber is not a specific compound. Rather, fiber consists of several different types of compounds with a variety of complex carbohydrates being the largest fraction. Fiber can be visualized by considering its role in plants as the material that provides structural strength and form. Generally, the vegetative parts, especially the stems, have the highest fiber content. When viewed microscopically, fiber is the major constituent of the cell walls. Seed hulls and/or coats also often contain appreciable fiber that protects the interior of the seed. This fiber gives structural resistance against organisms and other destructive environmental forces (Van Soest, 1969).

If one considers the enormous differences in the physical appearances, and structural strengths of plants, it is easy to understand how the content and type of fiber differs markedly among plants. These differences not only occur between species, but among parts of the same plant, and with maturity.

8.1.a. Chemical Composition of Fibers

Most of the substances which make up fiber are classified as complex carbohydrates; but some fiber components, such as lignin, are not true carbohydrates. Others, including silica and cutin, are not closely related to carbohydrates.

Celluloses, the most abundant of fiber constituents in feeds, provide tensile strength to plants (Van Soest, 1973). Chemically, celluloses are classified as polysaccharides because they are large molecules composed of numerous glucose units condensed into long chains. Starches, which are not fiber, also consist of a large number of glucose molecules, but differ from celluloses in the type of chemical linkage between the glucose molecules. Starch has an α linkage and cellulose a β. This type of linkage is enormously important for two reasons. First, the cellulose β linkage has much more tensile strength. (Cotton is almost pure cellulose.) Second, the digestive tract of animals does not have enzymes that can break the cellulose type (B) chemical bond. In contrast, even nonruminant animals have enzymes for easily digesting starch to the glucose units. Cellulose is fermented by the microbes of the rumen (see Sections 1.1.c and 8.2).

Hemicelluloses generally are the second most abundant class of fiber components (Dehority, 1973). As the name implies, these compounds are similar to celluloses but differ by being less resistant to chemical agents but not necessarily more digestible than celluloses (Van Soest, 1967) (see Section 12.5 and Table 12.1). Unlike celluloses, hemicelluloses are a mixture of complex polysaccharides. The building units of hemicelluloses often are simple sugars, which frequently have 5 carbon atoms, and uronic acids.

Pentosans differ from celluloses by consisting of long chains of five carbon sugars (glucose is a 6-carbon sugar). Often the pentosans make up about 20% of the complex carbohydrates of hays. Likewise, they are a major component of corn cobs and oat hulls.

Pectins, gums, and mucilages are sometimes classified with the hemicelluloses but contain somewhat different building units including especially galacturonic acid. The pectins, that occur in cell walls and intracellular spaces, are highly digestible. Feeds such as citrus pulp and beet pulp have a high pectin content (Smart *et al.,* 1964).

Lignin plays a key role in plants, but lowers the value of feeds to dairy cattle. In plants, lignin acts very much like concrete in a building. Whereas cellulose provides tensile strength to plants, lignin imparts stiffness. This can be visualized if one considers that the highly flexible cotton fibers are almost pure cellulose, whereas wood is high in cellulose, but also has a very high lignin content. The higher the lignin content, the more woody or "woodlike" the plant becomes.

In addition to being virtually indigestible, lignin, "acting in a role similar to that of concrete" in reinforced concrete, prevents other nutrients from being attacked by rumen microbes; thus, digestibility is reduced more than the amount of lignin present (Dehority *et al.,* 1962). Increased lignin content is a major reason for reduced digestibility in more mature forages. The type of lignin apparently changes with maturity with that in mature forage lowering digestibility to a greater extent.

Silica, although an essential trace mineral element (see Section 5.30), seems to function in forages in a way similar to lignin. The importance of silica varies tremendously in different types of forages. A high silica content appears to be a key reason for low performance of dairy cattle fed certain subtropical forages such as Bermuda grass.

Cutins are major constituents in seed hulls and coats contributing to their impervious nature (Van Soest, 1969). Stems and leaves also have a thin layer of cutin which tends to give an impervious surface. Cutins are complex polymers of long chain fatty acids, alcohols, aldehydes, ketones, and paraffin hydrocarbons (Van Soest, 1969). They are very resistant to digestion by rumen microbes (Van Soest, 1969).

8.2. FIBER DIGESTION IN THE RUMEN

In cattle, most fiber digestion is performed by rumen microbes. Without these microbes, cattle could not utilize fiber as a principal source of energy. The type of fiber greatly influences the percentage digested. Fiber in young plants is far more digestible than that in more mature plants. As plants mature, the cell walls, which contain most of the fiber, become thicker and more resistant to microbial digestion. This process involves increasing the amount of lignin and/or silica which also renders other nutrients unavailable for digestion. Likewise, in forage plants other fiber components change with maturity. The increased lignification and stiffness accompanying maturity enables the plant to resist breaking or falling for a longer period of time, thereby facilitating distribution of seed over a wider area. Although lowering the value of the forage for the cow, lignification increases the survival possibility for plants in nature. In its effect on fiber digestibility and in giving stiffness to the plant, silica appears to function in a way similar to lignin.

The digestion of fiber by rumen microbes is a very complex process involving many biochemical reactions (Leng, 1970). Digestibility of the major types of fiber components varies enormously depending on the nature of the fiber. Cellulose digestibility, for example, varies from 90% in very high quality, immature grasses to as little as 0% in certain wood fibers (Van Soest, 1973). The digestibility of cellulose and other fiber components is a key factor in the practical quality and value of forages for dairy cattle. Not only is the percentage of fiber digested important, but the rate of the digestion has a major influence on the practical value of feeds (see Section 9.2.a).

8.3. PRODUCTS OF FIBER DIGESTION—VOLATILE FATTY ACIDS

Although some components of the fiber in feeds, such as lignin and silica, are almost totally indigestible, substantial amounts of other constituents including celluloses and hemicelluloses are digested by the rumen microbes. The end products of this rumen fermentation are not the original, simple carbohydrate or other building units constituting fiber (Leng, 1970). Rather, the major end products of fiber digestion are the volatile fatty acids (acetic, propionic, and butyric), methane, carbon dioxide, water, limited amounts of lactic acid, and heat (Leng, 1970). Some of these products are incorporated into the microbes themselves (Leng, 1970). Since these volatile fatty acids also exist in the salt form, the terms acetate, propionate, and butyrate are used interchangeably with acetic acid, propionic acid, and butyric acid.

The carbon dioxide and methane produced by the rumen microbes are disposed of as gasses. The practical importance is largely limited to the energy losses they represent and the danger of bloat occurring if these gasses are not eliminated as fast as they are formed (see Section 18.3).

8.3.a. Influence of Diet on Volatile Fatty Acids (VFA's) Formed

The VFA's, acetic, propionic, and butyric are the most important sources of energy for dairy cattle, except for young calves. Typically, these three compounds make up about two thirds of the total digestible energy in feeds (Annison and Armstrong, 1970; McCullough, 1973). When dairy cattle are fed normal diets containing substantial amounts of good quality fiber, acetate is formed in greater quantities than propionate or butyrate. The amounts of the VFA's are generally expressed in ratios. In terms of "molar ratios" with a normal diet, the proportions are about 65% acetate, 20% propionate, and 15% butyrate. Because the molecules are of different sizes, on a weight basis, the proportions are closer to 50% for acetate, 25% propionate, and 25% as butyrate.

The relative quantities of the VFA's are greatly affected by the diet. For instance, if a very high proportion of concentrates, that are primarily starches, and insufficient "good quality" fiber is fed, the proportion of acetate produced is reduced but propionate is increased (Annison and Armstrong, 1970).

8.3.b. Effects of Physical Form of Diet on Volatile Fatty Acids

When forage is finely ground, the proportion of acetate formed by the rumen microbes is reduced and that of propionate increased. This effect is similar to feeding too little fiber. The fineness of the grind is important. Chopping or coarse grinding has relatively little influence on the proportions of the VFA's produced. Likewise, the changes associated with grinding vary with different forages. The typical VFA changes are readily produced with alfalfa, whereas with coastal Bermuda grass, it is a comparatively more difficult task. Part of the difference is associated with the tougher nature of the coastal Bermuda grass including greater resistance to breaking up by grinding.

8.3.c. Absorption and Utilization of Volatile Fatty Acids

Most of the VFA's produced in the rumen are absorbed directly through the rumen wall with smaller amounts from the reticulum, omasum, and large intestine. After absorption, the metabolism and utilization of acetate, propionate, and butyrate are quite different.

In their tissues, cattle are able to convert propionate into glucose. Since the rumen microbes utilize and ferment any sugar of starch present, very little glucose is available for absorption from the feed after digestion. The transformation of propionate to glucose in the tissues is quite important as glucose is the starting material for the lactose in milk (Table 8.1). Further, glucose is not only a major source of energy in milk synthesis but also the precursor for the glycerol fraction of the milk fat molecule. Acetate is not readily converted to glucose, but is used as a precursor for fatty acids of milk by the mammary tissue. As much as 50% of the fatty acids of milk are derived from acetate. The fatty acids of milk coming from acetate are those with a chain length less than the 16 carbons of palmitic. Acetate, propionate, and butyrate, are all utilized as energy sources in numerous complex reactions in body tissues.

8.3.d. Effects of Volatile Fatty Acids on Milk and Body Composition

As mentioned earlier (see Sections 8.3.a and 8.3.b), feeding too little fiber or finely grinding the forage materially lowers the relative proportion of acetate while increasing that of propionate. Even before the effect on volatile fatty acids was discovered, lack of fiber or finely grinding the forage was known to reduce the fat content of milk. The percentage of fat in milk will remain comparatively

TABLE 8.1
Sources of Energy for the Synthesis of Milk[a]

	Amount per 100 lb of milk	
	lb	% of total energy used by udder in milk synthesis
Glucose	10[b]	44[b]
Acetate	3.2	12
Amino acids	4.1	22

[a] Adapted from Blaxter (1967). Based on amounts of substances removed from the blood. Other substances including long-chain fatty acids are important sources of energy for milk synthesis.

[b] About half of the glucose is used for conversion to lactose.

normal over a relatively wide range of forage to concentrate ratios. Thus, major changes occur only in fairly extreme situations.

Because the nature of fibers differs markedly, the minimum amount of "unground" fiber required for normal milk fat cannot be precisely defined. The quantity of fiber from coastal bermuda grass, for example, would be smaller than that needed from alfalfa hay to maintain milk fat percentage. The fiber in soybean hulls is highly digestible and very ineffective in maintaining milk fat content.

Although too little acetate relative to propionate lowers fat content of milk, no such effect occurs in body composition. Rather, a low-fiber diet generally results in more efficient fattening of cattle. Since dairy cattle are ultimately disposed of for beef, there are times when the lower acetate to propionate is desirable (see Chapter 17).

8.4. OTHER EFFECTS OF TOO LITTLE FIBER; AN ESSENTIAL NUTRIENT

In addition to the effect on volatile fatty acids and milk fat composition, dairy cattle fed insufficient fiber or finely ground fiber often develop other metabolic or digestive problems (Miller and O'Dell, 1969). These include "founder" or lactic acid acidosis (see Section 18.8), abscessed liver (see Section 18.9), rumen parakeratosis (see Section 18.9), displaced abomasums (see Section 18.6), and other disorders (National Research Council, 1978).

The dairy animal probably evolved through the ages as primarily a forage and/or whole plant consumer. Since dairy cattle often develop problems when fed mainly concentrates with too little fiber, fiber can be considered as an essential nutrient. About 17% crude fiber or 21% acid detergent fiber, on a dry basis, in most situations is enough to prevent adverse effects of a deficiency for lactating cows (see Appendix, Table 3) (NRC, 1978). Because fiber is a general term for a variety of substances with different chemical and nutritional properties, the fiber requirement as a percent of the total ration varies with the nature of the fiber and the overall diet (see Sections 8.1 and 8.1a) (see Chapters 9 and 12).

8.5. EFFECTS OF TOO MUCH FIBER

The usable energy (see Chapter 2) of feeds generally declines as the fiber level increases, especially above required levels. Thus, when diets contain excessive fiber, often it is not possible for dairy cattle to eat enough feed to meet their energy needs. The energy needed relative to the capacity to eat feed varies with the level of production. Accordingly, low-producing cows with smaller energy requirements can obtain sufficient energy from a higher-fiber diet than can high producers.

Because of the variable nature of fiber in different feeds, the amount of energy which a given dairy animal can obtain at a given fiber percentage varies. For instance, cows will consume more alfalfa hay than coastal bermuda grass hay even though their crude fiber content is similar (see Table 9.5 and Chapter 12). Accordingly, dairy cattle could meet their energy needs with a higher total amount of fiber content when alfalfa hay is a major feed ingredient than if coastal bermuda grass were the main feed. Except for lack of energy or other nutrients, there are no known metabolic abnormalities associated with excessive fiber consumption.

8.6. SOURCES OF FIBER IN DAIRY RATIONS

By far the largest source of fiber in dairy cattle feeds is forages (see Chapter 9). A comparatively high percentage of most forages is fiber. As will be discussed more fully in Chapter 12, historically, there have been many difficulties in measuring the amount of fiber in feeds. The content depends on the methods used to define and measure the components.

Concentrates generally contain much less fiber than forages or roughages. However, most classification systems for concentrates depends on the origin of the ingredient: in essence, ingredients derived from seeds are classed as concentrates. Some products derived from seeds, such as oat hulls, have a higher fiber content and a lower digestible energy content than many forages.

REFERENCES

Annison, E. F., and D. G. Armstrong (1970). *In* "Physiology of Digestion and Metabolism in the Ruminant" (A. T. Phillipson, ed.), pp. 422–437. Oriel Press, London.
Anonymous. (1975a). *Dairy Counc. Dig.* **46**(1), 1–4.
Anonymous. (1975b). *Dairy Counc. Dig.* **46**(5), 25–30.
Blaxter, K. L. (1967). "The Energy Metabolism of Ruminants." Thomas, Springfield, Illinois.
Dehority, B. A. (1973). *Fed. Proc., Fed. Am. Soc. Exp. Biol.* **32**, 1819–1825.
Dehority, B. A., R. R. Johnson, and H. R. Conrad (1962). *J. Dairy Sci.* **45**, 508–512.
Kritchevsky, D. (1974). *Nutr. Notes* **10**(4), 4–5.
Leng, R. A. (1970). *In* "Physiology of Digestion and Metabolism in the Ruminant" (A. T. Phillipson, ed.), pp. 406–421. Oriel Press, London.
McCullough, M. E. (1973). "Optimum Feeding of Dairy Animals for Meat and Milk." Univ. of Georgia Press, Athens.
Miller, W. J., and G. D. O'Dell (1969). *J. Dairy Sci.* **52**, 1144–1154.

National Research Council (1978). "Nutrient Requirements of Dairy Cattle," 5th rev. ed. Natl. Acad. Sci., Washington, D.C.

Smart, W. W. G., Jr., T. A. Bell, R. D. Mochrie, and N. W. Stanley (1964). *J. Dairy Sci.* **47,** 1220–1223.

Van Soest, P. J. (1967). *J. Anim. Sci.* **26,** 119–128.

Van Soest, P. J. (1969). *Proc. Cornell Nutr. Conf. Feed Manuf.* pp. 17–21.

Van Soest, P. J. (1973). *Fed. Proc., Fed. Am. Soc. Exp. Biol.* **32,** 1804–1808.

9

Forages and Roughages for Dairy Cattle

9.0. INTRODUCTION

In feeding dairy cattle, careful attention should be given to the ingredients used to furnish the required nutrients. Most feeds are classified as forages (and/or roughages), concentrates, by-products, or special supplements. Forages are the vegetative parts of plants, including leaves and stems. Concentrates are primarily seeds and products derived from seeds. Large amounts of dairy cattle feeds are the by-products from processing various materials for other purposes, especially human food. Differences in nutrient composition and availability are enormous and of great practical importance in feeding dairy cattle.

9.1. FORAGES AND ROUGHAGES, A MAJOR SOURCE OF NUTRIENTS FOR DAIRY CATTLE

Historically, cattle lived on forages and roughages during most of the year with some seeds during limited periods. Thus, they are especially well adapted to utilization of forage nutrients (see Chapter 1). Typically forages make up the majority of the feed given dairy cattle, but the relative proportion of forages to other feeds varies greatly on dairy farms. As discussed in Chapter 8, dairy cattle require a substantial amount of fiber for good health and optimum performance. Although there are other sources, most of the fiber consumed by dairy cattle comes from forages.

In addition to fiber, dairy cattle obtain large amounts of energy, protein, minerals, vitamins, lipids, and water from forages. Usually forages are the most economical source of nutrients, especially energy but the cost relationships vary enormously in different areas, times, and individual situations. While utilization of the maximum amounts of forage nutrients generally is the most economical way to feed dairy cattle, there are exceptions (see Chapters 13, 17, and especially 19).

221

9.2. WIDE VARIATIONS IN NUTRIENT CONTENT AND FEEDING VALUE OF FORAGES

The great variation of individual nutrients in forages was mentioned frequently in earlier chapters. Major factors affecting nutrient content and feeding value of forages are (1) the species, variety, and strain of the plant; (2) the parts of the plant such as leaf or stem; (3) the stage of maturity at harvest; (4) the fertilization and soil characteristics; (5) climates, weather and season; and (6) changes during preservation, processing, and/or storage. Since these factors have many practical implications and interact in various crucial ways, an understanding of each is important in dairy cattle feeding and nutrition.

9.2.a. Effect of Species, Variety and Strain of Plants on Nutrient Content and Feeding Value

The genetics of a forage plant have a major influence on its composition and nutritional value. Forages fed to dairy cattle can be divided into legumes and grasses. Contrary to popular opinion, legumes are not consistently superior nutritionally in every respect to all grasses, even under the same conditions. For example, at the same stage of maturity often the digestible energy value of grasses such as timothy or ryegrass is at least equal to that of alfalfa. The legume, *Sericea lespedeza,* has a lower overall feeding value than most desirable grasses. Certain characteristic differences between legumes and grasses tend to occur. For instance, even with low nitrogen fertilization, legumes seldom have an extremely low protein content. On an average, legumes contain much more calcium and higher levels of potassium, magnesium, sulfur, iron, zinc, copper, and cobalt but less silica and manganese than grasses (see Chapter 5) (Underwood, 1966).

Alfalfa and *various clovers* are the legumes used in largest amounts for dairy cattle feed. Many other legumes are also important on dairy farms in limited localities throughout the world. Numerous grasses, either in combination with legumes or as "pure" stands, are widely used forages for dairy cattle. These include timothy, orchard grass, fescues, bluegrass, ryegrass, brome, bermuda grasses, oats, rye, wheat, sudan grass, millets, johnson grass, various hybrids of sorghums and johnson grass, *Phalaris* grass and many others (see Appendix, Table 4). Botanically, corn is classified as a grass. Using this system, the largest amount of grass forage comes from the nongrain part of the corn plant, primarily as silage. Certainly it is appropriate to view the corn plant as a grass, although a very specialized one in which seed production is a key aspect.

Some grasses such as ryegrass and bluegrass, are of very high nutritional value, especially at immature growth stages. However, in the practical feeding of dairy cattle, yields and other factors affecting cost must be considered along with nutritional value.

9.2.b. Different Nutritional Values of Forage Plant Parts

Generally, leaves of a forage plant have a higher content of digestible energy, protein, and essential mineral elements but less fiber than stems. Thus, the overall nutritional value usually is greater for leaves than for stems. Since dairy cattle ordinarily will eat more forage with a high leaf content, a greater proportion of the total nutrients can be supplied with leafy forage. The decreased percentage of leaves relative to stems is one of the changes usually occurring with advanced maturity and a factor in lowering nutritional value.

9.2.c. Stage of Maturity—The Key to Forage Quality

The stage of maturity at harvesting forages often is the most important factor affecting success in practical dairy cattle feeding (Hibbs and Conrad, 1975). As forage plants mature, several changes occur. A high proportion of the early growth is leaves. In the young forage plants even the stems have thin cell walls and usually are highly digestible. As most forage plants mature, the cell walls of stems thicken and become lignified. With most grasses and legumes, the content of digestible energy, protein, and essential mineral elements decline with advancing maturity (Tables 9.1, 9.2, 9.3, and 9.4). Likewise, the amount eaten decreases (Table 9.1). With young forages, a greater proportion of the total re-

TABLE 9.1
Digestibility, Dry Matter Intake, and Milk Production of Holstein and Jersey Cows Fed Green Chopped Alfalfa–Brome Forage at Various Stages of Maturity[a]

Maturity stage	Date	Dry matter Digestibility (%)	Intake (lb/day)[b]	Milk production (lb/day)	Grain needed (lb/day)[c]
Prebud	May 17	67	34	43	4
Bud	May 24	65	33	40	6
Early bloom	May 31	63	32	34	8
Mid-bloom	June 7	61	31	31	11
Full bloom	June 14	59	29	27	14
Late bloom	June 21	58	28	23	16
Mature	June 28	56	26	20	18

[a] Adapted from Hibbs and Conrad (1975). Cows fed forage free choice with 3–5 lb of grain per day.

[b] Dry matter intake per 1000 lb of body weight.

[c] Grain needed to maintain initial milk production level (i.e., to maintain an average dry matter digestibility of 67%) (see Section 2.6).

TABLE 9.2

Average Chemical Composition and Yield of Well-Fertilized Coastal Bermuda Grass Harvested at a 3-, 5-, or 7-Week Frequency throughout the Year[a]

Harvest frequency	Chemical composition (% of dry matter)			Yields (lb/acre/year)	
	Protein	Crude fiber	Nitrogen free extract	Dry matter	TDN
3-week	15.3[b]	28.0	47.6	12,953	8,730
5-week	11.8	30.2	49.9	13,069	8,050
7-week	10.8	30.9	50.4	18,548	10,609

[a] Adapted from Brooks et al. (1968).

[b] The protein content of the 3-week forage ranged from a high of 16.4% to a low of 11.6% among harvests.

quired nutrients can be supplied by forages with less needed from concentrates.

The higher feeding value of immature forage is partially offset by lower yields and therefore higher cost per pound (Tables 9.2, 9.4). In practical feeding, it is usually most profitable to compromise and harvest forage at a sufficiently early stage of maturity to obtain relatively high nutritional value but after considerable yield has been achieved. The optimum stage of maturity of harvest also is affected by the relative cost of feed nutrients from forages versus concentrates.

In temperate climates, often the most desirable stage of maturity for the first harvest can be approximated by calendar dates (Fig. 9.1) (Reid, 1961). In tropical and subtropical areas, dates are not as meaningful. With plants such as corn

TABLE 9.3

Digestibility of Various Components of Pelleted Coastal Bermuda Grass as Affected by Frequency of Harvest[a]

Component	3-week	5-week	7-week
Dry matter	65.9	60.7	56.6
Organic matter	66.0	60.7	56.8
TDN	67.4	61.6	57.2
Gross energy	63.1	56.8	53.2
Protein	67.0	58.8	51.7
Crude fiber	66.2	59.1	54.2
Ether extract	68.4	69.4	61.7
Nitrogen-free extract	65.3	61.7	59.2
ENE (calculated)[b]	59.2	51.2	45.0

[a] Adapted from Brooks et al. (1968).

[b] Megacalories (Mcal) per 100 lb.

TABLE 9.4
Effect of Stage of Maturity of Certain Forages on Dry Matter Yield and Crude Protein Content[a]

Stage of maturity	Alfalfa[b]		Smooth bromegrass[c]		Sudan grass[d]		Oats[e]	
	DM (tons/acre)	Protein (%)	DM (tons/acre)	Protein (%)	DM (tons/acre)	Protein (%)	DM (tons/acre)	Protein (%)
A	0.9	26.5	0.5	22.9	2.2	18.4	1.1	17.0
B	1.5	23.3	1.4	16.4	2.8	15.2	1.9	11.6
C	2.3	17.9	1.9	14.4	3.5	11.6	2.2	12.3
D	2.4	15.8	2.6	8.6	6.0	5.8	2.9	8.4

[a] Adapted from Hutjens and Martin (1976).
[b] 1st cut: (A) prebud, (B) bud to midbud, (C) first flower to one-tenth bloom, (D) full bloom, (Wisconsin).
[c] 1st cut: (A) vegetative, (B) 1st flower, (C) full flower, (D) 2 weeks after full bloom, (Minnesota).
[d] Season average (A) 5 cuts at 18 inches, (B) 3 cuts at 36 inches, (C) 2 cuts at 54 inches, (D) 1 cut at hard dough, (Iowa).
[e] (A) boot, (B) flower, (C) milk, (D) dough, (Minnesota).

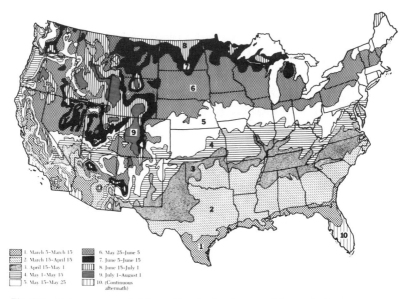

1. March 5–March 15   6. May 25–June 5
2. March 15–April 15   7. June 5–June 15
3. April 15–May 1   8. June 15–July 1
4. May 1–May 15   9. July 1–August 1
5. May 15–May 25   10. (Continuous aftermath)

Fig. 9.1   Dates recommended for making the first harvest of forages in the United States (see Reid, 1961). First date is the time to start cutting in the southern part of area; second date is the time to start cutting in the northern part of the area. The goal is to finish haying in approximately 15 days.

that produce large amounts of grain, the most desirable stage of maturity for harvest also involves the grain yield (see Section 9.7.c).

9.2.d. Effect of Soil Characteristics and Fertilization on Feeding Value of Forages

Fertilization and soil characteristics affect the nutritional value of forages by altering the composition of each plant and by influencing the plant species which grow. For instance, alfalfa requires a fertile soil, adequate amounts of several fertilizer elements, and a fairly high soil pH. In contrast, some legumes such as *lespedezas* will grow at a lower pH on soils having less fertility. Legumes generally are more sensitive to soil pH than grasses.

Although the composition of alfalfa and other legumes is influenced by fertilization and soil characteristics, usually they are affected much less than grasses. For example, with changing nitrogen fertilization, the protein content varies much more in bermuda grass or timothy than in alfalfa. Often bermuda grass forage contains less than 9% crude protein (see Appendix, Table 4), but with high nitrogen fertilization, a protein content of 16% or more can be attained readily at immature growth stages (Table 9.2). With high nitrogen fertilization, the crude protein content of some immature grasses exceeds 20%. In contrast, with low nitrogen fertilization at mature stages, values under 5% are common.

The mineral composition of forages, especially grasses, often are greatly affected by the minerals in the soil and fertilizer used. As discussed more fully in Chapter 5, many of the practical mineral problems of cattle are related to soil characteristics of the areas involved. Examples are iodine, cobalt, copper, phosphorus, and selenium deficiencies and toxicities of molybdenum, selenium, and fluorine.

9.2.e. Effects of Climate, Season, and Weather on Composition of Forages

Forages that can be successfully grown are largely determined by the climate, season, and weather. Thus, these factors profoundly influence nutritional value by affecting the species and varieties grown. For instance, the digestible energy content of perennial grasses grown in hot climates usually is substantially lower than with those grown in more temperate climates. In subtropical regions annual grasses with very high energy digestibility often thrive during the cool season. Highly nutritious clovers and ryegrass grow well in the southern United States during late winter and early spring. In northern sections of the United States, at the same stage of maturity, generally the digestible energy content of many forages is lower in the middle of the summer than at the first spring harvest (see Appendix, Table 4).

Often season, climate, and/or weather have a major influence on the mineral content of forages. For instance, the phosphorus content of forages is usually low with drought conditions. This was a key factor in the widespread phosphorus deficiencies that once were so widespread in South African cattle.

9.3. FORAGE EVALUATION—A KEY ROLE IN DAIRY CATTLE FEEDING

The enormous variability in the nutrient content and feeding value of forages was discussed in previous sections (see Section 9.2, and 9.2.a through 9.2.e). Because of these variations, forage evaluation is an important research area. Likewise, some evaluation of the specific forages used is needed in practical dairy cattle feeding (Chapter 12). Typically, it is desirable to make the maximum use of the forages which can be effectively grown or obtained locally. Generally, supplemental feed is needed for lactating cows and calves. When formulating the additional feed, a reasonable knowledge of the nutrients provided by the forages is essential (Crowley, 1976).

The term forage quality is widely used in dairy cattle feeding and nutrition. In the broadest sense this includes all aspects affecting the feeding and nutritional value of the forage. The first consideration is the ability of the forage to supply the energy needs of the animal which is determined by the content of usable energy and the quantity cattle will consume (see Chapter 2). Since dairy cattle generally are fed all the forage they will eat, the expression voluntary intake is used to describe the amount of forages eaten when offered "free choice."

9.3.a. Importance of Voluntary Intake in Determining Proportion of Forage That Can Be Utilized

The amount of forage dry matter dairy cattle will voluntarily consume often has more influence on the ability of a specific forage to supply the energy needs of dairy cattle than the usable energy content per pound. When the differences in digestible energy of two forages are caused by the stage of maturity, usually dairy cattle will voluntarily consume a greater amount of the more digestible forages (Table 9.1). Thus, harvesting most forages at a less mature stage increases both the digestible energy content and the pounds voluntarily consumed. With this double advantage, immature forages supply much greater proportions of the energy needs (Table 9.1).

Often dairy cattle will voluntarily consume greatly different amounts of forages having a similar digestible energy content. For instance, when the TDN contents are similar, dairy cattle usually will eat substantially more alfalfa hay

TABLE 9.5

Milk Production, Feed Consumption, Weight Gains, and Milk Composition of Cows Fed Alfalfa Hay, Pelleted Coastal Bermuda Grass Hay, and Baled Coastal Bermuda Grass Hay[a]

	Coastal pellets	Alfalfa hay	Coastal hay
	(lb/cow/day)		
Milk production	27.0	26.1	22.4
Forage consumed	37.6	32.2	23.6
Concentrates consumed	6.5	6.7	6.7
Weight gains	0.1	−0.1	−0.6
Fat in milk	4.27%	3.98%	3.78%
Solids-not-fat in milk	8.94%	8.58%	8.49%
Protein in milk	3.56%	3.37%	3.21%

[a] Adapted from Brooks et al. (1968).

than coastal bermuda grass hay (Table 9.5). Grinding and pelleting coastal bermuda grass does not materially change the TDN content, but greatly increases voluntary dry matter consumption (Table 9.5). With alfalfa, grinding and pelleting has much less effect on voluntary intake.

A major reason for the large differences in voluntary intake among forages is the varying rates of digestion, especially in rumen fermentation. With faster digestion, the material leaves the rumen more quickly making space for additional forage. Usually a more rapidly digested forage has a greater feeding value because it can be used in larger amounts with less concentrates needed. The amount of space occupied by the forage in the rumen also affects the amount eaten. The increased density appears to be an important reason for the greater intake of pelleted coastal bermuda grass relative to the long hay. Whereas tables of digestible energy (TDN, net energy, etc.) (see Chapter 2 and Appendix, Table 4) are generally available, no such tables are widely used for the amount of voluntary intake with different forages. Such tables would be of great value in feeding dairy cattle.

9.3.b. Variation in Protein, Mineral, and Vitamin Content of Forages

Although usable energy is the nutrient receiving most attention in evaluating forages, supplying the other essential nutrients is just as necessary. When designing a practical feeding program, usually maximum use should be made of the best forages readily and practically obtainable. Supplemental feeds should be formulated to supply the other required nutrients.

Even though there are large differences in the usable energy values among forages, the relative differences in protein, mineral, and vitamin contents are much larger (see Chapters 3, 5, 6, and Sections 9.2 to 9.2.e). Protein, mineral, and vitamin contents of individual forages available for feeding often are greatly different from average values in tables. For this reason, the content of many of these nutrients is determined in forage evaluation programs available to dairymen (see Section 12.15).

9.4. TYPES OF FORAGES AND ROUGHAGES

One of the major ways of classifying forages and roughages for dairy cattle is by use. The major forms are (1) pastures, (2) hays, and (3) silages. Others such as straws, various plant residues, and special forms including pellets, wafers, and harvested green forage are utilized in some situations. Likewise, intermediate forms such as haylage (between hay and silage) sometimes are widely used.

9.5. PASTURES, THE "OLDEST" FORM OF DAIRY CATTLE FEED

Although currently playing a somewhat smaller role in the feeding of dairy cattle than many years ago, pasture is still widely used and probably will continue to have an important role on numerous dairy farms (Crowley, 1975). Generally, pasture makes up a higher percentage of the forage for dry cows and young stock than for the milking herd. These nonmilking animals often graze land that is inconvenient to the milking barn. Frequently land unsuitable for machine harvesting of crops is used for pasture. The substantial amount of such land on numerous dairy farms is a major reason for pasture continuing to be important.

Pasture has the further advantage of requiring relatively little equipment and labor for harvesting. However, there are problems, disadvantages, and limitations in pasture utilization including seasonal production, selective grazing, and wastage due to trampling and fouling. Perhaps the most serious disadvantage of pasture is the very uneven growth throughout the year. Whereas the amount of nutrients needed by dairy cattle remains relatively uniform throughout the year, forage growth is seasonal. The uneven seasonal distribution can be partially overcome with plant species having different growth patterns, but these efforts may increase cost. Another approach to the seasonal growth problems, used in some areas such as Australia, is to field stockpile some of the pasture forage for grazing in the nongrowing seasons (Fig. 9.2). This method of "storing pasture on the stalk" works best when a dry season follows the growing period so that the forage is virtually "cured in place."

A great variety of plant species are used for dairy cattle pastures. Many are employed in "permanent" pastures, whereas others are used as temporary pastures. Some of the more widely utilized permanent pasture species are low growing, enabling relatively close grazing without serious damage to the plants (Fig. 9.3). For example, white clovers and grasses such as bluegrass, ryegrass, and fescue easily adapt to a low grazing pattern.

Many annual plants are used for temporary pastures, including various tall growing grasses such as sudan grass, millets, and sorghum—sudan grass hybrids. Grazing tall growing species too closely or continuously, may reduce yields and destroy stands. Thus, rotational grazing may be the most feasible management plan, but costs are increased.

Two problems with pastures that are interrelated with the seasonality of growth are selective grazing and low yields of forage per acre. Cattle generally closely graze areas or spots with more palatable forage, while largely skipping the tall or rougher plants. Since the very young forage is more nutritious and palatable, cattle may continue to graze and regraze the same spots.

Fig. 9.2 Cow in Queensland grazing forage which was grown in the rainy season for use in dry season. (Courtesy, W. J. Miller, Univ. of Georgia.)

Selective grazing often can be eliminated or greatly reduced by close grazing, periodic mowing the areas not being grazed, and/or "dragging" the pasture to spread manure. Pasture grazed close may not provide dairy cattle with enough nutrients, especially usable energy, for optimum performance. Frequently the mowed pasture herbage is wasted because preservation is not feasible due to poor quality and low yields.

Dry matter of forage plants is synthesized as a result of photosynthesis in the green forage. The photosynthesis process, which obtains its energy from the sun, requires a substantial amount of chlorophyll in green forage leaves. Thus, with too close grazing, leaf photosynthetic area is reduced and plant growth decreased resulting in lower yields per acre. The optimum grazing intensity depends on the individual species of forage. White clover and bahia grass can be grazed much more closely than alfalfa and sudan grass which should be rotationally grazed.

Pasture forage may be more or less nutritious than other forms of forage depending on how and when it is used. The digestible protein and available mineral content of young pasture forage are very high, but there is an exception. In the young, highly succulent pasture, mangesium availability is very low, often leading to grass tetany. The carotene content of young green forage usually is very high. When young pasture forage can be supplied economically, and the

Fig. 9.3 Pasture is a major source of nutrients in some areas. Cattle grazing on polder (reclaimed from the sea) land in the Netherlands. (Courtesy of J. Hartmans, NRLO-TNO, Wageningen.)

problems of seasonality of yields and low yields per acre satisfactorily solved, pasture may be a very useful part of a forage feeding program. However, to meet total forage needs, appreciable amounts of stored forages, as hay or silage, usually are needed.

9.6. HAYS—IMPORTANT ON MANY DAIRY FARMS

Because dairy cattle must be fed daily and forage grows seasonally, it is necessary to store forage for periods when little or no growth occurs. Historically, forage has been stored as hay much longer than in any other form.

9.6.a. Good Hay: Requirements, Problems

The first requirement for good hay is a nutritious forage; as the hay will be no better than the forages from which it was made (see Sections 9.2 through 9.2.e). Although numerous species of legumes and grasses are widely used as hay crops, in the United States, alfalfa and alfalfa–grass mixtures make up about 58% of the total (Cullison, 1975).

Drying is a most critical aspect of hay making, as too much moisture will cause molding or heating. When hay is moldy the feeding value is reduced and in extreme cases may become essentially worthless. Nutritional losses can be reduced or eliminated by propionic acid added as a mold inhibitor. Excess moisture results in heating that causes brown or black hay and a greatly reduced protein digestibility (see Section 3.1). The excessive heating also may lead to spontaneous combustion, possibly causing a fire and the loss of a barn.

When hay is overdried, leaves become brittle and shatter resulting in loss of considerable nutrients (Fig. 9.4). The hay which remains has fewer leaves and a lower nutritional value (see Section 9.2.b). The extent of the leaf loss through shattering depends on the nature of the forage as well as the degree of excess drying. For instance, the loss of both weight and quality, due to excessive drying, are greater with alfalfa than with coastal bermuda grass because leaves of alfalfa and most other legumes shatter more easily than those of grasses.

In addition to the physical losses of nutrients during hay curing, there are chemical losses resulting from respiration and/or fermentation of the partially dried forages (see Section 9.7.b). Because of these losses, rapid drying is desirable. When there is rain, nutrients also may be lost through leaching. On newly mown hay, rain causes relatively little damage, but the losses are much greater on forage almost dry enough to bale.

The ideal moisture content for storing hay depends on several factors, but often falls within the range of 18–22% (Cullison, 1975). With loosely pressed bales, a somewhat higher moisture content is more satisfactory than with denser bales. If there is considerable air movement around the bale, a higher moisture content is acceptable. Thus frequently, hay can be baled when the moisture

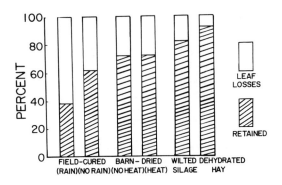

Fig. 9.4 Loss of alfalfa leaves during harvest by different methods. (Adapted from Shepherd *et al.*, 1954.)

Fig. 9.5   Hay making and storage involve many different procedures, equipment, and facilities in different parts of the world. This shows examples of the haying operations. A, Hay drying barn in Maine. B, Alfalfa cubes in California. C, Special storage building for chopped hay in the Netherlands. (A courtesy of C. A. Walker, Univ. of Maine. B and C courtesy of W. J. Miller, Univ. of Georgia.)

Fig. 9.5 Continued

content is about 25% as some further drying will occur during the handling process.

Often the most serious problem in making good hay is the lack of suitable weather when forage is at the optimum stage of maturity (Fig. 9.5). Drying is not difficult in arid regions. In many humid areas of eastern United States and Northern Europe, where temperatures are low and clouds are frequent, field curing of high quality hay often is not a practical way to preserve forage. If the weather is unsuitable for the field curing of hay, drying can be achieved artificially, either with or without supplemental heat (Fig. 9.5). Especially when heat is used, this type of drying substantially increases the cost.

9.6.b. Storage and Feeding of Hay

In earlier years, hay was stored and fed loose. On most dairy farms, other forms of storage, handling, and feeding are employed with small bales of some 30–100 lb being widely used. Most hay of commerce has been in this form (Fig. 9.6).

In recent years, large bales (and stacks) of 500–2000 lb have come into relatively wide usage (Fig. 9.7). Whereas, the small, rectangular bales require shelter, the large bale often is left unprotected. In some situations, storing the large bales without shelter may only result in nutrient losses near the surface, but with other conditions the losses are so large that shelter is highly desirable or imperative.

Hay can be stored as chopped, dry forage, but generally a lower moisture

Fig. 9.6 The small bale is the most widely used form of hay, especially for that handled commercially. View of California dairy farm. (Courtesy of W. J. Miller, Univ. of Georgia.)

content is required for satisfactory results (Fig. 9.5). Excessive dustiness can be a serious problem for the workers involved with chopped hay.

A limited amount of hay is made into pellets, cubes, and wafers. Generally these methods require sophisticated equipment and a greater energy cost during processing. However, they have the advantage of compressing the hay into a far smaller volume for transportation, storage and feeding. These forms are readily adapted to mechanization. Pelleting, cubing, and wafering affect the feeding value and often improve production efficiency. For example, the voluntary dry matter consumption of coastal bermuda grass is greatly increased when the forage is ground and pelleted (Table 9.5) (see Section 9.3.a).

With dry storage, the overall hay quality decreases very slowly but loss of carotene proceeds quite rapidly when hay is exposed to light. The amount of carotene in hay when fed is only a small fraction of that in the green forage at harvest (see Section 6.2.c). In spite of the relatively slow loss of nutrients during good storage, it is generally thought that hay more than one year old is less desirable than a "new-crop hay."

9.6.c. Other Advantages and Disadvantages of Hay as a Way to Preserve Forages

One of the key advantages of preserving forage as hay is the comparative simplicity of the process. Once dry, storage is relatively easy, requiring only that it be kept dry. Almost any shape or type of simple structure is adequate. Dry

Fig. 9.7 In recent years, large, round bales and stacks have come into relatively wide use in hay making. A, Bales. B, Stacks. (Courtesy of W. J. Miller, Univ. of Georgia.)

matter yields of hay generally exceed those of pasture and with suitable weather, harvesting hay at the optimum stage of maturity is relatively easy.

Hay also has key disadvantages and limitations. Unless expensive artificial drying is used, satisfactory hay making is totally dependent on the weather. Often good hay-growing weather and hay-curing weather do not occur at the same seasons. Unless pelleted, cubed, or wafered, hay is a very bulky material requiring considerable storage space. Also transportation for substantial distances is costly. Handling and feeding of hay is not easily mechanized, often resulting in a substantial labor cost. In addition, considerable care is required to avoid excessive wastage during feeding.

In common with most other forages, hay quality is difficult to standardize. Since there are huge differences in the feeding value of hay, this makes it difficult to accurately assess its commercial value. However, some success has been achieved in limited areas of the world. For example, in central and southern California and surrounding areas considerable alfalfa hay is grown, transported, and fed to dairy cattle using some form of quality evaluation.

9.7. SILAGES—THE MAJOR FEED ON MANY DAIRY FARMS

In recent years the use of silage on large numbers of dairy farms has increased greatly and in many areas has become the major feed, especially for lactating cows. Much of this increase centers around corn silage and its advantages over hay and other forms of silage as a way to preserve forage for dairy cattle. A key advantage of silages is the opportunity to harvest at an ideal stage of maturity in weather that is unsuited for making hay. Further, silage making and feeding is more easily mechanized, especially with large operations, than is haying. In humid areas, nutrient losses often are lower with silage than with hay.

Relative to hay, silage making has some disadvantages and limitations, including a more complex and involved technology. Likewise, the equipment needs and the structural requirements for storage are somewhat greater and more exacting. Efficient silage making is not readily adapted to small operations, and is difficult to commercialize or to feed at long distances from the place where the silage is produced.

9.7.a. Types of Silage Crops

In the United States corn is the "king of silage crops" with far more corn silage used than all other types of silages combined (Fig. 9.8) (Cullison, 1975). Most corn silage is made from the whole corn plant which results in high dry matter yields per acre. When corn is harvested for grain, the forage parts generally are not efficiently utilized, if at all. The whole corn plant gives a very high yield of nutrients per acre, and has excellent ensiling characteristics. Even so, when feeding corn silage, it is crucial to balance the ration for the essential nutrients. Corn silage is quite low in protein, very deficient in calcium, and generally inadequate in sulfur for lactating cows. Other essential mineral elements also may be below required levels.

Numerous crops other than corn are used for silage with many classified under the broad heading of "grass" or "hay crop silage." Sorghum silages are sometimes used. Although grass or hay-crop silages are made from about the same forages as those used for hay, often legumes are more difficult to ensile successfully than grasses (see Section 9.7.c). Thus, compared with hay, more crops such as oats and wheat are made into silage relative to alfalfa (see Section 9.6.a).

Fig. 9.8   In much of the United States as well as in many other countries, corn is the single most important feed crop for dairy cattle. (Courtesy of W. J. Miller, Univ. of Georgia.)

9.7.b. The Silage Forming Process

The key chemical and microbiological reactions occurring when forages are chopped and put in a silo have been divided into various phases by Barnett (1954) and discussed more recently by McCullough (1973). Immediately after being chopped, respiration proceeds very rapidly under the influence of the enzymes from the plant cells. This process rapidly uses the oxygen trapped in the forage, and forms carbon dioxide along with considerable heat. The "using up" of oxygen and replacement with carbon dioxide produces an anaerobic environment favorable for microbial fermentation that becomes the dominant phase within a few hours after chopped forage is placed in the silo.

The microbial fermentation produces organic acids that reduces the pH of the ensiling mass of material. Initially, acetic acid appears to be the major acid formed but if fermentation proceeds in a desirable way within 2 to 4 days considerable lactic acid formation begins. A substantial concentration of lactic acid is desirable as it lowers the pH much more than acetic acid. The lower pH is needed to prevent subsequent types of microbial activity that would produce undesirable products and a poorer quality silage.

A pH of around 4.2 with high-moisture material, to 4.5 with low-moisture forages, or below is needed to preserve silage in a desirable condition. If the pH is much above this, the fermentation may not cease after the lactic acid formation phase is complete. Any subsequent fermentation is undesirable as butyric acid and protein breakdown products are formed. Depending on the degree of such unwanted fermentaion, the silage may be unpalatable, stinky, and undergo considerable spoilage.

9.7.c. Good Silage

The first requirement for making good silage is to begin with a suitable and nutritious crop (see Sections 9.2 through 9.2.e). With corn and similar high-grain crops, the most desirable harvest stage generally is strongly influenced by the maturity of the grain. Opinions of scientists vary relative to the ideal maturity, but it usually is when the grain is somewhere between the early dough and hard dent stages. In recent years the maturity is more frequently described relative to dry matter percentage. Because of the time involved in practical operations, often parts of the crop will need to be ensiled before and some after the ideal stage. Usually good silage can be made over a relatively wide range of maturities. When the corn is substantially too immature, the yield of dry matter and nutrients is greatly reduced. If too mature, molding and spoilage are more difficult to avoid, and often voluntary dry matter consumption is decreased.

In addition to being highly nutritious, for good silage formation, the crop should have certain other characteristics. Sufficient, easily-fermentable carbohydrates are needed to serve as a substrate for the fermenting microbes that form the preserving acids, especially lactic acid. McCullough (1973) has concluded that at least 6–8% of ensiled dry matter should be "readily-fermentable" carbohydrate for a good fermentation. High-protein crops do not ensile as well as those of lower or moderate protein content. Difficulties are often encountered when ensiling alfalfa and soybean forage. Although somewhat less desirable for hay than alfalfa, oat forage generally makes better silage.

Exclusion of air is a key aspect of good silage making and preservation. On a practical basis, air exclusion is not absolute. Oxygen from air which is trapped in the forage or subsequently enters must be converted to carbon dioxide or other substances to prevent molding or spoilage. Enclosing the silage with materials such as plastic film and concrete permit only a very slow rate of oxygen penetration and greatly reduces the air problems.

Dry matter content of the ensiled material has a major influence on the nature of the silage fermentation. Excessive moisture results in considerable loss of nutrients by seepage. Of even greater concern, too high a moisture content adversely affects the fermentation, often resulting in a high pH and formation of considerable butyric acid. The effect may be a "stinky" unpalatable silage with low voluntary dry matter intake by cattle. When ensiled forage has less than about 28% dry matter, the undesirable clostridial-type fermentation also results in more protein being degraded to nonprotein nitrogen (Waldo, 1977b). Likewise, extremely high moisture in silage is more damaging to concrete or metal silos. In contrast, when the forage dry matter content is excessively high, exclusion of oxygen becomes much more difficult. Also, digestibility of the protein may be reduced due to heat damage (see Section 3.1).

Generally, a dry matter content of 28–40% is most desirable for good silage making, but this depends on other variables, especially the type of silo. In horizontal silos, a somewhat lower dry matter content is needed for good silage than in tower silos. An acceptable dry matter content of ensiled material is one sufficiently high to avoid excessive seepage but low enough to avoid high temperatures which increase losses and cause heat damage. The taller the tower, the narrower the acceptable range of dry matter contents (Waldo, 1977b).

When many hay-crop forages are sufficiently immature to make highly digestible silage, they are too low in dry matter for good ensiling. For instance, in one experiment, silage was made from highly nutritious ryegrass and crimson clover containing only 12% dry matter (Miller et al., 1961). If the dry matter content is too low, wilting the forage prior to chopping is beneficial. Although this requires an additional field operation, frequently it gives better and more economical feed than making silage with excessive moisture or adding sufficient dry feed to achieve a desirable dry matter content. Attaining the correct amount of wilting under variable weather conditions is often very difficult.

Relatively fine chopping of forage being ensiled usually gives best results. Normally a theoretical cut 1/4 and 3/8 inches in length with sharp cutter knives is most desirable. Inadequate chopping makes exclusion of air more difficult, and often increases fermentation losses and/or spoilage. Excessively fine chopping increases the power requirements and may adversely affect fat composition of milk (see Section 8.3.d).

Adequate packing is important, especially in horizontal silos. Packing removes entrapped air and slows the rate of air movements through the ensiled mass, greatly reducing the introduction of new air. A uniform distribution of the ensiled material is desirable, especially in tower silos.

A rapid rate of filling is beneficial to good silage making. With fast filling, the ensiling material is exposed to air for a shorter time before sealing. For instance, leaving chopped forage on wagons for long periods, such as overnight, is undesirable since the oxygen which has been used up by the early respiration and fermentation is reintroduced when the material is blown into the silo. More rapid filling reduces nutrient losses and often results in a better quality silage (Table 9.6) (Miller *et al.*, 1962).

Early, thorough, and permanent sealing of the ensiled material reduces losses

TABLE 9.6

Effect of Fast and Slow Filling of Silos on Nutrient Losses, Chemical Composition, and Performance of Lactating Cows[a,b]

	Rate of filling	
	Fast	Slow
Ensiled dry matter		
recovered (% of ensiled nutrients)		
As good silage	83.0	77.0
As seepage	8.3	8.7
Fermentation loss	8.7	14.3
Composition (% of dry matter)		
Lactic acid	9.2	6.0
Acetic acid	2.4	3.6
Butyric acid	1.1	2.2
pH	4.9	5.1
Milk production (lb/cow/day)	28.7	27.6

[a] Adapted from Miller *et al.* (1962) and Miller and Clifton (1962).

[b] Silage made in 8 by 24 ft experimental tower silos from oats, ryegrass, and crimson clover. The fast fill was done in one day compared with 5 days for the slow fill.

and improves silage quality. As soon as possible after filling is completed, the silo should be sealed with plastic film or other material to prevent the entrance of air (Hight *et al.*, 1975) (Table 9.7).

During storage silage should be protected from weather damage, especially air and water. Rainwater will leach out high quality, soluble nutrients and preserving acids lowering the nutritional quality as well as adversely affecting the preservation.

Often considerable emphasis has been attached to a suitable temperature in silage fermentation. Temperatures, over 95°F–100°F, often accompany high fermentation losses, and may result in brown, black, or "charred" silage (Fig. 9.9). The digestibility of the protein in overheated brown or black silage is substantially reduced (see Section 3.11). Heat damage appears to be closely related to the number of days the silage temperature exceeds 95°F (Waldo, 1977b). It has been suggested that achieving a wet density of 30 lb per cubic foot within 3 days after filling is essential to keep the temperature low enough to avoid excessive protein damage (Waldo, 1977b). The time required to achieve the needed density is reduced with lower dry matter content, more rapid filling, and finer chopping.

When feeding silage, it is important to remove it from the silo sufficiently fast to prevent spoilage or damage. The rate required for good results goes up with ambient temperature. As a general rule, a minimum of 2 inches per day in winter and 3 inches in summer should be removed from the silage surface. Faster rates are desirable. Likewise, it is important to avoid disturbing the silage not removed.

TABLE 9.7
Effect of Covering and Sealing Bunker Silos on Losses[a]

	Corn		Oats and vetch	
	Covered	Not covered	Covered	Not covered
Forage ensiled				
Fresh wt (lb)	45,655	47,990	27,450	26,190
Dry matter (lb)	11,551	12,141	10,452	10,314
Preserved silage				
Dry matter (lb)	8,568	7,056	7,336	4,911
Dry matter losses (%)	25.8	41.9	29.9	52.4

[a] Adapted from Hight *et al.* (1975). Covers were 6 mil polyethylene weighted with discarded auto tires. The experimental silos were 16 ft long, 10 ft wide, and 5 ft high.

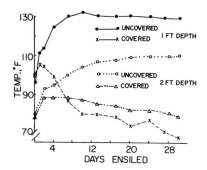

Fig. 9.9 When a silo is not sealed to prevent entrance of air, the temperature is elevated indicating increased fermentation. This increases dry matter losses and reduces quality by various processes including heat damage to protein. (Adapted from Hight *et al.,* 1975.)

9.7.d. Use of Preservatives and Additives in Silage Making: Fads, Fashions, and Facts

Because of technical problems in silage making, numerous substances have been investigated and used as silage preservatives (Waldo, 1977a). Some examples include phosphoric acid, hydrochloric acid, mixtures of acids, formic acid, formaldehyde, *p*-formaldehyde, propionic acid, carbon dioxide, sulfur dioxide, sodium metabisulfite, enzymes, molasses, ground grains, beet pulp, citrus pulp, straws, and common salt. No additive is needed for making excellent silage from the whole corn plant. The use of additives to improve silage preservation has been largely restricted to grass or hay-crop silages. Supplemental acids reduce the silage pH resulting in preservation with less lactic acid fermentation, and thereby reduce the need for readily fermentable carbohydrate in the forage. Use of the comparatively strong acids, such as phosphoric and sulfuric, have resulted in good silage preservation. However, objectionable working conditions and adverse effect on equipment usually cause farmers to abandon the practices. Some of the organic acids, that are weak acids, such as propionic, are effective in inhibiting mold growth in comparatively dry forages. Most of the organic acids are expensive.

In forage crops with low-fermentable carbohydrates, adding molasses, or ground feed grains can improve the fermentation and silage quality. However, when the additional cost including added silo capacity, extra nutrient losses, added filling costs, and the cost of the inventory are considered, the overall advantage often disappears.

Using large amounts of dry grains, citrus pulp, beet pulp, straw, or other materials to increase the dry matter content of a forage that is too high in moisture, generally is uneconomical. A key problem is the very large amount of additive dry matter needed to increase the dry matter content of the ensiled mixture sufficiently to obtain a material benefit. For instance, using the example cited earlier (Miller *et al.*, 1961) where the forage contained only 12% dry matter, to obtain a mixture with 30% dry matter would require adding more than twice as much grain dry matter as forage dry matter. Obviously, such an approach generally is not feasible because of increased fermentation losses, added silo capacity required, additional capital, and other factors.

Historically, numerous exotic silage additives and preservatives have been developed, promoted, used, and studied. In time, they usually prove to be only a passing fashion or fad. It is tempting to suggest that the parade of new fashions and fads will continue indefinitely. Real sucess in silage making depends on an understanding of the key principles and comes to those who carefully execute the important details.

9.7.e. Addition of Urea and Other Nutrients to Corn Silage

As discussed in Section 9.7.a, corn silage is easily preserved without additives, but is low in protein and several minerals. To partially correct the protein deficiency, addition of urea at the time of filling has become a widely researched, recommended, and utilized practice (see Section 4.4). Adding urea to corn at ensiling has several advantages including a way to feed urea more uniformly throughout the day than when it is fed with concentrates. Likewise, the corn silage supplies the readily-fermentable carbohydrates needed for good urea utilization and silage preservation. Thus, with good conditions the tendency of urea to adversely affect the ensiling process is minimized.

When urea is added to corn silage at the time of filling, several important aspects should be followed. Generally, the recommended rate is about 10 pounds per ton. It is essential to obtain good mixing of the urea with the silage prior to the time of feeding. Urea additions are not recommended when the dry matter is either too low or too high. With a dry matter content below about 30%, much of the highly soluble urea may be lost in seepage. If the dry matter of the silage is too high, above about 40%, the urea may substantially decrease silage palatability resulting in lower feed intake and milk production. When urea is added to the corn silage, somewhat less urea should be added to the concentrate, but usually a greater total amount can be effectively utilized. Nitrogen sources other than urea, including ammonia, are sometimes added to corn silage to increase the crude protein content.

Ground limestone has been added to corn silage at the time of ensiling, especially when the silage was to be fed to finishing beef cattle. The limestone may increase the amount of lactic acid formed, improve starch digestion (Wheeler and Noller, 1976), and in some instances increase milk production.

Addition of ground limestone is not consistently beneficial when the corn silage is to be fed to lactating dairy cows receiving sufficient calcium.

9.7.f. Types of Silos

Most of the many types of structures used for silos can be classified into one of three groups: (1) conventional tower silos, (2) horizontal silos, and (3) gas tight silos. On dairy farms probably the conventional tower silo has been most widely used. These are usually constructed of concrete staves, special tile brick, galvanized steel, wood staves, or other suitable materials. When properly constructed, maintained and used, good silage can be readily made in these conventional tower silos.

Horizontal silos come in many forms and are made from many different materials (Hendrix and Miller, 1964). Often, horizontal silos were not well managed and acquired a bad reputation among dairymen. With satisfactory attention to fundamental principles and essential details, horizontal silos have major

TABLE 9.8
Approximate Capacities of Tower Silos for Corn Silage (Whole Plant) Expressed as Tons of Dry Matter[a]

Size of silo (ft)		Percentage of DM				
Diameter	Height	30	35	40	45	50
12	40	30	32	32	29	22
12	50	38	41	40	36	28
14	40	41	45	44	40	31
14	50	53	57	56	51	39
16	50	71	76	75	68	53
16	70	103	111	110	98	75
18	50	93	99	98	88	69
18	70	135	145	143	127	98
20	50	116	124	123	111	87
20	70	170	183	181	162	125
24	60	212	227	226	204	159
24	70	253	272	270	243	188
24	90	337	362	360	323	249
30	80	472	507	508	464	363
30	100	604	649	654	600	469

[a] Adapted from Waldo (1977b). The tons of silage would be much larger. For example, the silo with a 16 ft diameter and 50 ft height would have a capacity of 71 tons of dry matter at 30% D.M. This would be 71 tons/0.30 = 237 tons of silage.

economic and operational advantages over other types of silos (Fig. 9.10). These include a lower initial cost, easier filling, and reduced danger from poisonous gas (Hendrix and Miller, 1964). The reduced investment needed is especially advantageous where capital is in short supply, if there is a possibility that the silage enterprise will not be continued for a long number of years, or in situations where a high return on invested capital is a key objective. Horizontal silos are especially desirable in large, efficient operations using large amounts of silage. In constructing and using the horizontal silo, it is critical that greater attention be paid to important details such as site location, preventing water damage, fine chopping, avoiding too high a dry matter content, permitting seepage, good packing, rapid filling, and maintaining a good seal (Hendrix and Miller, 1964).

The gas-tight, glass-lined silo, when properly used, will give good silage preservation. However, the initial investment and therefore annual capital cost is huge compared to other types of silos. Often maintenance cost is substantial. Thus relative to conventional tower silos properly used, the small advantages in reduced losses and possibly better preservation in the gas tight silos generally are not economical (McCullough, 1973).

With the continuing trend toward more cows per farm and greater economic efficiency, horizontal silos probably will continue to gain in use relative to other types of silos. Because of limitations on acceptable dry matter percentages (Waldo, 1977b), the need for rapid filling, power cost for very tall silos, and other technical problems, excessively large tower silos are undesirable.

9.7.g. Nutrient Losses in Silage Preservation

Considerable research and attention has been devoted to the nutrients lost during the making and preservation of silages. These losses can be divided into (1) field losses, (2) respiration and fermentation losses, (3) seepage, (4) spoilage, and (5) losses during feeding.

With good procedures the field losses are comparatively small and far lower than in hay making when losses of nutrients in the field may become sizable. The entire crop may be lost in hay making when adverse weather continues for a long period. Wilting forage prior to ensiling increases field losses.

Respiration and fermentation losses, sometimes called invisible losses, depend on the conditions involved. However, even with the best of conditions, some energy loss is a part of the processes required to achieve good preservation. The heat created in the process represents some loss of digestible energy. Factors in minimizing nutrient losses through respiration and fermentation include rapid

---

Fig. 9.10 On many farms properly designed, well managed, horizontal structures are the lowest in cost and most practical way to store silage. (Courtesy of W. J. Miller, Univ. of Georgia.)

filling, proper chopping, packing, good and quick sealing, and protection from air and water damage.

Seepage losses depend primarily on the amount of excess moisture in the forage, although they are modified somewhat by other factors, especially the height of the silo (Fig. 9.11) (Miller and Clifton, 1965). For example, in horizontal or low tower silos, losses through seepage will nearly cease when the dry matter exceeds 30%. In very tall towers, as much as 35% dry matter may be necessary to completely eliminate seepage (Waldo, 1977b). With a forage containing only 12% dry matter, 14% of the dry matter was lost as seepage from a small silo (Miller *et al.*, 1961). Sealing the silo in such a way that seepage cannot readily escape results in an unpalatable silage and should be avoided.

Spoilage or moldy silage develops when air is not excluded. The amount of nutrients lost from spoilage often is quite misleading. For instance, it is possible to spoil the entire contents of a silo and never have a bad layer more than a few inches thick. As the silage spoils, fermentation and microbial action convert the material into carbon dioxide, water vapor, and other volatile materials that evaporate into the air leaving only the mineral elements.

Losses during silage feeding are of two types. A variable amount may be wasted as it is spilled from the trough and in the feeding operations. If silage is allowed to remain in the trough for extended periods, some fermentation and spoilage may also occur. When the removal rate from the exposed surface of the silo is too slow, the exposed silage may deteriorate. Especially in horizontal silos, care must be taken that "loose" silage is fed fairly soon; otherwise it may damage or become worthless. Often self-feeding from horizontal silos results in considerable wastage.

With a good situation and careful attention to details, the total loss of utilizable energy can be limited to about 6–12% of that in the harvested forage (McCullough, 1973; Waldo, 1977b). In contrast, when conditions are poor, the losses can be many times higher.

9.7.h. Deadly Gases Evolved When Making Silage

Two types of gases which may form during silage fermentation can be lethal (Oleskie and Wright, 1974). Nitrogen dioxide which apparently develops from accumulated nitrates, especially in corn, leaves a yellow stain on materials touched. This gas is a deadly poison. The other major dangerous gas is carbon dioxide which forms as the respiration and fermentation of ensiling material utilizes the oxygen of the air (see Section 9.7.b). The main danger from carbon dioxide comes from the absence of oxygen that is essential to life.

Since nitrogen dioxide and carbon dioxide are heavier than air, they settle or accumulate in areas where there is inadequate air drainage. It is especially important to avoid entering silos, that have been filled from a few hours to several

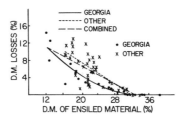

Fig. 9.11 Seepage losses are closely related to the dry matter content of the material ensiled. Each point represents one silo. (Adapted from Miller and Clifton, 1965.)

days, below the point of open doors. If one must enter the silo, operating the blower for at least several minutes before usually removes most of the dangerous gases. The carbon dioxide danger can be tested by using a small open flame which will not burn without oxygen.

9.8. CHOPPED FRESH FORAGE: GREEN CHOP, SOILAGE

On many dairy farms fresh green forage is harvested and fed to lactating cows for short periods during the year. In earlier years this practice, formerly called "soilage" and sometimes "zero grazing," involved a great deal of hand labor. The advent of the field chopper greatly reduced the labor needs.

Feeding freshly chopped green forage has some important advantages and some serious limitations. This method permits harvesting and feeding the forage at a near ideal stage of maturity without storage cost and with little field losses. Likewise, it avoids wastage through the trampling, fouling, and selective grazing that go with grazing a tall growing pasture crop. Also forage can be utilized from areas which might not be accessible for grazing. Feeding freshly chopped green forage, however, presents some problems. Maintaining the correct amount of forage at a suitable stage of maturity with variable and usually somewhat unpredictable growing conditions is difficult or impossible. The apparent saving by eliminating preservation and storage may be more than offset by the necessity for frequently harvesting a small amount of forage in all kinds of weather—the cows must be fed even when it rains and whether or not the fields are wet and muddy.

In feeding freshly chopped green forage, it is helpful to recall the rapid silage-forming process (see Section 9.7.b). When chopped forage stays on a wagon or in a feed trough, respiration and fermentation begin the initial stages of silage formation. Often as the forage heats it becomes unpalatable and the cows will not consume an adequate amount. The length of time the freshly chopped forage can be kept in wagons or in the feed trough varies considerably depending on temperature, the particular crop involved, and other factors. However, it is crucial that chopping and feeding of forage be done often enough to provide adequate forage in a palatable state to meet the needs of the cows.

9.9. HAYLAGE

Haylage, sometimes called low-moisture silage, is a form of preserved forage with characteristics between those of hay and silage. Although the exact points in dry matter content where the term haylage fits is not clearly defined, it is in the general area of 40–60% dry matter.

As with other forms of forages, haylages have important advantages but some serious limitations. For instance, if forage is mowed with the intention of making hay and the weather becomes unfavorable for drying, the partially dried forage can be made into haylage. Well preserved haylage may be more palatable and consumed in larger amounts than higher moisture grass silages.

With haylage, fine chopping, good packing, and complete sealing against air entrance become much more critical than with silage. Greater amounts of air are trapped and new air moves through the less densely packed haylage more rapidly. Thus molding and spoilage are major problems. Some additives such as propionic acid will reduce mold formation (Thomas, 1976; Waldo, 1977a). The danger of excessive heating which lowers protein digestibility is more acute in haylages than silages.

Haylage, originally limited to gas-tight silos, is now made successfully in good, conventional-type tower silos. Although horizontal silos have been used experimentally for storing haylage, considerable skill is required.

9.10. OTHER ROUGHAGES SOMETIMES FED TO DAIRY CATTLE

Although pastures, hays, and silages usually are the primary forages fed to dairy cattle, in some situations other roughages are important (Cullison, 1975). Large quantities of corn cobs, husks, and stover are grown in the United States as by-products of grain production. When more desirable forages are not readily available, these roughages can be utilized effectively as an energy and fiber source in a well-balanced ration. However, they are quite deficient in protein, minerals, and vitamins and have a relatively low level of digestible energy per pound. Accordingly, more concentrate will be required for efficient production than if a higher quality forage is fed.

When complete rations, sometimes called all-in-one rations, are fed to dairy cows, it is necessary to have fiber in a form that can be mixed with the concentrate, readily transported, and commercialized. In such rations, roughages such as cottonseed hulls are often used quite successfully.

Oat, barley, and wheat straws are sometimes fed to dairy cattle. As with corn cobs, stover, and husks, these straws are high in fiber and quite low in digestible energy, protein, minerals, and vitamins. Oat hulls and soybean hulls are fed to dairy cattle in substantial quantities. Although quite high in crude fiber content, the type of fiber in soybean hulls is highly digestible. Thus, substantial amounts of soybean hulls can be fed in lieu of concentrates. Oat hulls are much less nutritious. When peanut hulls and other roughage by-products of crops that are sprayed with pesticides are fed, it is important to be sure that the feed ingredient does not contain undesirable residues. Other roughage sources sometimes used include rice hulls and sugarcane bagasse. The latter has a very low digestible energy value and may cause mechanical injury to the digestive tract.

9.11. FORAGE PROCESSING: SPECIAL CONSIDERATIONS AND FUTURE POSSIBILITIES

In a sense, other than pasture, every forage is processed. However, in "dairy cattle nutrition circles," it is convenient to think of forage processings as the unusual types.

Grinding hay, other dry forages, and roughages together comprise one of the oldest forms of processing. These processes may reduce waste by decreasing the ability of the cattle to select only the best parts of the hay while refusing the stemmy, less desirable portions. Although grinding usually does not improve the digestibility of hay, sometimes there is an increase in the amount voluntarily consumed. When forage is too finely ground, the volatile fatty acids produced in the rumen may be altered and the fat content of milk reduced (see Section 8.3.b and 8.3.d).

Grinding and pelleting hay substantially increases the density and voluntary dry matter consumption of forages such as coastal bermuda grass hay (Brooks *et al.*, 1968). This higher consumption usually is largest in those forages which have a low intake when fed in the unground and pelleted form. Grinding and pelleting alfalfa hay affects voluntary intake less than with coastal bermuda grass. The influence of ground and pelleted hay on rumen volatile fatty acids and milk fat percentage are similar to that of ground hay.

In experimental studies, excellent results were obtained by grinding and pelleting the whole corn plant after it had been field dried (Beaty *et al.*, 1966). In many areas this practice appears to offer considerable potential as a major source of feed nutrients for dairy cattle. The whole-corn plant pellets utilize most of the nutrients grown. Likewise, they are a dense product readily adaptable to mechanization, commercialization, and transportation over substantial distances.

Hay cubes and wafers have found some use in limited areas especially in California (Dobie and Curley, 1969; Dobie et al., 1966) (Fig. 9.5.b). These primarily involve compression of unground hay into high density forms. They have the advantages of high bulk density, and are readily adaptable to commercialization and mechanization. The future of hay cubes and wafers appears to depend primarily on cost relationships for production relative to the advantages involved. Since the hay is not ground, cubes and wafers do not have an adverse effect on milk fat content.

The key difference between forages which have high quality and those of poor quality usually is associated with changes in the energy digestibility of the complex carbohydrates. Most of these carbohydrates are combinations and polymers of simple, readily digestible units. Accordingly, scientists have long been intrigued by the possibility of finding easy ways of processing poor quality forage or even such products as wood into high quality feeds. Experimentally, in pilot plants molasses and other digestible feeds have been made from wood. Total or gross energy of poor quality forages, roughages, and related materials is similar to that of high quality forages. Likewise, the basic chemical building units are similar with the differences primarily related to how the building units are combined. Rumen microbes cannot rapidly and thoroughly convert the poorly digested materials into usable products.

One of the major changes occurring when forage plants mature is cell-wall thickening and lignification (see Section 8.1.a). Because lignin is readily attacked by alkalis, many of the procedures attempted for improving digestibility of poor quality forages involved an alkali. In the future, practical combinations of fermentation, heat, pressure, and chemicals may be developed making it economical to convert poor quality fibrous materials into high quality dairy cattle feed (Miller and O'Dell, 1969).

REFERENCES

Barnett, A. J. G. (1954). "Silage Fermentation." Academic Press, New York.
Beaty, E. R., W. J. Miller, O. L. Brooks, and C. M. Clifton (1966). Agron. J. **58**, 424–426.
Brooks, O. L., W. J. Miller, E. R. Beaty, and C. M. Clifton (1968). Ga., Agric. Exp. Stn., Res. Bull. **27**.
Crowley, J. W. (1975). Hoard's Dairyman **120**, No. 11, 694–695.
Crowley, J. W. (1976). Hoard's Dairyman **121**, No. 10, 673.
Cullison, A. (1975). "Feeds and Feeding." Reston Publ. Co., Reston, Virginia.
Dobie, J. B., and R. G. Curley (1969). Calif. Agric. Exp. Stn., Circ. **550**.
Dobie, J. B., R. G. Curley, M. Ronning, and P. S. Parsons (1966). Agric. Eng. **47**(7), 378–380.
Hendrix, A. T., and W. J. Miller (1964). Ga., Agric. Exp. Stn., Bull. [N.S.] **113**.
Hibbs, J. W., and H. R. Conrad (1975). Hoard's Dairyman **120**, No. 11, 686 and 720.

Hight, W. B., D. L. Bath, and D. Miller (1975). *Calif. Agric.* **29**, (No. 4), 10-12.

Hutjens, M., and N. Martin (1976). *Hoard's Dairyman* **121**, No. 8, 533 and 552-553.

McCullough, M. E. (1973). "Optimum Feeding of Dairy Animals for Meat and Milk." Univ. of Georgia Press, Athens.

Miller, W. J., and C. M. Clifton (1962). *Ga. Agric. Res.* **3**(No.4), 4-6.

Miller, W. J., and C. M. Clifton (1965). *J. Dairy Sci.* **48**, 917-923.

Miller, W. J., and G. D. O'Dell (1969). *J. Dairy Sci.* **52**, 1144-1154.

Miller, W. J., H. L. Dalton, and J. K. Miller (1961). *J. Dairy Sci.* **44**, 1921-1927.

Miller, W. J., C. M. Clifton, and N. W. Cameron (1962). *J. Dairy Sci.* **45**, 403-407.

Oleskie, E. T., and F. A. Wright (1974). *Rutgers Univ. Coop. Ext. Serv., Leafl.* **504.**

Reid, J. T. (1961). *Hoard's Dairyman* **106**, No. 8, 405, 417, and 436-439.

Shepherd, J. B., H. G. Wiseman, R. E. Ely, C. G. Melin, W. J. Sweetman, C. H. Gordon, L. G. Schoenleber, R. E. Wagner, L. E. Campbell, G. D. Roane, and W. H. Hosterman (1954). *U.S. Dep. Agric., Tech. Bull.* **1079.**

Thomas, J. W. (1976). *J. Dairy Sci.* **59**, 1104-1109.

Underwood, E. J. (1966). "The Mineral Nutrition of Livestock." Central Press, Aberdeen.

Waldo, D. R. (1977a). *J. Dairy Sci.* **60**, 306-326.

Waldo, D. R. (1977b). *Proc. Res. Ind. Conf. Am. Forage Grassl. Counc., 10th, 1977* pp. 69-91.

Wheeler, W. E., and C. H. Noller (1976). *J. Dairy Sci.* **59**, 1788-1793.

10

Concentrates, By-Products, and Other Supplements for Dairy Cattle

10.0. INTRODUCTION

Most dairy cattle feed ingredients can be classified as either forages and roughages or concentrates. Generally, concentrates, many of which are grains, protein supplements, or by-product feeds, have less fiber and more digestible energy. The average nutrient composition of the most important ones is presented in Table 4 of the Appendix.

10.1. REASONS FOR FEEDING CONCENTRATES TO DAIRY CATTLE

10.1.a. Balancing Rations to Supply Needed Nutrients not in Forages

Because forages usually are a more economical source of nutrients than concentrates, the typical dairy cattle feeding program is built around the maximum use of forages (Chapter 9). On most United States dairy farms the best available forages do not provide all the nutrients needed for the most profitable feeding. Accordingly, concentrates are fed to make up deficiencies. As discussed in Chapter 2, usable energy is the nutrient required in largest amounts. Depending on price relationships between feed and milk, often it is not most profitable to feed sufficient energy for maximum milk production. Likewise, maximum rates of gain in growing dairy cattle are not desirable or most profitable (see Chapters 14 and 15).

With the typical dairy cattle operation in the United States, the practical approach is to feed sufficient concentrates to provide a usable energy level that results in the most profitable production. When this has been determined, the composition of the concentrate should be formulated to supply the other required nutrients not furnished by the forages. Mainly, concentrates are designed to balance the diet for the needed protein, minerals, and vitamins not in the forages.

10.1.b. Concentrate Feeding (When Concentrates Are More Economical Sources of Nutrients than Forages)

Occasionally forages are in short supply or so expensive that concentrates are a more economical source of nutrients. In such situations, it is desirable to make the maximum effective use of them. Except for inadequacies in some needed nutrients, especially energy, feeding dairy cattle only forages does not cause problems. In contrast, serious digestive and health abnormalities often result when dairy cattle are fed only concentrates (see Section 8.4) (Miller and O'Dell, 1969). Some of the problems associated with feeding excessive amounts of concentrates are due to insufficient intakes of unground fibrous feed.

With too high a proportion of concentrates the overall digestibility of the ration is reduced (Schmidt and Van Vleck, 1974; Wagner and Loosli, 1967). The highest, practical ratio of concentrates to forages, when they are more economical sources of nutrients than forages, depends on a number of variables. When it is fibrous, a smaller proportion of forage will suffice. Likewise, some concentrates such as beet pulp or citrus pulp contain appreciable fiber.

As a general rule, for best performance, lactating cows should not be fed more than about 60% of their total feed dry matter as concentrates; or less than about 40% as forage and roughages (Schmidt and Van Vleck, 1974; Ronning and Laben, 1966). Typically feeding more than 60% concentrates tends to decrease total milk yield, reduce fat percentage in the milk, and cause digestive and metabolic disturbances. In most situations, 17% crude fiber or 21% acid detergent fiber, from unground feeds, in the total ration dry matter is adequate even though the percentage of concentrates exceeds 60% (National Research Council, 1978).

Most of the above discussion concerns concentrate feeding for best performance. If the price relationships become distorted to a sufficient degree, it may be profitable to sacrifice some reduction in performance and accept some digestive problems, especially for short periods. Experimentally, cows have been fed rations containing 100% concentrates (Ronning and Laben, 1966).

With excessive levels of concentrates a number of additives, especially certain buffers, have at least partially alleviated the reduced milk fat percentage. Sodium and potassium bicarbonate and magnesium oxide appear to have some effect in such situations (Harris, 1970; Schultz, 1969; Thomas and Emery, 1969). Bentonite, a special type of clay which swells in water to many times its original size, also partially alleviates the low fat test exhibited with too high a proportion of concentrates (Bringe and Schultz, 1969; Rindsig *et al.*, 1969). Apparently none of the buffers will increase milk fat percentage in cows fed a "normal" diet.

10.2. GRAINS FOR DAIRY CATTLE

Grains are an important part of many dairy concentrates. The relative amounts of various grains fed to dairy cattle depend primarily on the quantities of different grains produced and price relationships.

10.2.a. Major Grains

Of the grains, corn (maize) is by far the most important in the United States. Corn grain is high in usable energy, fairly high in fat, very low in fiber, and relatively low in protein (see Appendix, Table 4). Also, corn is deficient in some minerals, especially manganese, calcium, and often sulfur. Yellow corn grain contains a vitamin A precursor, but white corn does not (see Section 6.2.b). Substantial amounts of corn are fed as ground grain, or in some situations as ground corn and cob meal, or as ground snapped corn. Likewise, the comparatively high percentage of grain is a key reason for the high usable energy content of corn silage (see Section 9.7).

Although most corn grain which does not become a part of corn silage is fed as dry grain, some is utilized in other forms. Especially in areas having a short growing season, corn is often preserved as high-moisture shelled corn or as high-moisture ear corn either by ensiling or by treatment with a preservative such as propionic acid. When ensiled, the corn grain should contain about 30% moisture (Crowley, 1976). If preserved with propionic acid, the lower the moisture content, the better. Properly preserved high-moisture corn will give satisfactory results in feeding dairy cattle, but the technology and economics of preservation are quite complex (Clark, 1976).

Oats, although used in much smaller amounts than corn, have been a traditional grain on United States dairy farms. Compared with corn, oats have a higher and much more variable fiber content, more protein but a lower usable energy value. The energy value is quite dependent on the "test" weight of the oats per unit volume and is inversely related to the fiber content. "Heavy" oats are a much more valuable feed per pound than light oats.

Sorghum grain occurs in various types and varieties, including hegari grain, milo grain, grain sorghums and kafir grain (Cullison, 1975). These grains are grown more extensively where drought may adversely affect corn growth. Generally the sorghum grains contain somewhat more protein than corn, a low level of fiber but have little vitamin A value. Normally, it is more important to grind or otherwise process sorghum grain than corn.

Wheat is widely grown in many countries, but because of its use for human food, usually much less is fed to dairy cattle. Except for a higher protein content, wheat has a feeding value similar to corn grain. Excessive amounts of wheat should not be fed, especially to unadapted cattle, as it is prone to cause acute indigestion and founder (see Section 18.8).

Barley grain is intermediate between corn and oats in protein, fiber, and usable energy value. It is a good dairy concentrate ingredient, but total barley production in the United States is comparatively small.

Rye grain generally is fed to dairy cattle in only very limited amounts. Although the usable energy and protein content are similar to other grains, rye grain is less palatable. Likewise, rye may be contaminated with the toxin, ergot. One of the major advantages of rye relative to other small grains is its wide geographical adaptability and comparatively vigorous growth under adverse conditions.

10.2.b. Processing of Grains

When whole grains are fed to dairy cattle, other than calves, often chewing is incomplete. Thus, a substantial percentage may pass through the digestive tract undigested. Accordingly, most grains receive some form of processing prior to feeding.

Grinding and/or cracking are used most to process grains (Williamson, 1973). Excessively fine grinding may increase dustiness and reduce palatability of the grain. In addition, too fine a grind may tend to reduce milk fat percentage and increase the probability of digestive problems. The most desirable fineness depends somewhat on how the grain is to be fed. More finely ground material will carry a larger amount of wet ingredients such as molasses and may "pellet" more effectively.

A substantial amount of the concentrates sold commercially are pelleted. The major advantages of pellets revolve around better handling characteristics and easier mechanization both before and after the feed reaches the farm.

Another popular form of grain processing is rolling. Rolled oats have long been a popular dairy feed ingredient. Often the rolling process involves the use of steam and therefore is a heat and moisture treatment as well as a mechanical one.

Many cooking, heating, popping, roasting, and flaking processes have been investigated (Hale, 1975). Generally if the process is only a mechanical one, the effects on lactating cows is similar to that of grinding. When the process involves cooking of the starches, the type of rumen fermentation may be altered, and may result in increased rumen production of propionic acid. Often these changes are accompanied by a lower milk fat percentage (see Section 8.3). The same changes may increase body fat deposition as well as efficiency of weight gain and therefore be beneficial for cattle being finished for slaughter (Hale, 1975). The benefits vary considerably with different grains and are quite dependent on quality control in processing (Hale, 1975).

10.3. PROTEIN SUPPLEMENTS

With many types of forages, especially corn silage, a supplement consisting only of grains is deficient in protein. Accordingly, additional protein often is required for optimum performance, particularly for lactating cows and young calves.

The best known type of high-protein feeds used for dairy cattle are the "oilseed meals." Of these, soybean oil meal, usually called soybean meal (Smith, 1977) is used in greatest amounts. Others include cottonseed meal, linseed meal, peanut meal, sesame meal, safflower meal, and sunflower meal. Generally these high-protein supplements are very palatable and contain from 35–50% crude protein depending on the processing methods involved (Cullison, 1975). The oil from the oil seeds are extracted either with solvents, leaving a product known as solvent-processed meal, or with large presses providing expeller-processed meal. Solvent-extracted meals contain less oil and therefore less usable energy, but have a higher protein percentage. Likewise, for several of the oil meals, the protein content is considerably influenced by the hulls content. Especially in using peanut meal, it is important to make sure that the oil remaining in the meal has not become rancid. Often this can be readily detected by smell.

Other frequently used medium or high-protein feeds for dairy cattle include corn gluten meal, corn gluten feed, copra meal from coconuts after oil is extracted, distillers dried grains, brewers dried grains, and nonprotein nitrogen supplements. These are further discussed in the by-product section below. In some circumstances, especially when vegetable oil prices are low, it is economical to feed whole cottonseed or ground whole soybeans as protein supplements. The unextracted seeds are lower in protein and higher in fat than the meals. Whole cottonseeds are high in crude fiber (18%) and not especially high in protein (25%), but whole soybeans are higher in protein, 41.7%. When properly used, these whole seeds are a good dairy cattle feed. Ground, whole soybeans should be fed soon, within 7 days, after grinding, and if raw should never be fed with urea or poultry litter, as they contain the enzyme urease which decomposes urea.

Most of the high-protein supplements, also, are comparatively high in usable energy content. However, usually it is not economical to feed larger amounts of these supplements than needed for the protein requirements. Several of the high-protein feeds are good sources of some of the essential mineral elements.

Except for their phosphorus content, generally the added monetary value of the minerals is quite small, either because of their low cost per pound or the very small amounts required.

10.4. BY-PRODUCT FEED INGREDIENTS

Large amounts of by-products are used in dairy cattle concentrates. In some areas of the world, very little grain suitable for human consumption is fed to dairy cattle. Nearly all the concentrates may come from by-products with these situations. Even in certain areas of the United States, most of the feed ingredients in concentrates are by-products from materials processed for other purposes. Historically, many of these by-products were wasted or poorly utilized. Often they were developed into feeds because of problems in finding acceptable disposal methods. Feeds formulated with large amounts of by-products may be fully equal in value to those using grain even though by-products usually are more economical nutrient sources.

As discussed in the previous section, because of their protein content, oilseed meals such as cottonseed meal, linseed meal, copra meal, and peanut meal are secondary products from materials primarily produced for other purposes. Soybean meal is the product left after the extraction of the oil, but because of the comparatively higher value of the protein meal, usually it is classified as a main product. Often protein suitable for dairy cattle can be obtained from other by-products at a lower cost than from soybean or other oilseed meals.

10.4.a. Citrus Pulp

In the processing of citrus fruit for juice, large amounts of citrus peel and pulp are produced; most of which is utilized as dried citrus pulp. Often in areas close to production points, very high levels are fed to dairy cattle. Dried citrus pulp is low in protein, comparatively high in usable energy, highly palatable, and is an excellent feed ingredient. Since dried citrus pulp is quite bulky and contains more fiber than most grains, in addition to being fed as a primary concentrate ingredient, frequently it is used as a substitute for part of the roughage. The fiber in citrus pulp is of high quality (see Section 12.5). Because of the processing methods employed, dried citrus pulp is very high in calcium.

10.4.b. Beet Pulp

In certain areas large amounts of beet pulp are produced as a by-product of the sugar beet industry. This excellent feed ingredient is used in a way similar to that of citrus pulp. Beet pulp contains more fiber than most grains, but the fiber has a high digestibility. Partially because of its excellent palatability in earlier years, large amounts of soaked, dried beet pulp were fed to "test cows." This increased total feed consumption and more fully met the energy needs of high-producing cows in early lactation without causing digestive problems which often accompanied very high grain feeding. Although a value of about 0.8 ppm is widely cited in tables, beet pulp is not extremely low in zinc. Rather the zinc content of dried beet pulp generally is around 10 ppm (Miller and Miller, 1963).

10.4.c. Wheat Bran and Other Wheat Milling By-Products

Wheat bran consists largely of the seed coat that is removed as wheat is processed into flour for human consumption. This excellent feed ingredient is highly palatable, slightly laxative, a good source of phosphorus, contains an intermediate amount of protein and usable energy, and has substantial amounts of several other essential minerals.

In various processes of milling wheat for flour several types of shorts, middlings, and related by-products are available for cow feed. These ingredients generally contain less fiber and more usable energy than wheat bran. Often they furnish nutrients at a lower cost than grains and oilseed meals. In addition, they can "carry" large amounts of liquid feeds, such as molasses, and improve the pelleting characteristic of many feed mixtures.

10.4.d. Molasses

Molasses is a very popular feed ingredient for dairy cattle. Three major types or sources of molasses are widely used, each having several aspects in common with the others but some important differences. The best-known type is that obtained from the refining of sugar cane juice into sugar. In a similar way beet molasses is obtained from sugar beets and citrus molasses from the processing of citrus fruit.

Molasses has a very low protein content, is a comparatively good source of energy and in small amounts is noted for increasing the palatability of feeds. Small amounts, less than 5–10% of the feed, of molasses improve efficiency of rumen fermentation, especially in feeds having little readily fermentable carbohydrates or those inadequate in some essential mineral elements. Large amounts of molasses may depress fiber digestion in the rumen. It is a good source of many essential mineral elements, but the content varies substantially with molasses from different sources. Cane molasses is an excellent example of feeding the highly nutritious portion to cattle while taking the less nutritionally desirable part for human consumption.

10.4.e. By-Products of Corn Processing

In processing corn grain for starch and oil, several products are obtained and used in dairy cattle feeds. Corn gluten meal and corn gluten feed were mentioned in Section 10.3 because of their protein content. For nonruminants the amino acid balance of corn protein is less desirable than that of soybean meal. However, corn protein is less easily degraded by rumen microbes (see Sections 3.10 and 3.11). Accordingly, in some situations, corn gluten meal and corn gluten feed may have a very desirable type of protein, especially for cows having a high protein requirement.

Hominy feed is a by-product from the manufacture of hominy, grits, and cornmeal. It is similar to corn grain in nutrient value but usually contains a little more protein, fat, and fiber. Rancidity may become a problem if hominy feed is stored too long in a hot, humid area with little or no air circulation.

10.4.f. Distillers and Brewers By-Product Feeds

Substantial amounts of brewers and distillers grains are produced from the spent grains in the production of alcoholic beverages. Basically, these are the products left after the starches and sugars have been converted into alcohol and removed. Some of the products are fed in the wet form to dairy cattle within reasonable shipping distances of the point of production. Special attention must be given to the dry matter content because they often have a very high water content. Large amounts of the brewers and distillers grains are fed after drying.

Because most of the sugar and starch have been removed, brewers and distillers grains are substantially higher in fiber, protein, and some minerals than the original grains. Generally they are a somewhat better source of protein than of energy. Due to the relatively high fiber content, most of the products are fed to ruminants rather than to swine or poultry. Palatability varies substantially with different products.

10.4.g. Other Important By-Product Feeds

Rice bran, obtained from rice processing for humans, has a very high fat content (15.8%) (Cullison, 1975), a substantial amount of fiber, and an intermediate level of protein. When properly used in limited amounts, it is a satisfactory ingredient. However, it is less palatable than wheat bran and rancidity may easily become a problem with the high fat content.

Oat mill feed and oat hulls are sometimes used in dairy feeds. However, because of the high content of poorly digested fiber, these products have a low energy value.

Soybean hulls and soybean mill feed, although having a fiber content as high as that of forages and roughages, have a digestible energy content equivalent to that of concentrates (Macgregor *et al.*, 1976). This is largely due to the unusual nature of the fiber which makes it highly digestible (see Chapter 8).

Many by-products of animal origin are used as feed ingredients including fish meal, tankage, meat and bone scrap, bone meal, blood meal, feather meal, poultry by-product meal, dried skim milk, whey, dried whey, and dried buttermilk. Most of these animal by-products are more valuable, because of the good balance of amino acids, as ingredients in feed for nonruminant animals such as swine and poultry than for cattle feed. The milk by-products are often used as the major ingredients in milk replacers for baby calves. Due to its high lactose content, whey is quite laxative, making it necessary to limit the amount fed. Fish meal has been used successfully in research studies as a source of protein for milk replacers. Some bone meal is used as a source of phosphorus for dairy cattle. However, it is important that bone meal be steamed to prevent diseases, especially anthrax.

When the prices are competitive, small amounts of by-product fats both of animal and/or vegetable origin are sometimes fed to dairy cows. A much higher percentage of added fat is generally used in milk replacers. Often this fat is especially processed to improve performance of the calves (see Chapter 14).

10.5. SPECIALTY AND MISCELLANEOUS FEEDS AND INGREDIENTS

10.5.a. Urea

Substantial amounts of urea and other nonprotein nitrogen compounds are fed to dairy cattle. This is the subject of Chapter 4: the addition of urea to corn silages is discussed in Section 9.7.e.

10.5.b. Liquid Feed Supplements (LFS) and Liquid Protein Supplements (LPS)

Except for the young calf, traditionally most of the feeds given dairy cattle have been in dry or solid form. However, in recent years feeds known generally as liquid protein or liquid feed supplements have come into relatively wide use.

Although liquid feed supplements are more widely used for beef cattle, often they are fed to dairy cattle. One of the popular ways of feeding is free choice in a tank with a wheel that rotates as the animal licks the sticky liquid. Also, liquid feed supplements may be mixed with either the concentrate, the forage or with a complete feed. Regardless of the method of feeding, the general purpose is as a supplement to balance the deficiencies of the remainder of the ration.

Because of the nonprotein nitrogen used (see Chapter 4), excessive intake of liquid free supplements in free-choice feeding can produce toxicities. Also, the rations frequently can be better balanced when the liquid feed supplement is mixed with the forage or the concentrate. With cattle on pasture, free-choice feeding may be more convenient.

Whether a feed is liquid or dry does not materially affect the nutritional or practical feeding value. Rather the most desirable form is the one that is the most effective and economical way to provide the needed nutrients. There are several reasons for the large increase in the use of liquid feed supplements in comparatively recent years. Largely, these center around the convenience in feeding and the ability to use some low-cost ingredients in a liquid form that are not nearly as easily utilized in a dry feed.

Although a multitude of different formulations are used in liquid feed supplements, certain characteristics are fairly common. Most include urea and/or ammonium compounds as sources of nonprotein nitrogen. Likewise, molasses is used as a highly palatable liquid carrier which tends to mask unpalatable ingredients. The molasses provides many useful nutrients including sugar, which in limited amounts stimulates rumen microbial activity, and many trace elements which might be deficient in some forages and roughages.

Generally liquid feed supplements are designed to supplement diets having large amounts of forages. Accordingly, most contain substantial amounts of phosphorus either from phosphoric acid or ammonium polyphosphates or both. The phosphoric acid gives the liquid a lower pH and better flow characteristics but is very corrosive to equipment. In contrast, use of ammonium polyphosphates results in a less acid pH, a more viscous liquid, and is less corrosive to equipment. Ammonium polyphosphates are good sequestering agents for many of the essential minerals and trace elements (Miller and Stake, 1972). Thus, with this ingredient, these elements do not settle out as readily. However, the more viscous liquid is more difficult to pump, especially in cold weather.

Because of the low solubility of calcium, and the large amounts required, most liquid feed supplements do not provide a substantial percentage of the calcium needed by dairy cattle (Miller and Stake, 1972). Often liquid feed supplements contain vitamin and/or antibiotic supplements. The degree of stability of these materials depends on the forms used and on the total formulation involved.

10.5.c. Sprouted Grains (Hydroponic "Grass")

Periodically, spectacular claims are made for the merits of hydroponically producing grass for dairy cattle. Research, basic science, and cost relationships always show this to be a very expensive way to provide needed nutrients (Schmidt and Van Vleck, 1974; Thomas and Reddy, 1962). The hydroponic production of grass or sprouting of grains is called by several names and varies in procedural and equipment details. However, the essentials of the procedures and the economics are similar for the various approaches.

In essence, grains, often oats, are sprouted or grown for a few days in special chambers with controlled temperatures, humidity and often light. Supplemental minerals and fertilizer nutrients may or may not be added. Under these growing conditions, the grains absorb large amounts of water and greatly increase in total weight. However, the total amount of dry matter and energy decrease as the energy produced by photosynthesis is less than that expended by the rapidly metabolizing young grain plants. In Michigan studies, hydroponically sprouted oats contained only 77% as much TDN as the oats put into the chambers (Thomas and Reddy, 1962). The composition of the dry matter changes as starch in the grains is converted to fiber and other products in the young plant. If the young sprout develops a green color, the amount of carotene, precursor of vitamin A, may increase greatly. Likewise, if the cultural liquid medium used contains the needed elements and a usable nitrogen source, the mineral and protein percentages of the young plant usually increase.

When the sprouting and/or growing period is completed, the total product contains a smaller total amount of digestible energy than the grain used. Likewise, due to the substantial investment cost in the equipment as well as labor and other expenses, the nutrients produced become quite expensive. If the added minerals, protein, and vitamin A value in the hydroponically produced grass are needed by dairy cattle, they can be provided at a much lower cost in other ways. Hydroponically produced grass does not appear to have any special milk stimulating properties.

10.5.d. Roots and Tubers

Although not used in substantial amounts in the United States, in some areas of the world, particularly Northern Europe, tubers and various roots are frequently fed to dairy cattle. Some of the crops used include fodder beets, turnips, cull potatoes (Hoover *et al.*, 1976), carrots, and mangolds (or mangels). Many of these feeds are quite palatable and provide considerable nutritional value.

10.5.e. Animal Waste, Excreta, or Manure

In recent years, great interest has developed in using recycled feces and urine from various types of animals as a cattle feed ingredient. Generally this interest has been greater with beef than with dairy cattle. Usually recycled excreta has a relatively low digestible energy value. The amount depends somewhat on the nature of bedding material used and on the species of animals providing the excreta. Often a substantial quantity of the nitrogen in urine and feces is in a form which can be utilized by rumen microbes. Generally the minerals can be used, but excesses may pose a problem. Depending on the metabolic characteristics of the mineral, the excreta may contain much higher levels than did the feed dry matter. Thus, in some situations, precautions must be taken to avoid a buildup of certain mineral elements when excreta is recycled repeatedly.

When feeding animal excreta to dairy cattle, key considerations other than nutritional value are involved. Poultry excreta, especially, may contain undesirable drugs. Likewise, special treatment to avoid disease spread may be required, and various government regulations must be considered.

10.5.f. Other Ingredients

Many substances in feeds are included for a variety of other reasons. Some are used primarily for other purposes but also may have nutritional value. An example is lignin sulfonates used as a pellet binder.

REFERENCES

Bringe, A. N., and L. H. Schultz (1969). *J. Dairy Sci.* **52**, 465–471.

Clark, J. H. (1976). *Hoard's Dairyman* **121**, No. 16; 935, 944, and 945.

Crowley, J. W. (1976). *Hoard's Dairyman* **121**, No. 19, 1124–1125.

Cullison, A. (1975). "Feeds and Feeding." Reston Publ. Co., Reston, Virginia.

Hale, W. H. (1975). *Proc. Ga. Nutr. Conf. Feed Ind.* pp. 131–141.

Harris, B., Jr. (1970). *Hoard's Dairyman* **115**, No. 14, 777 and 793.

Hoover, W. H., C. J. Sniffen, and E. E. Wildman (1976). *J. Dairy Sci.* **59**, 1286–1292.

Macgregor, C. A., F. G. Owen, and L. D. McGill (1976). *J. Dairy Sci.* **59**, 682–689.

Miller, W. J., and J. K. Miller (1963). *J. Dairy Sci.* **46**, 581–583.

Miller, W. J., and G. D. O'Dell (1969). *J. Dairy Sci.* **52**, 1144–1154.

Miller, W. J., and P. E. Stake (1972). *Proc., AFMA Liq. Feed Symp., 1971* pp. 54–64.

National Research Council (NRC) (1978). "Nutrient Requirements of Dairy Cattle," 5th rev. ed. Natl. Acad. Sci., Washington, D.C.

Rindsig, R. B., L. H. Schultz, and G. E. Shook (1969). *J. Dairy Sci.* **52**, 1770–1775.

Ronning, M., and R. C. Laben (1966). *J. Dairy Sci.* **49**, 1080–1085.

Schmidt, G. H., and L. D. Van Vleck (1974). "Principles of Dairy Science." Freeman, San Francisco, California.

Schultz, L. H. (1969). *Hoard's Dairyman* **114**, No. 04, 209 and 243.

Smith, K. J. (1977). *Feedstuffs* **49**, No. 3, 22–25.

Thomas, J. W., and R. S. Emery (1969). *J. Dairy Sci.* **52**, 60–63.

Thomas, J. W., and B. S. Reddy (1962). *Mich., Agric. Exp. Stn., Q. Bull.* **44**, 654–665.

Wagner, D. G., and J. K. Loosli (1967). *N.Y., Agric. Exp. Stn., Mem.* **400**, Ithaca.

Williamson, J. L. (1973). *In* "Effect of Processing on the Nutritional Value of Feeds," pp. 349–355. Natl. Acad. Sci., Washington, D.C.

11

Nonnutritive Additives and Constituents

11.0. INTRODUCTION

Many different types of substances that do not provide required nutrients are sometimes included in dairy cattle feeds for a variety of purposes. Likewise, other nonnutritive materials in feeds often have a substantial effect on the health and performance of the animals or on other aspects of practical value. Some of the more important nonnutritive additives and naturally occurring constituents are discussed in this chapter.

11.1. ANTIBIOTICS

Frequently, antibiotics are included in the feed of young dairy calves. Especially when raised with unfavorable sanitation, housing, or disease conditions, young calves benefit from some antibiotics such as chlortetracycline (Aureomycin is one brand) or oxytetracycline (Terramycin is one brand) (Lassiter, 1955; National Research Council, 1978). The improved responses of calves to antibiotics are associated with increased feed consumption, more rapid growth, and decreased diarrhea. Best results are obtained early in the life of the calf during the milk or milk-replacer feeding period (NRC, 1978). Thus it is desirable to begin feeding the antibiotics as soon as possible after birth (Bartley *et al.*, 1954). Usually little if any benefit occurs after the calf is four months of age (NRC, 1978).

Generally, for the promotion of growth, 20–40 ppm (9–18 mg/lb) in the dry milk replacer or the equivalent quantity of whole milk is suggested (NRC, 1978). The recommended level in the dry starter feed is about half that of the dry milk replacer. The amount needed for control of diarrhea is about two and one-half times (50–100 ppm in dry milk replacer) that indicated for growth promotion (NRC, 1978). Although slight increases in milk production have been noted from some studies when antibiotics were fed to lactating dairy cows, other results are conflicting (NRC, 1978). The continuous feeding of antibiotics to lactating cows generally is not recommended (NRC, 1978) due to cost and the possibility of antibiotics in milk. Some antibiotics can have adverse effects on rumen microbes. The feeding of antibiotics to dairy cattle, especially near the time of slaughter, and to lactating cows is closely regulated in many countries. Although such regulations may not always be justified by scientific knowledge, careful attention should be paid to local regulations when and if antibiotics are to be fed.

11.2. HORMONES

Many substances with hormone activity have been investigated for possible beneficial effects on the performance of dairy cattle. None have given sufficient promise to be generally used commercially, and few have been extensively studied. Thyroprotein, a substance with thryoxine activity, is sometimes sold to increase milk production. This controversial substance is discussed more fully in Section 11.3. Although apparently never used commercially, growth hormone has increased milk production in research studies.

Hormone-like substances to increase performance of cattle for beef production have been more widely used than those for improving milk production. Since some dairy cattle are produced for beef, a limited discussion of such products is appropriate here. Diethylstilbestrol (DES), a synthetic estrogen, for many years has been widely used in finishing steers. This hormone increases rate of gain and feed efficiency in steers by approximately 10–15%. Knowledgeable scientists are convinced that the FDA efforts to withdraw approval for DES use is an unjustified political decision rather than one based on scientific merit. Such decisions should be made on a scientific basis. Unfortunately the pattern of putting them in the emotional and political realm is increasing to the detriment of the best public interest. The increased production costs from being unable to use the most efficient, safe methods elevate the price of food to consumers.

Some information indicates that Melengestrol acetate (MGA), a substance with progesterone-like actions, will increase rate of gain in heifers (Lowrey *et al.*, 1974). Likewise, substances having testosterone male sex hormone activity, often have increased growth rate in heifers and steers.

11.3. THYROPROTEIN

Thyroxine, the hormone produced by the thyroid gland, regulates the basal metabolic rate in animals. A synthetic thyroxine called thyroprotein is produced by iodinating casein. Some of the tyrosine in casein combines with iodine to form thyroxine. It is sold under one or more trade names. Thyroprotein has been used experimentally and commercially for several decades. Even so, its practical use for dairy cows is still controversial and generally not recommended (NRC, 1978). Although the response is quite variable among individual cows, feeding thyroprotein often materially increases milk production, especially for a limited period of time (Schmidt et al., 1971).

Thyroxine production in the normal animal is quite closely regulated. Thus, if thyroprotein is fed, the added thyroxine activity suppresses the usual formation of the hormone by the thyroid gland. Accordingly, when thyroprotein is withdrawn, the cow is suddenly left with less than a normal amount until the thyroid gland can adjust back to a normal production of this key hormone. The subnormal level of thyroxine results in a sudden marked drop in milk production when thyroprotein feeding is discontinued, and usually to a lower level than that of controls never fed thyroprotein. A gradual withdrawal of thyroprotein feeding has not consistently proven beneficial in preventing the sharp decrease in milk production. Because of the decrease in milk after withdrawal, total milk production for a lactation may or may not be as high or higher than that of cows not given this hormone feed. When thyroprotein is fed, generally there is a transitory increase in the milk fat content.

Since thyroxine regulates basal metabolic rate, feeding thyroprotein increases metabolic, heart, and respiration rates, excitability, as well as body temperature and susceptibility to stress from heat or other factors. The energy requirement for maintenance is elevated. Accordingly, when thyroprotein is fed, larger total amounts of feed and concentrates are needed to supply the extra energy needed. If more feed is not provided, there may be a serious loss of weight. With the same level of feed intake, cows fed thyroprotein, generally gain less or lose more weight.

Because of the increase in milk production which may be as much as 20% for limited periods in some cows, considerable effort has been devoted toward finding ways of using thyroprotein profitably in practical dairy operations. Some of these have been associated with very sophisticated marketing programs for the commercial products. Since dairy cows, especially high producers, often are unable to consume sufficient feed to meet their energy needs, generally suggested programs for thyroprotein feeding include waiting until the cow is past the peak of lactation. Also, economic benefits are greater if milk produced during a given season has a higher value than at other times. For instance, in some situations milk sold during a base-building period may have greater effective value. This would be especially true if the thyroprotein withdrawal period coincides with a surplus milk season or with a dry period. Often programs for use of thyroprotein suggest removing any cow which does not show a positive response within several days.

Although considerable effort has been devoted to research and marketing of thyroprotein over many years, use of this product has never achieved a widespread and long-continued use on large numbers of dairy farms. In addition to the problems caused by the withdrawal and the added expense of the extra feed needed, other adverse side effects have been observed. The extent of these side effects, sometimes termed hyperthyroidism, are quite variable among cows. Generally the degree of hyperthyroidism increases with the amount of thyroprotein fed. Other undesirable effects including increased services per conception (Schmidt *et al.*, 1971), possibly increased injuries due to more excitability, weak calves, failure to exhibit estrus, and longer calving intervals sometimes occur. Likewise, muscular weakness and coughing have been observed.

Another factor limiting the feeding of thyroprotein is its prohibition for cows on DHI (Dairy herd improvement) or DHIR (Dairy herd improvement registry) testing programs. Thus records of cows fed this hormone cannot be used in USDA sire evaluation programs. Cows on Owner-Sampler or other unofficial record keeping programs can be fed thyroprotein.

11.4. ENZYMES

Enzymes play a key role in almost all digestive, metabolic, and other life processes. Because of this great importance, attempts have been made to find practical uses for adding supplemental enzymes to the feeds of dairy cattle. However, no important application has been clearly established. Efforts to improve rumen development and fermentation or silage preservation by adding enzymes apparently have not resulted in great success.

11.5. SPECIAL DRUGS AND CHEMICALS

Considerable interest and attention have been devoted to finding special drugs and chemicals to improve the performance and/or feed efficiency of dairy cattle. Only a very few have shown real promise.

11.5.a. Monensin

A complex chemical, monensin, an anticoccidial drug for poultry, increases the production of propionic acid by decreasing methane. Since methane represents an energy loss, feed efficiency is improved in growing and finishing beef cattle (Anonymous, 1975; Perry *et al.*, 1976; Utley, 1976). A major advantage from feeding monensin occurs in animals fed a high proportion of forage and/or roughages. An adverse effect on the fat content of milk might be expected in some situations, because of the lower acetate to propionate ratio when monensin is fed. Since the product is comparatively new, future research should provide more definitive answers on monensin's practical value for lactating cows.

11.5.b. Buffers and Additives to Alleviate Low Milk-Fat Content

See Section 10.1.b.

11.5.c. Feed Flavors

Often special flavoring compounds have been added to dairy cattle feeds in attempts to increase the palatability of the feeds. However, relatively little research has been conducted by university scientists to determine whether these additives were needed or if they achieved their purpose. For instance, at one time various anise oils were used in calf feeds. Generally these compounds provided a desirable aroma for humans examining the feeds. In controlled experiments, however, the addition of anise oils to calf feeds decreased the palatability (Miller *et al.,* 1958).

Unless the amount of feed consumed is unsatisfactory due to poor palatability, there is little justification for the addition of special flavoring compounds. Since humans and cattle do not respond to flavoring compounds in the same way, objective research should establish the beneficial effects of the flavor additive before their use is justified. Cattle on the "show circuit" often are given chlorinated water that is unpalatable, resulting in decreased water intake. Popular reports indicate that flavor compounds, such as Jello, added to the water may alleviate the problem.

11.5.d. Antioxidants, Feed Preservatives

Some feeds and ingredients, especially those containing a high level of unsaturated fats, oxidize and/or become rancid quite easily. Likewise, unless stabilized, fat-soluble substances such as vitamin A may lose potency rapidly in some feeds. To prevent or retard deterioration, frequently antioxidants and/or other preservatives are added to feeds. Antioxidants that have proven effective include: ethoxyquin, butylated hydroxyanisole (BHA), butylated hydroxytoluene (BHT), and propyl gallate (PG). Frequently a level of about 0.01% of such antioxidants is sufficient.

11.5.e. Special Drugs in Feeds

In some instances, nonnutritional specialty drugs are administered in the feed. An example is the use of special chemicals, including stirofos, for fly control

(Miller, 1976). Likewise, the essential nutrient, iodine, in organic form may be included in the feed for a nonnutritional purpose such as control of foot rot.

11.6. MYCOTOXINS AND MOLDS

Molds and closely related organisms in feeds are a very old phenomenon. However, they have become of much greater concern and practical importance in recent years (Wyatt, 1976). In some situations the growing molds produce highly toxic and often very stable by-products called mycotoxins that, when eaten, can adversely affect dairy cattle. It is estimated that more than 100 different mycotoxins can occur in animal feeds (Wyatt, 1976).

Molds, when growing on feeds, use part of the nutrients, especially the highly digestible carbohydrates. However, the adverse effects of the mycotoxins on the performance and/or health of dairy cattle can be of much greater importance. Some of the more important mycotoxins include aflatoxins, T–2 toxin, ochratoxin, and zearalenone (Wyatt, 1976). Even though this area is poorly understood, the adverse effects of the different mycotoxins vary greatly. The toxic symptoms range from lower feed intake, decreased growth or milk production and lethargy, to respiratory problems, reproductive problems including abortions, liver damage, increased susceptibility to stress, and even death. Within the same herd the range in effects often is quite variable among individuals. Young animals, especially calves, are much more susceptible to aflatoxin toxicity than older cattle. Relatively, cattle are more tolerant than swine and poultry. Aflatoxins have been studied more than any other mycotoxins. The most important one of this group of compounds is aflatoxin B_1 (AFB_1). Some aflatoxin may be secreted into the milk. At a dietary level much lower than required to cause toxic effects, approximately 0.1–0.4% of the dietary aflatoxin (AFB_1) is found in milk as aflatoxin M_1 for which there is a zero tolerance (Bodine *et al.*, 1976). As little as 15 ppb (parts per billion) of the aflatoxin B_1 in the diet can be detected in milk. Fortunately, the aflatoxins are not stored in the body; thus the milk is free soon after intake ceases.

The growth of molds and therefore production of mycotoxins is determined primarily by the environmental conditions during growth. These organisms require air, moisture, and a source of nutrients, including readily digestible carbohydrates. Some molds, including those producing aflatoxins, multiply far more rapidly with a warm temperature. Others require a low temperature. Mold growth can be prevented in any one of several ways. When hay, grain, or mixed feeds are dry, molds will not grow. Likewise, they will not grow in silage that has no free oxygen. Aflatoxins are much more troublesome when grains are damaged by drought, insects, or cracked during harvesting or handling; as the seed coat normally provides protection against mold invasion. Aflatoxins can be produced in the field prior to harvest, especially when the grain is damaged. Silages which have too high a dry matter content, are poorly sealed, not finely chopped, or not well packed, are much more likely to mold. Likewise, when the silage pH is not sufficiently low, molding is much more likely. Generally aflatoxins are not produced by molding silage. However, aflatoxins in the corn grain at the time of ensiling probably will not be destroyed by the process.

Hay with a moisture content above 15% may mold. During feeding, silage will mold when left in the feed trough too long, depending on the temperature and other factors, including the pH. Silage near the feeding surface of a silo may mold if feeding rate is too slow or if it is "loosened" during feeding.

One approach to controlling mold growth and mycotoxins is the addition of mold inhibitors. Propionic acid, when properly used, is effective in reducing mold growth on grains, haylages, hays, silages, and high-moisture grains. These and other preservatives, including formic acid and heavy metals such as copper, have been used for preservation of high-moisture corn grain (see Section 10.2.a) (Britt and Huber, 1976). None of the inhibitors decrease the toxicity of compounds such as aflatoxins after they are produced.

REFERENCES

Anonymous (1975). "Rumensin (Monensin Sodium)," Technical Manual. Elanco Products Co., Eli Lilly Co., Indianapolis.

Bartley, E. E., F. W. Atkeson, H. C. Fryer, and F. C. Fountaine (1954). *J. Dairy Sci.* **37**, 259–268.

Bodine, A. B., G. D. O'Dell, and J. J. Janzen (1976). *S.C., Agric. Exp. Stn., Dairy Res. Ser.* No. 65.

Britt, D. G., and J. T. Huber (1976). *J. Diary Sci.* **59**, 668–674.

Lassiter, C. A. (1955). *J. Dairy Sci.* **38**, 1102–1138.

Lowrey, R. S., W. J. Casey, and H. C. McCampbell (1974). *Proc. Ga. Nutr. Conf. Feed Ind.* pp. 44–53.

Miller, R. W. (1976). *Hoard's Dairyman* **121**, No. 9, 585.

Miller, W. J., J. L. Carmon, and H. L. Dalton (1958). *J. Dairy Sci.* **41**, 1262–1266.

National Research Council (NRC) (1978). "Nutrient Requirements of Dairy Cattle," 5th rev. ed. Natl. Acad. Sci., Washington, D.C.

Perry, T. W., W. M. Beeson, and M. T. Mohler (1976). *J. Anim. Sci.* **42**, 761–765.

Schmidt, G. H., R. G. Warner, H. F. Tyrrell, and W. Hansel (1971). *J. Dairy Sci.* **54**, 481–492.

Utley, P. R. (1976). *Proc. Ga. Nutr. Conf. Feed Ind.* pp. 123–126.

Wyatt, R. D. (1976). *Feedstuffs* **48**, No. 15, 22–23.

12

Evaluation of Feeds for Dairy Cattle

12.0. INTRODUCTION

The nutritive content of different forages, even of the same species, vary enormously (see Chapter 9). Although the variability generally is less among grains and other concentrates, large differences often occur in these ingredients. Because of these variations in nutrient content and the importance of providing the needed nutrients without excessive wastage of expensive ingredients, feed evaluation is crucial to practical dairy cattle feeding.

The great variability in the nature of different feeds and forages makes it impossible to accurately evaluate every feed for all nutrients with only a few, simple determinations. Since the type of analyses needed varies with the situation and kinds of feeds involved, understanding the major types of procedures used to evaluate feeds is helpful. Most can be classified as either chemical analyses or as biological evaluations with many of the more useful measures, especially for research work, involving both types.

12.1. CHEMICAL ANALYSES OF FEEDS FOR GROSS COMPOSITION

Because of the complex chemistry of plant materials and other feed ingredients, some of the most useful chemical analyses are empirical in nature. With these, a specific substance or compound is not measured but rather a prescribed procedure is employed to determine compounds of a general type or those with similar characteristics. An empirical analysis often is especially useful when many compounds can serve the same nutritional need: energy-providing material is a good example. Due to the essentiality of many specific compounds or elements, several chemical analyses are for specific nutrients such as for individual mineral elements.

12.2. PROXIMATE ANALYSIS

The most widely used system of evaluating feeds by chemical analyses is known as proximate analysis. This system was a remarkable accomplishment when developed at the Weende Experiment Station in Germany more than a century ago. The basic concept was to separate feeds into the nutrient components needed by the animal including dividing the carbohydrate into readily utilized and poorly used fractions. Although the proximate analysis system is still widely used, nutritional discoveries since its development have shown that the basic objectives were only partially accomplished.

Proximate analysis of feeds divides the total weight into (1) water, (2) crude protein, (3) crude fat or ether extract, (4) ash or mineral matter, (5) crude fiber, and (6) nitrogen-free extract. Each of the fractions is empirically determined by an exact procedure. These procedures are presented in detail in publications of the Association of Official Agricultural Chemists and briefly outlined by Cullison (1975).

12.2.a. Water, Moisture, or Dry Matter

Although water is essential for dairy cattle (see Chapter 7), generally it is not considered to have any special value in feeds. The water (moisture) content of feeds varies greatly, making it necessary to know the content. Often in tables of feed composition, the nutrients are presented on a dry matter basis (see Appendix, Table 4). This is quite helpful in comparing feeds of widely differing dry matter contents. For instance, silage, on an average, will contain only about one-third as much dry matter per pound as hay, with considerable variation among silages. The amount of dry matter consumed by the cow is about the same with varying dry matter contents.

The methods used for measuring dry matter content of feeds varies with the type of feed. For many feeds a small quantity is dried in an oven under prescribed conditions with the weight loss in drying being water. With some feeds such as silages, oven drying removes substantial amounts of the usable nutrients that volatize with the heat. Similarly determining other silage constituents on a dry sample underestimates the amounts of energy and crude protein (Waldo, 1977). The dry matter content of silages should be determined by a method which measures the water directly. Likewise, protein and energy analyses should be made on undried silage samples.

12.2.b. Crude Protein

In the proximate analysis system, crude protein (see Chapter 3) is calculated from the amount of nitrogen in the feed which usually is determined by the

Kjeldahl procedure. Since proteins on an average contain 16% nitrogen, crude protein is calculated by multiplying total nitrogen by 6.25 (100%/16% = 6.25). A different multiplier is sometimes used for special proteins known to contain another percentage nitrogen (see Section 3.0). Crude protein still is a key analysis in the evaluation of feeds (see Chapter 3 for further discussion).

12.2.c. Crude Fat (Ether Extract)

In the proximate analysis system, crude fat is determined by extracting the feed with ether. Fat is soluble in ether. One of the reasons for developing the crude fat (either extract) determination is the much higher energy value of fat relative to carbohydrates and proteins. Fat has about 2.25 times as much energy per pound as carbohydrates. Unfortunately, from the standpoint of feed evaluation, ether also extracts many substances other than true fats, some of which have a much lower energy value. For instance, in pasture herbage and other forages much of the pigment is extracted and shown in the proximate analysis system as crude fat. These extracted pigments do not have the high energy value of true fat. In silages, substantial amounts of acids such as acetic, propionic, lactic, and butyric appear in the ether extract fraction. Again, as with pigments, these organic acids do not have the high energy content of true fats.

12.2.d. Ash (Mineral Matter)

The total mineral content of a feed is determined by burning it to ashes. Although the ash value is still used, it is not especially useful in evaluating the content of the essential minerals. In some situations where a feed contains a great deal of dirt, the ash content may be extremely high. Often this dilutes the other nutrients but does little or no direct harm.

12.2.e. Crude Fiber

In determination of crude fiber, a sample is boiled (refluxed) in dilute acid to remove soluble material. This is followed by the same procedure with a dilute alkali. The carbohydrate material not removed by the acid and alkali treatments is called the crude fiber. That which was dissolved is the nitrogen-free extract (NFE).

A major objective of the proximate analysis system of feed evaluation was division of the carbohydrate into the readily digestible and indigestible fractions. Great volumes of research have shown that a substantial amount of the crude fiber is digested by dairy cattle (see Chapter 8). Likewise, in some feeds much of the dissolved carbohydrate (NFE) is not digested. Thus, the developers of the crude-fiber determination did not achieve the objective of a measure of the indigestible carbohydrate. Of course, in many nonruminants, the digestibility of crude fiber usually is quite low.

12.2.f. Nitrogen-Free Extract (NFE)

In the proximate analysis system, the nitrogen-free extract (NFE) was intended to be the highly digestible fraction of the carbohydrate. NFE, which is determined by the difference, is the dry matter remaining when the crude fat, crude protein, ash, and crude fiber are subtracted from 100%. Although this fraction contains any errors in determination of the other fractions, basically it is the carbohydrate and carbohydrate related materials removed by the weak acid and weak alkali in the crude-fiber procedure. It includes sugars, starches, pectins, hemicelluloses, and some of the lignin (Table 12.1). The weak alkali removes a substantial portion of the lignin, one of the most indigestible components in feeds (see Section 8.1).

12.2.g. Calculation of Total Digestible Nutrients (TDN)

The TDN (total digestible nutrients) content of feeds is determined by a combination of the proximate analysis of feeds with a digestion trial. In essence, both the feed and the feces of animals receiving the feed are analyzed by the proximate analysis (see Section 12.2) system. The digestibility of the crude protein, crude fiber, crude fat and NFE of the feed are calculated by subtracting the amount of each of these components excreted in the feces from that fed. By definition, digestibility is the portion which the animal consumes but does not excrete in feces (see Section 2.1.a).

The TDN is calculated by adding digestible crude protein, crude fiber and NFE plus the digestible ether extract times 2.25. The 2.25 factor for digestible ether extract is used because digestible "true" fat has that much more digestible energy. TDN as a measure of energy value of feeds is discussed more fully in Section 2.1.d.

12.3. GROSS ENERGY; BOMB CALORIMETRY

The gross or total energy content of a feed is determined by burning a sample in a suitable container and measuring the heat released. The procedure known as "bomb calorimetry" has been used for a long time and is a very accurate measurement. Total energy content of feed and various fractions lost during the digestive and metabolic processes of animals is a key part of several measures of usable energy value (see Chapter 2). Determining the energy content of samples is much simpler than the other aspects of the balance experiments involved in energy evaluations (see Section 12.9).

12.4. LIGNIN

As discussed earlier (see Section 12.2), the proximate analysis system fails to achieve the objectives of its developers, relative to division of the carbohydrate fraction into the highly digestible and poorly digested fractions. Thus for several decades nutritionists have attempted to develop chemical procedures that would more accurately reflect the true energy value of feeds for animals. Because of the almost total indigestibility of lignin and its effect in decreasing the digestibility of other nutrients, considerable research has been devoted to this fraction (see Section 8.1.a) including developing several empirical methods of analyses. Since lignin is not a specific substance, the content is influenced by the method used (see Section 12.5).

12.5. THE DETERGENT OR VAN SOEST METHODS OF ANALYSES

A new system of chemical analyses of forages was developed with the intent of replacing the crude-fiber and nitrogen-free extract analyses of the proximate analysis system (Goering and Van Soest, 1970; Van Soest, 1964, 1966, 1967). This system developed primarily in the USDA Laboratory at Beltsville, Maryland, appears to achieve much more closely a division of the carbohydrate fraction of the forage into the readily digestible and less well-digested fractions (Goering and Van Soest, 1970).

The Van Soest system of forage analyses separates the forage dry matter into the cell wall constituents and the soluble contents of cells (Goering and Van Soest, 1970). The cell walls are further broken down into subfractions that substantially improve measurement of the nutritive value for dairy cattle. Details of the Van Soest procedures for analyses of the forages are presented in a USDA handbook (Goering and Van Soest, 1970). A feed sample is boiled in water containing a neutral detergent to dissolve the cell contents. The cell contents designated as neutral detergent solubles (NDS) includes sugars, starches, most of the protein, nonprotein nitrogen, lipids, and other soluble substances (Table 12.1). The cellular contents (NDS) have a true digestibility of about 98% in many different forages (Van Soest, 1971). The material not dissolved by boiling in neutral detergent water is designated as neutral detergent fiber (NDF) or as the cell wall fraction. This cell wall fraction includes cellulose, hemicellulose, lignin, and some silica (Van Soest, 1971).

TABLE 12.1

Nutritive Characteristics and Distribution of Forage Components by Proximate and Van Soest System Analyses

Forage component	Proximate analysis fraction	Van Soest system	True digestibility
Sugars, starches, soluble carbohydrates, pectins	NFE[a]	Cell contents (NDS)[a]	Complete
Protein, NPN	Crude protein	Cell contents (NDS)	High
Fats, lipids, pigments	Ether extract	Cell contents (NDS)	High
Minerals, dirt, soluble	Ash	Cell contents (NDS)	Variable
Insoluble (including silica)	Ash	Divided[b]	Variable
Hemicelluloses	NFE	Cell wall constituents (NDF)[c,d]	Variable
Cellulose	Crude fiber	Cell wall constituents (NDF)(ADF)[e]	Variable
Lignin	Crude fiber and NFE	Cell wall constituents (NDF)(ADF)	Indigestible
Heat-damaged protein	Crude protein	(ADF)	Indigestible

[a] NDS, Neutral detergent solubles; NFE, nitrogen-free extract.

[b] Some silica goes with the cell contents and some with the cell wall fraction.

[c] NDF (Neutral detergent fiber) is cell wall constituents.

[d] NDF includes total cell wall constituents, whereas ADF excludes the hemicelluloses but includes some protein.

[e] ADF, Acid detergent fiber.

The cell wall fraction (NDF) is digested only by the microbes in the digestive tract, primarily in the rumen. Thus most nonruminants obtain little energy from this fraction. The structural portion, cell walls, of the forage determines the amount of space occupied in the digestive tract of dairy cattle (Van Soest, 1971). For most forages, the amount of space used is the primary determinant of the amount which the animal will voluntarily consume.

Some of the advantages of determining the cell wall fraction compared to crude fiber is illustrated by the values obtained for alfalfa relative to bermuda grass (Table 12.2). Although alfalfa hay and bermuda grass had similar amounts of crude fiber, the bermuda grass hay contained almost twice as high a percentage of cell wall constituents. This correlates well with the recognized much higher voluntary intake and superior animal performance with unground alfalfa hay when these feeds make up the main sources of feed nutrients for dairy cattle (see Section 9.3.a and Table 9.5). The digestibility of the cell wall constituents by ruminants varies greatly depending on the makeup of the fractions. Accordingly, procedures have been devised to measure some of the components having the greatest effect on digestibility of this fraction (Goering and Van Soest, 1970).

As a preparatory step for the lignin analysis, the ADF fraction is removed and determined by boiling the sample in an acid detergent solution. Acid detergent fiber consists of cellulose, lignin, cutin, and acid-insoluble ash that is mostly silica (Table 12.1). The difference between the NDF and ADF fractions is an estimate of hemicellulose, but also contains some of the protein from the cell walls (Goering and Van Soest, 1970). The ADF fraction also includes heat damaged protein.

Acid detergent lignin is determined by digesting the ADF in 72% sulfuric acid to remove cellulose. The insoluble residue contains lignin, cutin and silica. Silica remains when the sample is ashed. Cutin is determined by an acid detergent cutin procedure (Goering and Van Soest, 1970). The cutin content may be quite high and important in many seed hulls.

Another method of determining lignin is the permanganate lignin method. This procedure has important advantages over the acid detergent procedure in many situations but also some limitations (Goering and Van Soest, 1970). The most suitable lignin procedure depends on the nature of the forage being analyzed. One important application of lignin values is as a ratio to the percentage of ADF. This ratio is very useful in estimating digestibility of the cell wall fraction (Table 12.3).

In lignin determination, avoiding artifacts is quite important. For instance, especially in very succulent, high-protein forages, heating or drying forages above 110°–120°F can greatly increase the nitrogen content of the lignin and the total amount of lignin determined by the analysis. Procedures have been devised for determining the amount of artifact lignin (Goering and Van Soest, 1970).

Since a high silica content in many grasses is a major depressant of cell wall digestibility, often a measure of this component is very useful. The silica content varies widely between species of forages and within the same forage grown in different locations (Van Soest, 1971). In evaluating the effects of silica, it is

TABLE 12.2
Comparison of Certain Temperate and Subtropical Forages[a]

Forage	Digestibility	Crude fiber	Cell walls	Cellulose	NFE[b]	Soluble CHO[c]	Crude protein
Temperate							
Alfalfa	60	30	40	32	43	33	17
Corn silage	70	24	45	26	61	40	9
Young orchard grass	70	27	55	28	49	21	15
Subtropical							
Bermuda grass	50	30	77	35	56	8	9
Pangola grass	54	30	70	34	50	10	11

[a] Adapted from Van Soest (1971). Data given as % of dry matter.
[b] NFE, nitrogen-free extract.
[c] CHO, carbohydrates.

TABLE 12.3
Estimation of Cell Wall Digestibility from Percentage of Lignin in Acid Detergent Fiber (ADF) of Forages[a]

Lignin[b] (% in ADF)	Estimated true digestibility of cell walls[c] (%)
6	85
8	76
10	68
12	62
14	57
16	52
18	48
21	43
25	37
30	30
40	21
50	13

[a] Adapted from Goering and Van Soest (1970).
[b] Lignin by 72% acid method. Conversion values also available for lignin from permanganate method (Goering and Van Soest, 1970).
[c] If silica exceeds 2%, values are decreased 3.0% for each 1% silica. Silica from sand or soil is excluded.

important to distinguish between silica grown into the cell walls and that from dirt or other foreign material. Even a high percentage of silica as dirt and dust will not reduce the digestibility of the cell wall components.

Scientists, who have been closely associated with the Van Soest procedures for chemical evaluation of feeds, feel that this system is far superior and should replace the crude-fiber and nitrogen-free extract parts of the proximate analysis system.

12.5.a. Estimating Digestibility and Feeding Value from Van Soest Analyses

The purpose of chemical analyses of feeds is to estimate the feeding value of the feeds and to aid in balancing rations and planning feeding programs. The Van Soest procedures (see Section 12.5) can be used with suitable formulas to estimate the digestible energy value of a feed. These formulas have been developed from data obtained in digestibility studies with feeds analyzed by the procedures. The true digestibility of the cell contents, neutral detergent solubles, is about

98% (Goering and Van Soest, 1970). The digestibility value generally measured and used is apparent digestibility (see Section 2.1.a). Apparent digestibility is lower than true digestibility by the amount of the endogenous or metabolically excreted nutrients. On an average this endogenous component is about 12.9% of the dry matter consumed (Goering and Van Soest, 1970). Thus the apparent digestibility is about 85.1% for the dry matter of cell contents. There is no endogenous excretion of fiber.

The true digestibility of the cell wall fraction is calculated from the ratio of the lignin to acid detergent fiber as illustrated in Table 12.3 (Goering and Van Soest, 1970). Further research should result in improvements in the formulas used to estimate digestible dry matter or other constituents from chemical analyses of feeds.

12.6. CHEMICAL AND MICROBIOLOGICAL ANALYSES OF INDIVIDUAL NUTRIENTS

Dairy cattle require a substantial number of specific nutrients in the diet, several of which may be deficient in some practical feeding situations. Likewise, a few specific constituents, sometimes found in feeds have adverse effects on dairy cattle. Good methods of analyses for both the essential nutrients and those causing toxicity are important in feed evaluation.

12.6.a. Mineral Elements

Satisfactory procedures for determining most of the required major mineral elements (see Chapter 5) have been in use for many years. Often the sulfur analysis has been less satisfactory. In earlier years, frequently only calcium and phosphorus were determined in feed ingredients.

The trace mineral elements obtained the designation *trace* from the fact that earlier methods of analyses were not quantitative. Rather they could only be measured on a qualitative basis indicating that an indeterminate amount was present. Since the late 1920's and early 1930's as their importance in animal nutrition became appreciated, gradually procedures for the quantitative measurement of the trace elements were developed and improved. The early chemical methods were laborious and tedious, generally requiring considerable analytical skill. For example, in 1935 a single duplicate analysis for cobalt required five days (Underwood, 1970). Because of the difficulty involved, until recently, the amount of information on the trace mineral content of feeds was very small (Miller *et al.*, 1972).

Beginning largely in the 1960's there has been a virtual revolution in methodology for determining trace mineral elements. The sophisticated instrumentation now available makes it possible to quickly and relatively easily determine the amount of most trace elements in feeds. Other improvements are still being made. The analytical method of choice varies for different trace elements. Likewise, the concentration in the feed, and other factors such as the precision needed and the number of samples to be analyzed, become considerations. Although the new sophisticated instruments are vastly superior to the older procedures, appreciable skill, knowledge, and expertise are needed to obtain uniformly reliable and useful values.

Two of the most useful methods for determining trace elements are *atomic absorption spectroscopy* and *direct reading emission spectroscopy*. With the direct reading emission spectrograph, many elements can be determined in the same sample at the same time. Atomic absorption spectroscopy is a very useful, sensitive and comparatively accurate analytical method for several mineral elements.

Other examples of newer methodology which are useful in some situations include neutron activation analysis, plasma emission spectroscopy, spark source mass spectrometry, and x-ray fluorescence. Neutron activation analyses appears to be especially promising with elements such as vanadium, cobalt, arsenic, selenium, and molybdenum at very low concentrations. In spite of the great advances, some of the elements such as selenium still present difficult analytical problems in dairy cattle feeds.

12.6.b. Vitamins

Although huge differences exist in vitamin content of feeds, in the practical feeding of dairy cattle analyses of feeds for vitamins is largely limited to research work. Even in research most of the analyses are for vitamin A activity, i.e., carotene. Frequently vitamin A is added to dairy cattle feeds. Since the types of feeds where vitamin A activity would likely be deficient are fairly well characterized (see Chapter 6), a reasonable evaluation for the need or lack of vitamin A supplementation can be made without analytical work.

12.6.c. Other Specific Constituents

A great many harmful substances are sometimes found in dairy cattle feed. These include natural substances such as nitrates, prussic acid (hydrogen cyanide, HCN), ergot, aflatoxins (see Section 11.6), or such materials as antibiotics, pesticides, and drugs. The analyses for nitrates and prussic acid are fairly simple, when suspected forages are involved (see Chapter 18). Although analytical determination for many of the pesticides, antibiotics, and drugs is quite complex, because of the great emphasis on this type of problem, many laboratories are able to make such analyses.

12.7. BIOLOGICAL AND COMBINATION EVALUATION METHODS

The purpose of evaluating feeds is to determine their ability to meet the nutritional needs of animals. Although most of the discussion up to now in this chapter has concerned chemical analyses, the real value of a feed is biological in nature. The usefulness of the chemical analyses depends on their utility in predicting biological performance of animals. Since most biological methods for evaluating feeds are rather complex and expensive, they are mainly research measures. Because the data are frequently and widely used, some understanding of the more important methods is useful in practical dairy cattle feeding.

12.8. DIGESTIBILITY

A major biological method of evaluating feeds is digestibility, which is simply the difference between the amount an animal eats and the amount excreted in the feces (Schneider and Flatt, 1977). Thus, digestibility can be measured for the dry matter, the energy content, or any combination of nutrients.

In the typical "digestion trial" a group of animals are fed a given feed for at least several days in an adjustment period to accustom the animals to the feed. As weighed amounts of the experimental feed are continued for several additional days, the total quantity of feces excreted is collected. Both the feed and the feces are analyzed for the nutrient(s) being studied and the difference between the amount eaten and that in the feces is calculated. Usually the answer is expressed as a percent of the nutrient in the feed. The values obtained from the typical digestion experiment are generally referred to simply as the digestibility (Schneider and Flatt, 1977). However, these values are apparent digestibility values and not true digestibility. The true digestibility is the amount of any nutrient absorbed by the animal.

In two important respects, apparent and true digestibility are different (Schneider and Flatt, 1977). Often a substantial amount of many nutrients is reexcreted into the digestive tract for elimination via feces after absorption. The reexcretion usually is of endogenous (metabolic) origin (see Section 5.7.c). Another source of difference between true and apparent digestibility is the conversion of compounds from a form that appears in one fraction in the feed but in a different fraction in the feces. This occurs especially with empirical procedures. For example, a portion of the crude fiber of feed may be converted into substances that are determined as NFE or ether extract in feces.

12.8.a. Indirect Ways of Measuring Digestibility

Determination of digestibility in conventional total fecal collection digestion experiments is comparatively expensive and often restricted to animals fed in stalls. Thus the measurement can only be made on a limited number of feeds. Accordingly, simpler methods, including various indicator procedures have been used to estimate apparent digestibility. Most of these involve an indigestibility marker such as lignin, chromic oxide, or polyethylene glycol. With these methods, only samples of feces are needed, eliminating the requirement for total fecal collections. Unfortunately, even the indicator methods of digestibility determination require considerable analytical work and expense.

12.8.b. Digestibility of Feeds in Nylon Bags

Since much of the digestion of forages occurs in the rumen, scientists have devised systems of measuring the disappearance of dry matter from feed samples in the rumen. In a typical procedure, a substantial number of experimental forage samples in suitable bags are placed in the rumen for a definite time period. In many situations the amount of dry matter disappearing from the nylon bag is a useful index of the digestible energy value of the forage to dairy cattle (Neathery, 1972). Since such procedures are much simpler than the conventional digestibility determination, more samples can be evaluated. Although results are not directly translatable into digestion values, often they are a useful tool in research designed to improve forages.

12.8.c. Test Tube Digestibility Studies *(in Vitro)*

In a manner somewhat similar to that of determining digestibility in nylon bags, the digestibility of forage in test tubes or other glass containers using microbes from the rumen of cattle is sometimes measured. These procedures have applications, advantages and limitations similar to those of nylon bag digestion procedures.

12.9. BALANCE STUDIES

When more information is desired than just the digestibility value of feeds, often nutritional balance studies are used. There are many types and variations of balance techniques and experimental methods. The digestibility measurement is a partial balance in that the amount of any nutrient eaten but not appearing in the feces is determined. Nitrogen balance is a frequently used measure. In addition to the fecal nitrogen, that excreted in the urine and going into milk, if lactating

Fig. 12.1 Measuring energy metabolism and balance of dairy cattle in calorimetric studies is exceedingly complex and sophisticated work. In earlier years, this was often done with huge amounts of skilled labor. This shows two views from the USDA Energy Laboatory at Beltsville, Maryland in which much of the labor has been replaced with complex equipment, including computers to make the voluminous number of calculations. (Courtesy of W. P. Flatt, USDA and Univ. of Georgia.)

cows, are measured. Although there are other minor losses such as in hair, skin, and hoof sloughing, the nitrogen not excreted in the feces and urine or secreted into milk provides a reasonable indication of the nitrogen retained by the animal or the nitrogen balance.

The nitrogen balance often gives a better indication of the ability of feed to support growth or maintenance than does weight changes. This is especially true for short periods of time as the weights of cattle can change substantially due to differences in "fill," independently of true body tissues. The primary reason for large differences in fill are the variable amounts of material in the rumen. Often fill weight is closely related to the type of feed consumed. Although the nitrogen balance is a most useful way of evaluating adequacy of dietary protein for meeting the needs of animals, results are easily misinterpreted. For instance, many nutrient defiencies can reduce the amount of nitrogen retained. As illustrations, inadequate energy intake or a zinc deficiency will lower the nitrogen balance even though there is more than sufficient protein for the needs of the animal.

One of the most important types of balance studies is that related to energy. In Chapter 2 the various ways of measuring energy including gross energy, digestible energy, metabolizable energy, and net energy were discussed. These values are obtained from specialized balance and calorimetry studies. A number of highly sophisticated laboratories have been established throughout the world to study energy metabolism including the ARS-USDA Beltsville installation (Figure 12.1).

Balance studies for certain individual minerals are sometimes very useful in evaluating the adequacy of the mineral in the diet. This can be especially true if large amounts of the element are stored in the body tissue in forms that can be withdrawn to meet the needs for extended periods (see Section 5.7.d). Iron and calcium fit in this category. Unfortunately, technical difficulties often limit the usefulness of balances established by differences. In such situations, determining the content in the tissues of the animals or in representative samples can be very beneficial. For example, changes in the ash content of the bones are very helpful in evaluating the adequacy of calcium and phosphorus intake of cattle.

12.10. FEEDING EXPERIMENTS AND ANIMAL PERFORMANCE

12.10.a. Growth, Milk Production, Reproduction, Milk Composition, and Body Composition

The value of a feed is determined by the effect it has on performance of the animal. All the other ways of evaluating feeds are but attempts to more accurately or more easily predict the effects, including long-range changes, of feeds on animal performance. With dairy cattle the most important performance aspects are milk production, growth, reproduction, milk composition, body tissue composition, and health of the animals. The various types of performance experiments make up a major part of the total dairy cattle feeding and nutrition research. Usually, the experimental feeds, forages or combinations are fed to dairy cattle and the response of the cattle determined in terms of one or more performance measures.

Different feeds do not have the same relative value for all functions. For instance, a ration causing a high propionate to acetate ratio in the rumen is relatively more effective in fattening cattle than in the production of milk of a normal composition (see Sections 8.3.a, 8.3.b, and 8.3.d).

The variations and combinations in types of diets is infinite. Thus, to be highly useful, results of feeding experiments need to be understood in terms of what causes the performance differences among feeds. For instance, if milk production of cows fed a specific forage is low relative to other forages, it is important to know why. The answer to this often lies in an understanding of the nutrient(s) or factor that is inadequate (see Chapters 2–8). In many situations the key factor is the amount of digestible energy voluntarily consumed by the cow.

Voluntary dry matter intake is one of the most important characteristics of many forages. The measure of voluntary dry matter intake is the amount an animal will consume when the forage is available all the time or almost all the time. One of the major advantages of the Van Soest methods of analyses is the positive relationship between the amount of cell walls in a forage and the amount voluntarily consumed (see Section 12.5). This readily explains some of the earlier observed major differences in performance of cows fed forages such as coastal bermuda grass and alfalfa (see Section 9.3.a and Table 9.5). Although the differences in cell wall contents aid substantially in our understanding, research has not yet advanced to the point where accurate predictions can be made on the performance of dairy cattle from these analyses alone.

12.11. HEALTH-RELATED MEASUREMENTS

In all animals, feeding and nutrition have a very key role in maintenance of normal health. Obviously, less than normal health decreases performance of dairy cattle and usually reduces profits. Accordingly, even though dairy cattle feeding and nutrition are centered around providing the needed nutrients, the relationship to health is always important. The feeds of dairy cattle can contribute to less than optimum health either through nutrient deficiencies, toxicities, or metabolic disorders. Various deficiencies and excesses are discussed in Chapters 2 through 8. Some of the more important metabolic disorders and a few common toxic substances found in feeds are the subject of Chapter 18. Others are discussed in Chapters 5 and 11.

12.12. OTHER BIOCHEMICAL MEASUREMENTS

Frequently when a feed or diet has deficiencies or other undesirable properties, biochemical changes occur before performance is affected. For instance, liver and blood copper values reflect both an excess or a deficiency before gross symptoms are evident (see Section 5.24). Thus, biochemical measurements are often used in evaluation of feeds, especially when certain types of problems might be suspected.

Although highly useful in many situations, biochemical measurements are not available for diagnosing other types of feeding problems. Likewise, interpretation of results often is exceedingly complex. In recent years, the series of biochemical measurements have been used in diagnostic programs for practical dairy feeding problems. These have different names including metabolic profiles. Often the metabolic profile programs are intended to anticipate or diagnose the possibility of nutritional problems so that feeding changes or corrections can be made before animal performance is seriously affected.

The analyses for metabolic profile studies generally are made primarily on blood plasma or serum samples, although milk or urine samples also may be used. Likewise, feed analysis normally is involved. Usually the biochemical measurements on blood samples are those established as useful in diagnosing nutritional and/or metabolic abnormalities. Examples include several mineral elements such as calcium, phosphorus, magnesium, and copper. Other metabolities such as blood glucose, ketones, blood urea nitrogen, and hemoglobin are frequently determined.

One of the major difficulties in using biochemical measurements in diagnosing practical feeding problems is in establishing normal values for the specific farm conditions. This is especially true where the objective is to anticipate problems before performance is seriously affected. Many biochemical changes can be readily demonstrated with specific deficiencies in carefully controlled experimental situations. Unfortunately, often the same biochemical measures are greatly affected by many other normal conditions such as climate, other feed ingredients, and physiological changes. To overcome the difficulty of establishing normal values, these are sometimes developed for certain specific areas, conditions, or even individual farms. Even so, the best of analytical programs leave much to be desired in characterizing the nutritional status of the dairy herd.

12.13. SPECIALTY AND MISCELLANEOUS FEED EVALUATION METHODS

A complete presentation of the useful methodology for evaluating feeds for dairy cattle is beyond the scope of this book. However, it is important to recognize that this is a constantly expanding field with research workers always seeking more effective and simpler ways of evaluating feed.

Useful methods for determining the value of feed for dairy cattle not mentioned above vary from the simple to the very complex. For instance, visual and sensory inspection of feed ingredients is most useful. Such aspects as deterioration due to weather or storage damage, often can be easily detected. Problems such as mold, browning or blackening of forages due to heat damage, loss of green color in hay, and others, frequently are easily observed. Likewise information on the stage of forage maturity, poorly developed grain, rancidity of some feeds, poorly preserved silage, and similar problems often are readily detected by careful inspection. Often dairy cattle nutrition researchers use very sophisticated methods to obtain information needed in evaluating feeds and nutrient requirements.

12.14. RADIOISOTOPES: POWERFUL RESEARCH TOOLS

One of the most effective and powerful tools available to nutrition research workers is the use of isotopes. Although there are innumerable other applications, the place of isotopes can be illustrated by their role in determining the value of certain mineral elements in feeds. With an element such as calcium, determining the apparent digestibility in a feed (see Section 12.8) has little meaning because so large a portion of the element in the feces is of endogenous origin and because the apparent absorption is so closely related to the needs of the animal (see Section 5.16). Without the use of isotopes, it is not readily possible to determine the amount of calcium in a feed which can be absorbed and utilized. In contrast, absorption of radioactive calcium can be easily determined experimentally after incorporating it into the feed ingredient. This is the most effective way of establishing the availability of several elements in feeds.

12.15. FORAGE AND FEED EVALUATING PROGRAMS

The ultimate objective in evaluating feeds for dairy cattle is to meet the nutrient requirements more effectively and/or at a lower cost. For this objective to be achieved, the results must be applied on individual farms. Many of the various procedures and methods discussed earlier in this chapter are research tools. The information obtained, however, has had and will continue to have considerable effect on dairy cattle feeding practices. Often these applications primarily affect how concentrate mixtures are formulated over wide areas, how

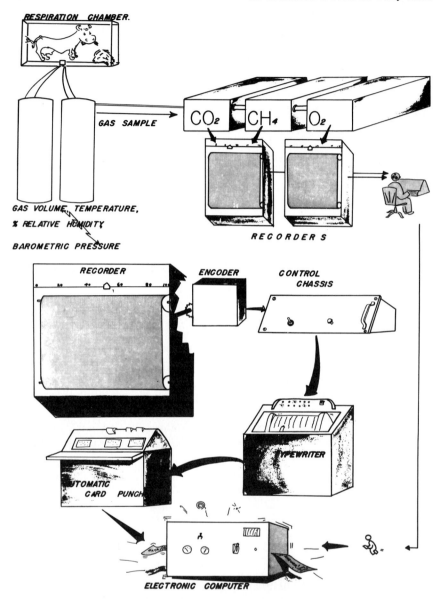

forage crops are produced and used, and influence other aspects resulting from general recommendations and information.

As discussed in Chapters 5, 9, and 10, often the variability in nutrient content of feeds, especially forages is enormous, even within the same species. Accordingly, dairy cattle nutritionists have long recognized the need for simple and effective ways of determining the nutrient content of individual lots of feeds.

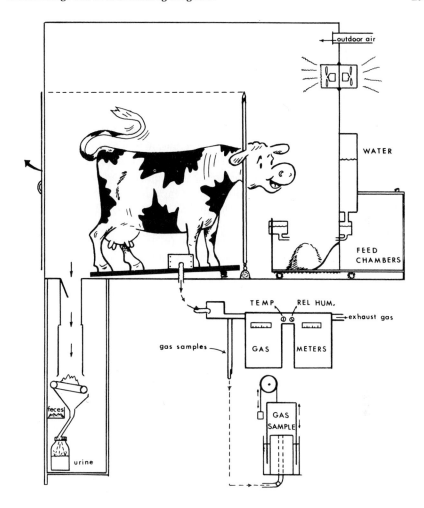

WATER

FEED
CHAMBERS

outdoor air

TEMP.   REL HUM.

exhaust gas

gas samples

GAS   METERS

GAS
SAMPLE

feces

urine

OPEN CIRCUIT RESPIRATION CHAMBER

Fig. 12.2 These two illustrations depict in simplified form the procedures used in energy metabolism studies at the USDA Energy Laboratory at Beltsville, Maryland. (Courtesy of W. P. Flatt, Univ. of Georgia.)

Beginning in the late 1950's, many states and other organizations have developed forage and feed testing programs with at least 48 of the states in the United States having them available (Coppock, 1976).

Although it is well recognized that the answers obtained from forage testing are not always highly precise, the practical benefits to dairymen have been considerable. In any forage or feed testing program, a representative sample is crucial for the results to have meaning. Thus, most of the analyses are on stored feeds such as hays or silages. Especially for silages, it is desirable to obtain samples as the forage is harvested. Otherwise, the material accessible for sampling may not be representative of that which will be fed when the results are received.

One of the primary purposes of forage testing is to more accurately supplement the forages with concentrates (see Section 13.2). But there are other important benefits. These programs are an excellent educational tool for many people including extension workers, feed company representatives and dairymen. Likewise, the data obtained are quite useful to research workers.

Answers desired from forage testing services include (1) the nutrient composition needed in the concentrate and (2) the amount which should be fed. Generally determining the composition needed is easier than establishing the ideal amount to feed: chemical analyses are less effective in predicting voluntary dry matter intake than dry matter digestibility. Fortunately, moderate deviations from the ideal amounts of concentrates to feed do not have a major impact on profits (see Section 13.2).

Because they are the major cost items, in forage testing programs, most emphasis has been devoted to protein and energy. Usually determining the crude protein content of a forage is fairly simple and reliable (see Section 12.2.b and Chapter 3). Because they are very closely related, usually digestible crude protein is easily and reliably calculated from crude protein determinations (Coppock, 1976). In most of the forage testing programs values have been expressed as digestible protein. In keeping with recommendations of the National Academy of Science (National Research Council, 1978), protein requirements are expressed in terms of total crude protein; thus no conversion is needed (see Sections 3.1 and 3.2). When proteins in forages are "heat damaged" the value is reduced (see Section 3.10). Thus this should be considered in practical feeding. Several states provide analyses for available protein using either acid detergent insoluble nitrogen or pepsin insoluble nitrogen analyses to indicate the degree of heat damage (Coppock, 1976) (see Section 3.11).

As discussed more fully earlier in this chapter, establishing the usable energy value of a feed from simple chemical analyses is much more difficult than for protein. No one ideal approach for all situations has been developed. Accordingly, a number of systems have been used. The Penn State Forage Testing Service was one of the first widely used systems of forage evaluation for application to individual farms. Initially, samples were analyzed for crude fiber with the values used in regression equations to calculate TDN. Although the regression formulas have varied considerably, often depending on the particular forages and local conditions, the crude fiber content has been widely used to calculate TDN, ENE (estimated net energy), or some other measure of usable energy value. For specific species and conditions, energy values obtained this way have been useful. However, as discussed in Sections 12.5 and 12.5.a, the Van Soest methods of dividing carbohydrates into cell contents and cell wall components appears to be a more effective way for use with many different types of forages. Several forage testing programs determine acid detergent fiber (Coppock, 1976). For first cutting of forages, especially in the more temperate parts of the United States, cutting date has been used as an effective way of estimating the energy value of forages (see Section 9.2.c).

In some of the forage testing programs, the mineral content of forages and/or other feeds are determined. These results reveal a tremendous variability in the mineral content of the same forage or grain on different farms (Adams, 1975; Miller, 1975). Because of its great importance, it is expected that research related to more effective forage testing procedures and applications will continue. Likewise, the procedures will continue to be changed, modified, improved, and refined.

REFERENCES

Adams, R. S. (1975). *J. Dairy Sci.* **58,** 1538–1548.
Coppock, C. E. (1976). *J. Dairy Sci.* **59,** 175–181.
Cullison, A. E. (1975). "Feeds and Feeding." Reston Publ. Co., Reston, Virginia.
Goering, H. K., and P. J. Van Soest (1970). *U.S., Dep. Agric., Agric. Handb.* **379.**
Miller, W. J. (1975). *J. Dairy Sci.* **58,** 1549–1560.
Miller, W. J., J. W. Lassiter, and J. B. Jones (1972). *Proc. Ga. Nutr. Conf.* pp. 94–106.
National Research Council (NRC) (1978). "Nutrient Requirements of Dairy Cattle," 5th rev. ed. Natl. Acad. Sci., Washington, D.C.
Neathery, M. W. (1972). *J. Anim. Sci.* **34,** 1075–1084.
Schneider, B. H., and W. P. Flatt (1977). "The Evaluation of Feeds through Digestibility Experiments." Academic Press, New York.
Underwood, E. J. (1970). *Trace Elem. Metab. Anim., Proc. WAAP/IBP Int. Symp., 1969* pp. 5–21.
Van Soest, P. J. (1964). *J. Anim. Sci.* **23,** 838–845.
Van Soest, P. J. (1966). *J. Assoc. Off. Anal. Chem.* **49,** 546–551.
Van Soest, P. J. (1967). *J. Anim. Sci.* **26,** 119–128.
Van Soest, P. J. (1971). *N.Y. Food Life Sci. Bull.* **4,** No. 4, 4–7.
Waldo, D. R. (1977). *Proc. Res. Ind. Conf. Am. Forage Grass. Counc., 10th, 1977* pp. 69–91.

13

Feeding the Milking Herd

13.0. AN OVERALL PLAN

In earlier chapters, consideration was given to how dairy cattle use nutrients, the needed nutrients, feeds as sources of nutrients, and the evaluation of feeds. All of this information is useful in planning practical feeding programs for different types of dairy cattle and varying farm conditions. More than two-thirds of the total feed used on dairy farms is for the milking herd. In several respects, feeding the lactating dairy cow is different from any other farm animal. Good nutrition for the milking herd is an indispensible part of successful dairy farming, and a comprehensive overall plan to provide the needed nutrients from feeds that are sufficiently palatable to obtain adequate consumption is important. The feeds are needed every day of the year, and the cost must be reasonable if the operation is to be adequately profitable.

The total amount of feeds required for a moderate sized dairy herd during a year is huge. A typical 1400-lb cow will eat about 45 lb of air-dry feed each day. For a 100-cow milking herd this is about 2.25 tons daily or 820 tons per year. If a part of the feed is silage, pasture, or green chopped forage, the total weight involved is even greater.

Most of the best practical feeding programs for milking herds are based on the use of large amounts of high quality forages (see Chapter 9). Because high-producing cows usually cannot consume sufficient forage to fully meet their energy requirements, supplemental concentrates are needed. Since forages often are deficient in one or more other needed nutrients, the concentrate mixture should be formulated to make up any deficiencies.

13.1. PROVIDING FORAGE FOR THE MILKING HERD

In most dairy areas of the United States providing an adequate supply of high quality forage at a reasonable cost is the most important single factor in the success of a dairy farming operation. The best overall way to supply high quality forages to the milking herd varies from region to region and on individual farms within a region (see Chapters 9 and 19). Since forages generally are a more economical source of nutrients than concentrates, usually they should be fed to the maximum feasible extent.

The degree to which lactating cows can utilize forages to meet their nutrient requirements varies with several factors. Higher quality, more digestible forages are voluntarily consumed in larger amounts and supply a higher proportion of the total nutrients needed (see Sections 9.2 and 12.10). The "quality" factor receiving first consideration in planning a forage program is the ability of the forage to meet the energy and protein needs of the cows (see Chapter 9).

Although many other nutrients are just as essential as energy and protein, the cost of providing the others in supplemental feeds is much smaller. Thus, usually the practical approach is to develop the forage system on the basis of the energy and protein quality factors along with cost considerations. The most practical system of providing forages for the milking herd often involves different types of forages during the year. Long range planning is needed to insure high quality forage at all times.

13.2. PROVIDING CONCENTRATES AND SUPPLEMENTS FOR THE MILKING HERD

After the forage program is established, concentrates and supplemental feeds should be formulated to furnish any nutrient inadequacies in the forages. Whereas, selecting forages for the milking herd usually should be a long-range program, both the composition and the amount of concentrates fed should be tailored to best fit short-range conditions. The short-range conditions refer specifically to the current milk production level plus the amount and quality of forage being fed (see Sections 19.3 and 19.3.d).

Since concentrates are formulated to supply the required nutrients not obtained from the forage, the composition needed is dependent on the forage. Thus two basic decisions are needed for the concentrate program. First, the amount to be fed should be established and then the composition of the concentrate determined. In the next several paragraphs, the logical steps by which these decisions are made are discussed. The general procedure is applicable to most situations starting from "scratch." An understanding of the approach also is useful in using various "thumb-rule" or "short-cut" methods which often are employed for making practical decisions.

13.2.a. General Approach for Concentrate Feeding Programs

The quantity of concentrate needed for maximum profits usually is determined by the amount needed to supply an optimum energy intake. This is influenced by the relative prices of concentrates, forages, and milk (Bath, 1975). For instance, when concentrates are quite expensive relative to milk, maximum profits may be attained with less concentrate than when the situation is reversed. The amount of concentrates necessary for optimum energy intake is greatly affected by the forage fed (see Chapter 9). Wide differences exist in the ability of different forages to supply the energy needs of milking cows with the amount voluntarily eaten varying more than the digestible energy content (see Chapter 9 and Section 12.10).

In determining the quantity of concentrates needed, the total usable energy obtained from the forage is subtracted from the total energy requirements of the cows. The difference is the amount needed from concentrates. When the amount of energy needed has been established, the number of pounds of concentrates involved can be determined from the energy content of the concentrate mixture involved. For nutrients other than energy, the needed composition of the concentrate mixture can be calculated by subtracting the amount of each nutrient in the forage from the total amount needed by the cow. The difference is the quantity of the nutrient needed in the concentrate. This can be readily translated into a percentage (or other suitable units such as ppm) by dividing the needed amount by the pounds of concentrates to be fed.

When a procedure similar to the above has been followed for each essential nutrient, the needed nutrient composition of the concentrate will have been established. Because many different combinations of concentrate ingredients would provide the needed nutrients, the formulator has considerable choice. The actual ingredients chosen from those available should minimize costs, but often are not least-cost rations. Aspects other than cost are important as discussed in subsequent sections of this chapter and in Chapter 19. A detailed discussion of basic procedures and calculations in balancing rations are presented in the book by Cullison (1975) and in the publication on Nutrient Requirements of Dairy Cattle (National Research Council, 1978).

It is not practical to formulate concentrate mixtures for individual cows. Rather only one or, in the case of large farms, only a very small number of mixtures are used. These are formulated after the forage available and the general level of concentrate feeding has been selected for the herd or the group.

13.2.b. Formulating Concentrates for Lactating Cows—Shortcuts

Generally, the procedures described in the previous section for determining the amount and composition of concentrates that should be fed are not followed in detail. In effect such calculations often are made on a background basis and converted into various shortcut methods for routine use. For example, the nutrient composition of the total feed dry matter for cows at various levels of production is given in the Appendix, Table 3 (NRC, 1978).

Routinely, except for the protein content, often the content of other essential nutrients in concentrates is formulated from background information to fit most situations with the major types of forage feeding programs for the area. The standards to be used may be developed by individual companies or may be obtained from general recommendations of professional dairy cattle nutritionists. Thus in practice, one of the major shortcuts is following a selected standard of composition for most of the essential nutrients. Often there are further shortcuts in formulation for many of the essential nutrients. For instance, frequently the supplemental trace element is provided by a trace mineralized salt or mineral mixture fed either free choice or included as part of the concentrate or complete feed. Such shortcut procedures give quite satisfactory results in many situations, but those not based on sound information may not.

When fed free choice, the amounts of mineral elements consumed are not closely related to the animals' needs (see Section 5.6.e) (Coppock *et al.,* 1976). Thus combining minerals with other feeds is desirable. If this is not feasible, feeding a mineral or trace mineral mixture is more satisfactory than feeding individual element sources free choice.

13.2.c. Determining the Amount of Concentrates to Feed—Complete Feeds vs. Separate Feeding of Forage and Concentrates

In the routine feeding of lactating dairy cows, several shortcut approaches can be used for determining the amount of concentrates needed. An approach with many advantages is to feed a mixture of concentrate and forage, especially silage, at a given ratio. It is sometimes called "complete feed," "optimum ration" or "optimum feeding" (Coppock, 1977; McCullough, 1973; Spahr, 1977) (see Sections 13.6.b and 19.4.b). This method is quite effective and has important

Fig. 13.1 In recent years, group feeding of a complete ration has been widely used on United States dairy farms. (Courtesy of W. J. Miller, Univ. of Georgia.)

advantages for silage feeding or other forms of forage that are readily mixed with concentrates at the time of feeding (Fig. 13.1 and Table 13.1). Usually it is not feasible with hay, pasture, or other forms of forage not easily mixed with concentrates. Complete feeds are best adapted to large herds divided into 3 or more groups of lactating cows (Table 13.1). More productive cows generally require a higher ratio of concentrates to forage to meet their energy needs. This can be achieved by having a minimum of two, and preferably three or four groups of cows receiving different ratios of forage to concentrates. Assignment to the groups should be based on stage of lactation as well as level of production with fresh cows included in the highest production group. Feeding forage and concentrates as a mixture can reduce the probability of digestive problems such as founder (see Section 18.8). Especially dry cows, late lactation cows, and low producers should be fed a smaller preparation of concentrates to forage than high producers. Otherwise some will be overfed and others underfed. Underfed animals produce less milk, whereas overfed ones become excessively fat (see Section 18.12). Although most complete feeds are composed of concentrates and silage, some involve a dry forage or roughage. These have been called all-in-one rations. Some high-fiber feed or a combination such as hay pellets, cottonseed hulls, oat hulls, or coarsely ground corn cobs are used as a source of fiber. Although high in fiber, soybean hulls will not meet the fiber requirement because of their high digestibility.

In earlier years, most recommended procedures for allotting concentrates to

TABLE 13.1

Advantages and Disadvantages of Complete Feeds vs. Separate Feeding of Concentrates and Forages[a]

Advantages of complete feeds	Disadvantages and limitations of complete feeds
Cows cannot select an unbalanced ration (all cows in a group receive the same diet)	Not readily adaptable to pasture or long hay feeding
NPN[b] and other nutrients are available throughout the day giving more even rumen fermentation	Probably not economically feasible in small herds
Can mask unpalatable flavors and ingredients	Well adapted only where herd is divided into groups
Easier to make rapid changes in formulations	Mixer wagons and other equipment are expensive
Fewer digestive upsets	Often not easily adapted to old barns and facilities
Free-choice minerals not needed	
Unnecessary to feed concentrate in milking parlor	
Often reduced labor needs due to increased mechanization	

[a] Adapted, in part, from Coppock (1977).
[b] Nonprotein nitrogen.

dairy cows were based on individually feeding each cow. With higher labor wage rates and larger herds, this has become much less practical. Group feeding of concentrates is much more widely used. Many small herds, and others for different reasons, are not readily adapted to use of complete feeds and grouping according to nutrient needs. In these situations variable amounts of concentrates can be fed in stanchion barns. To reduce labor and other costs, many dairy farms keep cows in loose housing systems which makes individual feeding impractical. If the lactating cows are kept in one herd, various magnetic or electronically controlled feeders can be used to give high-producing cows extra grain. Until recently, it was generally accepted that each cow should be fed the exact amount of concentrates needed. Careful analyses of milk production response curves for different levels of concentrates show that feeding somewhat more or somewhat less than an ideal amount has only a minor effect on net profits. Accordingly, a high degree of precision in allotting amounts of concentrates is not crucial to good dairy management.

Frequently, a specific ratio of concentrates is fed for a given quantity of milk. This can be done by some automatic systems of feeding in milking parlors in which concentrates are released mechanically as milk is removed from the udder. The biggest disadvantage of this approach is the cows eat only about 0.7 lb per minute and thus usually are through milking before they are through eating the needed concentrates. Where cows are fed individually, most of the decisions are made from the use of tables which consider the important factors affecting the concentrates needed. These factors are: (1) the forage quality, (2) fat content of the milk, (3) the level of milk production, (4) whether the cow needs additional concentrates for growth, and (5) the relative price of milk, forages, and concentrates. Essentially the same factors are involved if cows are to be group fed.

13.3. THE IMPORTANCE OF HIGH PRODUCTION

When evaluating the success of a dairy farming operation, one of the key measures is level of milk produced per cow. Although other factors play a role, many studies have demonstrated the close relationship between the milk yield per cow and the total profits from dairy farms (see Section 1.0.e). This high correlation is not accidental. When the cost items for milk production are reviewed, the reasons are apparent. A substantial percentage of the milk production costs is fixed and not influenced by the level of milk produced. For instance, the investment and interest costs in such items as land, buildings, milking and feeding equipment, waste disposal facilities, and cows are essentially the same regardless of milk production level. Labor costs, property taxes and depreciation of cows, buildings, and equipment are not greatly affected by the milk yield per cow. If only the nutrients needed for milk production are considered, the amounts used are almost proportional to the milk production level. But the total feed needs per pound of milk are much lower with higher production because of the nutrients required for maintenance and growing of replacements (see Sections 1.6.a and 2.2.a). For example, a 1400-lb cow producing 20 lb of 3.5% fat milk per day requires 9.82 lb TDN for maintenance and 6.08 lb for milk production or a total of 15.9 which equals 0.8 lb TDN per pound of milk. However, if she produces 100 lb of milk, the total TDN requirement is 40.2 lb (9.82 for maintenance and 30.4 for the milk) and equal to only 0.4 lb per pound of milk. When the total feed used on a dairy farm is considered, the proportion going directly for milk production is quite modest. Per pound of milk, much less total feed is required on farms having a high production per cow. Considering the much lower fixed cost in nonfeed items and the smaller amount of feed used per pound of milk, it is very evident that higher production per cow is a tremendous cost-reducing factor.

13.4. UNDERSTANDING THE LACTATION CURVE: MILK SECRETION UNDER HORMONAL CONTROL

Although this book gives only limited attention to the important nonnutritional factors in milk production, some understanding of other factors is necessary in planning a practical feeding program. Various hormones have a major influence on milk production. Normally lactation is initiated in cows with the birth of a calf. In the typical lactation, milk production will increase for approximately six weeks after calving when the peak level is reached. A slow but steady decline generally occurs until approximately two months prior to the next calving when the cow is turned dry. This normal lactation curve usually represents the maximum practical amount of milk attainable. If the level of milk production during the early part of the lactation is reduced materially below the normal potential, the yield during the remainder of the lactation will be adversely affected even though the reason is corrected.

Because of the adverse effects on subsequent milk production, generally it is not profitable to underfeed cows during early lactation, even when relative milk and feed prices temporarily may not justify liberal feeding of concentrates. The effect on subsequent yield in the lactation is a key reason why adequate nutrients are needed every day. Growing animals, through ''compensatory growth,'' can make up for short periods of underfeeding with little or no permanent adverse effects, but such is not true for lactating cows.

During early lactation, the hormonal stimulus to secrete milk is very strong. Thus, a cow that is underfed, especially on some nutrients such as energy (see Chapter 2), will use body reserves to produce milk. Even though a moderately inadequate feeding of energy may have a relatively small immediate effect on milk production, excessive use of body reserves can greatly lower milk production in later lactation.

Underfeeding during late lactation has much less effect than during early or mid-lactation. Only the remainder of the lactation would have to be affected as body reserves could be replenished during the dry period. Adequately feeding cows during the first few months of the lactation is much more demanding and important than later.

13.5. THE IMPORTANCE OF COST CONTROL

Although dairy farming has advantages not associated with many types of enterprises, milk production is basically a business conducted for a profit. It is a "commodity type" business which, among other economic characteristics, includes a high degree of price competition. To be successful, production costs must be competitive. Feed may represent 70% or more of the total costs of producing milk (Smith, 1976).

Income over feed cost is a frequently used index of efficiency in milk production (Bath, 1975). The level of production greatly affects income over feed cost (see Section 13.3).

13.5.a. Feeding the Needed Nutrients at Lowest Cost

Except for energy during some periods and to a limited extent for protein, generally it is most profitable to supply each required nutrient in amounts sufficient to permit maximum milk production. Because of the relatively small increase in milk produced per unit increase in usable energy when the maximum production is approached, especially in mid- and late lactation, usually it is not advisable to supply adequate usable energy for maximum production. Feeding sufficient energy for maximum production, in the short run in these periods, may result in excessively fat cows later. Overfat cows have higher maintenance requirements and more metabolic and health problems (Chapter 18). Since the cost of nutrients from different feed sources varies widely, the most practical and profitable feeding program depends on the specific situation and conditions.

13.5.b. Least-Cost Rations

The concept of least-cost rations has long been used as a way to provide the needed nutrients at lowest cost (Bath, 1975). Most often it is applied to the formulation of the concentrate portion of the ration to furnish the desired nutrient composition at the lowest total cost. Especially in large operations, such as commercial feed mills, the formulation may be altered frequently as ingredient prices change. Often the formula is calculated by computer (Bath, 1975) and is sometimes called linear programming (see Section 19.5.a). Although the computer is very useful, competent feed formulators have long achieved almost the same objective by simpler shortcut methods.

The least-cost approach also can be applied to the total feeding program (Bath, 1975) involving both the forage and concentrate and even the cost of physically feeding the animals. A part of the long-range objective is to provide the needed nutrients at the lowest total cost.

13.5.c. Maximum Profit

Maximum profits, the objective in feeding dairy cows, generally correlates quite well both with the highest production of milk per cow and with providing required nutrients at least cost. Since neither correlation is perfect, a compromise between highest milk production per cow and lowest cost of nutrients will achieve the largest net profit. Perhaps an illustration will aid in understanding the concept of maximum net profit. A part of the protein needs of dairy cattle can be met by the use of urea and grains as a substitute for plant proteins (Chapter 4). Results of many experiments suggest that usually there is a small reduction in the level of milk produced. Using the maximum net profit concept, if the cost savings from the use of urea is substantial and the reduction in milk produced small, net profits will be higher with the urea.

For the highly skilled practicing dairyman or dairy cattle nutritionist, many examples similar to that for urea are followed. Although any one of these choices may have only a small effect of net profits, with a substantial number, the total effect may be appreciable.

13.6. MANAGING THE FEEDING PROGRAM

Although most of this book is devoted to the nutritional aspect of the feeding programs on dairy farms, obtaining the feeds and physically making them available to the cattle is a huge task. As discussed in Section 13.0, a typical 100-cow dairy herd will consume about 820 tons of air-dry feed equivalent per year.

In earlier years, a tremendous amount of human labor was involved in obtaining feed and giving it to dairy cows. More recently there has been a constant increase in automating the handling of feeds for dairy cows. Developments continue to make new innovations practical. For instance, the mixing of silages and concentrates into a single complete feed or "optimum ration" would not have been feasible without considerable mechanization. In the overall management of the feeding program, buying and/or producing the needed feeds or ingredients, storing the feed until needed, physically feeding the animal, as well as the business aspects of arranging financing, facilities, etc., are important in the success of the dairy operation (see Chapter 19).

13.6.a. Feeding Facilities and Equipment

A detailed discussion of the different types of feeding equipment and facilities is beyond the scope of this book. Even so, it is important to recognize that physically giving the desired feed to dairy cattle is a key part of practical feeding and nutrition and that the effectiveness and costs affect net profits from the dairy operations.

Throughout most of the world, the cost of labor relative to milk and other items has advanced for many years along with vast improvements in mechanization. Consequently, feed handling has become much more mechanized with a great reduction in labor needs. With these changes, investment and capital costs have risen sharply. In some instances, such investments have not given satisfactory results due either to excessive costs, poor design and performance, or rapid obsolescence. In planning new feed storage and handling equipment and facilities, it is important to obtain highly competent advice and information. In many states, such assistance is available from Cooperative Extension Service personnel.

13.6.b. Feeding Procedures; Mechanics and Frequency of Feeding

Concentrates and forage can be fed separately, mixed, or some combination or modification. For instance, cows may be fed a mixture of silage and concentrates and a limited amount of pasture, or silage may be fed with concentrates put on top along with hay in a self-feeding rack. Whatever the mechanical procedure used to feed cows, a few basic aspects are needed for most efficient production. If the objective is to use maximum forage, the best intake is usually obtained when the forage is readily available to the cows throughout much of the 24-hour day or at least for several fairly long periods. Second, the feed should be protected from deterioration due to weather or other adverse factors until it is consumed. Third, adequate "feed bunker space" should be provided so that each cow can eat all the forage she cares to consume. Water should be convenient so that intake will not be limited; it should be available in both feeding and shelter areas.

Some research suggests that more efficient rumen fermentation may be achieved by mixing the concentrates and forages prior to feeding (McCullough, 1973). Although this results in more uniform rumen fermentation and definite advantages in some situations, in many instances separate feedings of concentrates appears to give just as good performance (see Sections 19.4.a, 19.4.b).

13.6.c. Making Changes in the Feeding Program

A key aspect of providing the nutrients for dairy cattle at a low cost and for maximum net profit is changing feed ingredients as availability and/or prices change. Although considerable detailed knowledge is helpful in obtaining most efficient overall production, a number of general principles are helpful. Generally, it is desirable to make changes gradually. Usually no metabolic problems result from a sudden increase in the amount of forages fed relative to concentrates; but underfeeding may occur with too little concentrate (Hernandez-Urdaneta *et al.*, 1976; Moseley *et al.*, 1976). In contrast, a sudden large increase in the proportion of concentrate in unadapted cows may cause serious metabolic problems. The extent of the problem is greatly influenced by the ingredients in the concentrates (see Section 18.8). Since nonprotein nitrogen is not well utilized in unadapted cattle, changes from plant proteins to urea should be made gradually. In forage feeding, sudden changes from pasture to stored feeds such as hay or silage usually does not cause serious problems. However, a sudden change from total stored forage to all pasture may lead to bloat, diarrhea, or other problems such as grass tetany. Thus it is advisable to make the change over at least a few days.

When substituting one ingredient for another in the concentrate mixture, a palatability problem may result if a very high percentage of the total concentrate is involved and/or if the new ingredient is not very palatable under the best of conditions. Many ingredients are not efficiently utilized when they make up a too high proportion of the total concentrate fed, especially when a high level of concentrates are fed. Molasses is a good example. Other ingredients such as citrus and beet pulp can be fed at a very high level with no adverse effect. If palatable high quality ingredients are used, generally no advantage in cow performance is achieved from a complex concentrate mixture relative to one composed of very few ingredients. Often the major advantage of the more complex mixture is the ability to utilize some lower cost ingredients without adverse effects on performance.

13.7. PREVENTING AND/OR CORRECTING FEEDING PROBLEMS—QUALITY CONTROL

In feeding the milking herd, it is more desirable and more profitable to prevent problems than to have them arise and need correcting. If problems develop, the sooner they are recognized, diagnosed, and corrected the smaller the adverse effects on profits.

In feeding the milking herd, problems may come from a variety of causes including nutrient deficiencies or excesses, palatability, toxic materials or substances, or deterioration of the feed before consumption. Some problems may be economic, managerial, or mechanical in nature. Examples would include excessive feed wastage, insufficient supply of needed ingredients, too high a cost of some ingredients or feeds, breakdown in the feeding equipment, or other aspects resulting in feed not being available to the cows when needed.

Careful planning should prevent most problems before they arise. Proper evaluation of the feeds used along with balancing of the rations to make up all deficiencies should avoid most of the nutritional and biological problems. Adequately planned, designed, and maintained equipment and facilities should avoid most of the mechanical problems. Similarly, economic planning, along with sound procurement, and/or storage of feeds or ingredients should avoid most of the economic problems. Adaptation of feeding plans to specific farm conditions including such questions as the growing of feeds versus purchase is discussed in more detail in Chapter 19.

13.7.a. Evaluation of Feeds for the Milking Herd

The evaluation of feeds for dairy cattle is discussed in some detail in Chapter 12. From the standpoint of the milking herd, to prevent or correct feeding problems, two general approaches are pertinent. The first need in planning the feeding program is a knowledge of the nutritional and general characteristics of the types of feeds and ingredients to be used. In choosing the feeds, both the nutritional and economic factors should be weighed to obtain the most profitable program. Second, evaluation of the nutrient content of the specific lots of feeds is important (see Sections 9.2 to 9.2.e). This is the basic purpose of forage and feed testing programs conducted by Cooperative Extension Services or other organizations for individual farms. If the nutrient content of the individual forage being fed on a farm is not considered in balancing the ration, problems often arise. For instance, protein deficiency may result if the content is below average when only book values are used in the planning. On the other hand, if the protein content of the specific forage is unusually high, failure to consider this may result in unnecessary expense for protein in the concentrate.

All the information in the earlier chapters concerning the nutrient requirements of lactating cows as well as that on feeds as sources of nutrients and on evaluation of feeds is pertinent to evaluating feeds for the milking herd. However, in any one situation only a few of the possible evaluation procedures would be indicated.

13.7.b. Balancing Rations to Prevent or Correct Problems

If a good forage program has been combined with concentrate feeding to provide any nutrients not supplied by the forage, nutritional deficiencies, excesses, or deficiencies should rarely occur. However, some nutritional or metabolic disorders may develop in well-executed programs (see Chapter 18). Likewise, with sound planning and good execution, problems sometimes arise because ingredients are unusually low or high in certain nutrients without being suspected. As discussed in several earlier sections, the knowledge of many aspects is quite incomplete. Thus much needed research remains to be done. When unusual health problems arise or when production is lower than expected, one of the first approaches is to recheck all aspects of the program for balancing the rations for essential nutrients (see Section 13.7.d).

13.7.c. Management Control

In preventing or correcting feeding problems, a good knowledge of results and effective managerial control are essential. Good records are important. Since one of the most sensitive indications of nutritional problems is reduced production, some form of "testing" cows for milk yield is essential. Likewise, good records on milk composition, reproduction, and other aspects of individual and herd health are needed. These records should be reviewed frequently to spot the possibility of problems before they seriously affect profits. Likewise, information on the availability and costs or prices of feed supplies should be followed so adjustments can be made as needed.

13.7.d. Trouble Shooting

Even with careful planning and good management, problems arise in feeding a milking herd. Some are relatively easily solved, but others are perplexing. Unfortunately, when a difficult problem arises in a dairy herd, "no one puts up a sign" indicating whether it is due to the feed or to some other cause such as an infection. In these situations the best of professional assistance is often needed. If the problem is especially difficult, none of the ablest specialists may find it easily. In such cases, each professional may suggest that it is not in his area of expertise. For instance, if the dairy cattle nutritionist cannot find a cause, he may feel that the problem probably is infectious. In contrast, when the veterinarian can find no other cause, he may suggest that something is wrong with the feed.

In the typical milking herd, the greatest losses due to less than the best feeding programs are caused by small unrecognized and unexpected reductions in performance. It is the old story that no one rings a bell to indicate that a problem is about to arise that will reduce performance. In most situations, the best trouble shooting requires a thorough knowledge and careful attention to important details. Obviously, a manager cannot be concerned with every possible problem at all times. Thus it is important to know what types of problems are likely with the specific feeding program being used.

REFERENCES

Bath, D. L. (1975). *J. Dairy Sci.* **58**, 226–230.

Coppock, C. E. (1977). *J. Dairy Sci.* **60**, 1327–1336.

Coppock, C. E., R. W. Everett and R. L. Belyea (1976). *J. Dairy Sci.* **59**, 571–580.

Cullison, A. E. (1975). "Feeds and Feeding." Reston Publ. Co., Reston, Virginia.

Hernandez-Urdaneta, A., C. E. Coppock, R. E. McDowell, D. Gianola, and N. E. Smith (1976). *J. Dairy Sci.* **59**, 695–707.

McCullough, M. E. (1973). "Optimum Feeding of Dairy Animals for Meat and Milk." Univ. of Georgia Press, Athens.

Moseley, J. E., C. E. Coppock, and G. B. Lake (1976). *J. Dairy Sci.* **59**, 1471–1483.

National Research Council (NRC) (1978). "Nutrient Requirements of Dairy Cattle," 5th rev. ed. Natl. Acad. Sci., Washington, D.C.

Smith, N. E. (1976). *J. Dairy Sci.* **59**, 1193–1199.

Spahr, S. L. (1977). *J. Dairy Sci.* **60**, 1337–1344.

14

Feeding and Raising the Young Dairy Calf

14.0. INTRODUCTION

Dairy cows stay in the herd, on an average, for only about four lactations. When death and other losses of calves are considered, about three-fourths of the heifer calves born must be raised to maintain equal cow numbers. Raising all the heifer calves gives more opportunity for culling on the basis of desirable characteristics and generally is a very profitable practice. In most dairy areas there are several advantages to raising herd replacements rather than purchasing them. These advantages include more opportunity for genetic improvement, better disease control, elimination of dependency on outside sources of cows with accompanying transportation and other added costs, and usually a lower cost for replacements.

Feeding young dairy calves during the first three months is a relatively small proportion of the total feed cost on a dairy farm. Because it does not have a functional rumen, the newborn calf has very different nutrient requirements from older dairy cattle (see Chapter 1). Its nutrient needs are more similar to those of simple stomached animals than to older cattle. The newborn calf is a baby which presents many nutritional and management problems including a much greater tendency to succumb to infections and other problems that may be fatal.

14.1. LOW MORTALITY: THE CENTRAL AND MOST IMPORTANT OBJECTIVE

Although a good calf feeding and raising program should accomplish several objectives, no program which results in high death losses is satisfactory. The percentage of calves lost during the first several weeks of life is extremely variable on dairy farms. With some excellent situations death losses may average less than 1% over a period of years. In others the mortality may exceed 25% of all calves born. The death rate among young calves, generally, is higher during the first week of life and rapidly decreases thereafter. To raise a healthy calf, adequate nutrition, establishing resistance to infections, prevention of overexposure to infectious organisms, and avoidance of excessive stresses are important.

Considerable research has been conducted on numerous aspects of different ways to feed and raise young dairy calves. An abundance of information exists on most of the important questions. With any program, taking care of the important details in a careful and skillful way is crucial to success. In most of the research with young calves, appreciable attention is given to the rate of growth. Moderate differences in growth rates during the first few weeks are not of major direct importance. Rather, a well-fed, well-grown calf is more likely to survive. On a statistical basis, underfed, slow-growing calves have a higher mortality. If the calf is healthy, substantial differences in growth rate during the first 6 weeks, are easily overcome later.

14.2. MILK OR MILK REPLACER

Being a baby, the newborn calf requires milk or a satisfactory substitute during the first few weeks of life. Most good milk substitutes contain a high percentage of milk constituents. Although calves can be raised with few deaths on a good milk replacer, often mortality may be somewhat higher than when whole milk is fed. Usually the feed cost with milk replacer is lower than with whole milk.

Many experiments have been conducted relative to the length of time whole milk or milk replacer should be fed. Although strong calves weaned at about three weeks of age have survived, a longer period is more satisfactory. Providing milk or milk replacer until the calf is at least four or five weeks old, often will decrease death losses and frequently is a practical routine program. For weak calves, slightly longer milk or milk-replacer feeding may be beneficial. The additional cost of feeding milk or milk replacer a week or two longer is small compared to the financial loss due to a death.

14.2.a. Colostrum: Importance for Survival

One of the most essential requirements of a good calf raising program is that the calf receive colostrum as its first feed. In addition to being an excellent source of essential nutrients, colostrum contains antibodies, which is nature's way of providing protection to the young calf against many disease producing organisms (Baumwart et al., 1977). The calf should receive the colostrum as soon after birth as possible, preferably within an hour; additional colostrum is desirable for 2 or 3 days after birth (McClurkin, 1977). To win the race against the infectious organisms which may reach critical numbers at a very early age, the colostral antibodies are needed. The ability of the calf to absorb antibodies from the digestive tract declines with age and is only effective for about 24 hours after birth (Roy, 1969). It is important that the young calf receive at least four pounds of the first colostrum produced from the cow. If several cows freshen at the same time, mixed colostrum will be higher in antibody protection than will colostrum from individual cows.

Colostrum is the name given the milk produced just before and after the calf is born. The first colostrum is very different from most of the milk produced during the lactation. It appears to be a mixture of the true milk and nondiffusible constituents of blood plasma concentrated some 10- to 15-fold in passage into the udder prior to calving (Roy, 1969). Initially, colostrum is higher in total solids, fat, vitamins, proteins, and minerals but lower in lactose (Table 14.1). The greatest difference in colostrum and later milk is the extremely high level of the protein, globulin, that contains the antibodies. This fraction declines quite rapidly with succeeding milkings after the first. Within three or four days after parturition, the milk has lost the special colostral properties and is quite similar to the usual or true milk.

14.2.b. Whole Milk

Limited whole milk feeding during the first few weeks of life is the standard against which other feeding programs are usually compared. To avoid diarrhea, the amount fed during the first few days should not be excessive. Traditionally, recommended whole milk or milk-replacer feeding programs included a complex

TABLE 14.1
Composition of Colostrum and Whole Milk[a]

Constituent	Colostrum	Whole milk
Fat (%)	3.6	3.5
Nonfat solids (%)	18.5	8.6
Lactose (%)	3.10	4.60
Ash (%)	0.97	0.75
Vitamin A (ppm in fat)	42–48	8
Choline (ppm)	370–690	130
Protein (%)	14.3	3.25
Casein (%)[b]	5.2	2.6
Albumin (%)[b]	1.5	0.47
Immunoglobulin (%)[b]	5.5–6.8	0.09

[a] Adapted from Roy, 1969.
[b] These constituents are a part of the protein.

scheme in which the amount fed to each calf was calculated as a percentage of birth weight. Initially the amount was low at about 6% of birth weight for a few days, then gradually increased to about 10% of birth weight, and after 2–3 weeks of age was reduced every few days until weaning. The complex schemes of determining the amount of whole milk or milk replacer require considerable labor and record keeping. However, they do not give better performance than a very simple system of feeding each calf of a given breed the same amount after it is taught to drink until weaning (Miller and Clifton, 1962). For Holsteins, about 7 lb of whole milk per day or a similar amount of nutrients in milk replacer usually gives satisfactory results in most situations. If the calf is weaned at 4 weeks of age, about 170 lb of whole milk would be fed after a colostrum feeding period of 4 days.

In feeding whole milk or milk replacer, good sanitation, especially including clean and sterilized utensils, is very important. Milk can be fed satisfactorily at a variety of temperatures but it should be consistent from feeding to feeding.

14.2.c. Colostrum: A Whole Milk Substitute

Although colostrum is quite nutritious, in most areas it cannot be sold legally for human food, especially with other milk. Except as a feed for calves, usually colostrum has little economic value on a dairy farm (Foley and Otterby, 1978). Accordingly, considerable research has been conducted on its use for feeding calves. Fresh colostrum is a very satisfactory feed, both for the calf at the time it is produced by the dam, and for older calves. Generally the amount fed should be reduced somewhat or diluted with 1 part water to about 2 or 3 parts colostrum to reflect the higher nutrient content of the colostrum (Jenny et al., 1977). It is highly variable in composition. Thus, feeding fresh colostrum often results in feeding a variable amount of ingredients. Lactose, a key constituent in milk, tends to cause diarrhea when excess whole milk is fed. Colostrum contains less lactose than whole milk (Table 14.1) and therefore is not as laxative.

Generally the amount of excess colostrum produced by a dairy cow is sufficient to meet at least half the total milk needs of a calf (Muller et al., 1975). Since the time of production does not totally coincide with the time it is needed, some method of preservation and storage is beneficial.

Frozen colostrum. When properly handled, colostrum can be frozen until needed and then thawed for feeding. This method has never been widely used on practical dairy farms, because of operational problems including considerable labor, freezer space, and freezer utensils.

Fermented colostrum. In recent years fermentation, sometimes called souring or pickling, as a preservation method has been studied and used widely (Fosgate and Harrington, 1977; Jenny et al., 1977; Muller et al., 1975; Foley and Otterby, 1978; Swannack, 1971). When careful attention is devoted to the essential details, fermented colostrum usually gives good results, but problems may arise especially in hot weather. Although it may appear quite "stinky" to some humans, properly fermented colostrum is readily consumed by calves.

In a manner similar to that of silage, fermented colostrum is preserved by developing sufficient acidity (low pH) which prevents undesirable putrefraction organisms from dominating. In some instances, especially in warm weather, a preservative such as 1% propionic acid may be desirable to bring the pH down quickly. A small portion of fermented colostrum may be used as a starter culture. Unlike well-preserved silage, fermented colostrum cannot be preserved easily for long periods.

Several rules and points appear important in using fermented colostrum as a milk replacer.

1. The colostrum should be milked and maintained under sanitary conditions to avoid infections. Storage containers should have a plastic liner.

2. Colostrum from cows treated with antibiotics for mastitis should not be used. Antibiotics may interfere with the desirable fermentation.

3. Fermented colostrum should be kept at reasonable room temperature between about 40°F and 80°F. If the temperature is too hot, it may spoil due to an undesirable type fermentation. When this occurs, the putrified product should be discarded. In some situations it may be necessary to discontinue feeding fermented colostrum in very hot weather. If too cold, fermented colostrum does not mix easily or may freeze.

4. If batches of colostrum do not develop a suitable type of fermentation, the "raunchy" material should be discarded. Such a product is likely to be refused by calves.

5. As a general rule, fermented colostrum should be kept no longer than 4 weeks.

6. Fermented colostrum should be stirred daily and when fresh material is added.

7. At the time of feeding, the colostrum should be stirred and diluted with water. About 1 part water to 2 to 3 parts colostrum is satisfactory.

8. The amount of colostrum fed should be about the same as for whole milk. When diluted with one part water to 2 or 3 parts colostrum, the total solids content is comparable to whole milk.

9. Fermented colostrum that is too cold should not be fed, as a thorough mixing of the fat is not easily achieved. In cool weather the temperature of the product to be fed can be increased by using warm water for the dilution.

10. Feeding fermented colostrum with a nipple pail seems to give better results than out of an open pail. Use of the nipple pail enables the calf to avoid much of the undesirable smell and therefore may improve acceptability.

11. If a calf refuses a good, properly fermented, product, it may be necessary to use whole milk or a milk powder based replacer.

14.2.d. Nutrient Composition of a Milk Replacer

The requirements for a good milk replacer are more rigorous than for any other type of feed normally used by dairy cattle. On a dry matter basis, a milk replacer should contain a minimum of 22% protein, 10% fat, 95% TDN, 1.90 Mcal of digestible energy per pound, 0.7% calcium, 0.5% phosphorus, 1720 IU of vitamin A per pound and other nutrients as shown in Appendix, Table 3 (National Research Council, 1978). In addition to the requirements in terms of nutrient composition, a good milk replacer should meet several other specifications. Many types of proteins, carbohydrates, and fats, which are readily utilized by older cattle, are not used efficiently by young calves. Thus the choice of suitable ingredients is quite limited. Ingredients used in milk replacers must be readily soluble so the powder can be dissolved quickly at the time of feeding.

14.2.e. Milk By-Product Based "Milk Replacers"

Because of the comparatively high cost of nutrients from whole milk relative to those from other feeds, using milk replacers for raising young dairy calves has received considerable attention for many years. In earlier times, when large amounts of milk were skimmed on farms for the cream, often the calves were fed fresh skimmed milk. Later with skimmed milk no longer readily available, another approach was required. Most successful substitutes for whole milk have been based largely on milk by-products. Early attempts to use dried skimmed milk as a milk replacer usually met with problems associated with difficulties of dissolving the powder. This problem has been largely solved with better skimmed milk powder. Dried skimmed milk makes an excellent base for many of the nutrients needed in a milk replacer. Frequently, other milk by-products are lower in cost providing an incentive to find a way to use them.

Dried whey, because of its high lactose content is not a satisfactory ingredient for the bulk of the milk replacer, but limited amounts have been used with satisfactory results. The maximum amount which can be used with good results is controversial due to conflicting research findings (Schingoethe, 1976). Partially delactosed whey can be used in larger amounts. Dried buttermilk is sometimes used as an important component of milk replacers. Less used but good products include casein and other whey proteins.

Dried skimmed milk and other milk by-products in which the fat has been removed, contain substantially less usable energy per pound of dry solids than whole milk. Accordingly, in most good milk replacers, another source of fat is used. To obtain good results, these added fats often are especially processed to give good mixing and utilization by the calf (see Section 14.2.j). Generally a milk replacer with a higher fat content will give faster gains than one with less fat. About 10% fat is quite satisfactory for calves being raised for herd replacements. For veal production (see Chapter 16) more fat is needed. The inclusion of vitamins, especially vitamin A, antibiotics, and possibly some minerals usually improves results with milk replacers.

14.2.f. Nonmilk Ingredients for Milk Replacers

Many attempts have been made to find suitable lower cost ingredients to replace milk by-products in milk replacers. Limited success has been achieved but there have been many failures.

14.2.g. Fish Protein

In experimental studies fish protein has replaced up to 35% of the milk protein in milk replacers with only a small reduction in calf performance (Huber, 1975). When higher levels of fish protein were fed, much poor growth occurred, especially at younger ages. Some of the reasons for the lower performance of calves fed high amounts of fish protein include lower protein digestibility, a less well balanced amino acid composition, lower fat digestibility, and possible problems associated with solvents used in preparing the fish soluble concentrate. Likewise, there is a possibility of a vitamin E deficiency with fish proteins caused by an increased requirement due to the large amount of polyunsaturated fats (see Sections 6.4.a; 7.1). Future developments may provide ways to obtain good calf performance when fish proteins are fed in large amounts.

14.2.h. Plant Proteins in Milk Replacers

Many attempts have been made to develop plant proteins, especially soybean protein, which would effectively replace that protein from milk in milk replacers (Ramsey and Willard, 1975; Roy, 1969). Obtaining a readily soluble protein is one problem. Another is poor utilization of the substantial amount of carbohydrate found with soy protein (Ramsey and Willard, 1975). When and to what degree the problems associated with soy protein as a substitute for milk protein will be solved, is uncertain.

14.2.i. Use of Grain Products in Milk Replacers

Many earlier milk replacers contained an appreciable amount of grain products especially starches. These were disasters. The newborn calf could not effectively digest starch, and thus obtained little usable energy from the grain products (Roy, 1969). Also, diarrhea often developed. In the newborn calf, utilization of the grain protein also probably leaves much to be desired.

14.2.j. Fats and Sugars in Milk Replacers

Generally the ability of the young calf to utilize nonmilk fat and sugar is better than for starches and plant proteins. However, for good results, these must be chosen with care. To be used effectively in milk replacers, fats should be readily soluble, highly stable, and easily utilized by the calf. Often special processing including homogenation and/or addition of emulsifying agents such as lecithin are required to achieve these goals (Roy, 1969). Usually animal fats have given better results than vegetable fats (see Section 16.1). Frequently lecithin is included in the formula (see Section 16.1). The young calf can effectively digest some simple sugars such as glucose (corn sugar), but it does not have the ability to effectively digest all sugars, including sucrose (Roy, 1969).

14.2.k. How to Feed Milk Replacer Powder

Generally milk replacer powders are mixed with about 7 parts water to 1 part powder by weight giving a mixture containing about 12% total solids comparable to whole milk. Due to the much lower fat content, most milk replacers provide less usable energy than whole milk. Typically, recommendations for milk replacer feeding suggest one 25-lb bag per calf. This would provide energy equivalent to about 150–180 lb of whole milk depending on the fat content of the milk and the replacer.

Usually after mixing with water, milk replacer is fed similarly to whole milk. With either system, it is desirable to give the calf colostrum for as long as it is available. The liquid milk replacer can then be fed in the same amounts as whole milk (see Section 14.2.b). Some dairymen feed either whole milk or colostrum during the first one, two, or three weeks after birth and then switch to milk replacer. Since the early period is more critical, this gives the calf the benefit of whole milk when it may be most beneficial. Likewise, some dairymen may switch any calf that develops health problems on milk replacer back to whole milk.

14.3. CALF STARTERS

Grains and plant proteins are much lower cost sources of nutrients than milk or ingredients suitable for use in milk replacers. Thus one of the objectives in feeding dairy calves is to utilize nutrients from the cheaper sources as soon as feasible. It is desirable to begin offering the young calf a high quality, calf-starter concentrate mixture by about one to two weeks of age. Initially, the calf will eat only a very small amount. Old feed should be replaced with fresh feed daily.

Consumption will increase quite rapidly with most calves eating enough grain, along with some forage, to meet their minimum nutrient needs for satisfactory growth by the time of weaning at 4 to 6 weeks of age. In addition to providing nutrients and adjusting the calf to concentrate consumption, dry feed (concentrates and hay) is very beneficial in stimulating normal and rapid rumen development.

A calf starter on a dry matter basis should contain a minimum of 16% protein, 80% TDN, 0.6% calcium, 0.42% phosphorus and other nutrients as shown in Appendix, Table 3 (NRC, 1978). Many studies have shown that the starter does not need to have a large number of ingredients (Miller et al., 1969). If high quality ingredients are used, there is no disadvantage to the large number of ingredients in a complex formula. Often a commercial feed company can formulate a complex calf starter at a lower cost than a simple one consisting of only two or three major ingredients.

A calf starter should be fed free choice to the young calf. As the calf becomes older, the amount consumed continues to increase. Accordingly, dairymen often limit the amount of starter to around 4 or 6 lb per day to keep feed cost down and to prevent the possibility of excessive fattening. By the time the calf is three months old, a less expensive concentrate containing less protein and other essential nutrients can be used.

14.4. HAY, SILAGE, PASTURE, OR ROUGHAGE

Generally hay is the preferred forage for dairy calves. Although initially the calf will consume even less hay than starter (Miller and Clifton, 1962), offering free choice, high quality hay beginning around two weeks of age is a good practice. The hay can be from any of many species of forages. It should be grown on land with adequate fertility and harvested at an early stage of maturity to ensure a high nutrient content (see Chapter 9). Likewise, the hay should be well preserved and palatable to encourage higher consumption.

Silage generally is less desirable for calves than hay. If grass silage is fed, dry matter intake and growth rates may be reduced. With corn silage special attention should be given to avoid deficiencies of nutrients such as protein, manganese, calcium, and sulfur.

Pasture can be utilized instead of hay with good results. If the pasture is of high quality and the supply is adequate, it will provide essential nutrients just as effectively as hay. Parasites are much more likely to be a serious problem when young calves are grazing than with older cattle (Herlich and Douvres, 1977). Adequate procedures to avoid excessive parasite damage are needed, including avoiding pasture areas recently used by other cattle. Best results are obtained when the pasture land has been plowed and reseeded since cattle had access to the area.

In some situations, it may not be feasible to include hay, pasture, or forage in the calf ration. If a complete feed or an all-in-one ration is fed, another source of fiber may be necessary. Good results can be obtained with such feeds when they are properly formulated. When roughages such as cottonseed hulls or corn cobs are used, it is important to be certain that adequate quantities of all the essential nutrients such as protein, minerals, and vitamins are supplied. Also, a sufficient amount of fiber is important (Miller *et al.*, 1969; Van Horn *et al.*, 1976) (see Chapter 8).

14.5. ANTIBIOTICS AND VITAMINS FOR CALVES

Considerable research has shown that low levels of antibiotics in calf feeds increase feed consumption, improve growth, reduce diarrhea, and lower death losses (see Section 11.1 for a more detailed discussion). A major portion of the benefits are obtained in the milk or milk-replacer feeding period with progressively less later and with little if any advantage after four months of age. Although antibiotic feeding usually results in a greater percentage improvement in performance with poorer management and sanitation, it is not a substitute for good sanitation and management. When management and care are poor, death losses and poor performance often are excessively high, even with antibiotics.

Vitamins. The newborn calf does not have a functional rumen and must obtain all the essential vitamins from other sources (see Chapter 6). Generally, the calf receives a substantial part of its vitamin needs from colostrum, milk, and/or milk replacer until the rumen is able to synthesize them. Usually supplemental vitamin A should be provided. Accordingly, good milk replacers contain vitamin A. The amount of vitamin A in colostrum and whole milk is quite dependent on the level of vitamin A and carotene in the cow's diet. Since milk sometimes does not contain sufficient vitamin A to meet the calf's needs, and because of the low cost, often vitamin A is routinely added to calf starters. During the first few days of life, frequently calves are given a special capsule, or other supplement, containing vitamins A, D, and sometimes E and perhaps an antibiotic.

When the calf is exposed to sufficient sunlight or is eating adequate suncured hay, no supplemental vitamin D is needed. Because many calves do not receive sufficient vitamin D from these sources, generally this vitamin should be included in the milk replacers and/or starter feeds. In special circumstances, such as low selenium areas and diets with considerable unsaturated fats, supplemental vitamin E may be needed in the practical feeding of calves (see Chapter 6 for details). Because of the expense involved and the apparent infrequent need, it is not believed practical to include supplemental vitamin E in calf feeds on a routine basis.

14.6. CALF DISEASE AND PARASITES

The newborn calf is especially susceptible to infectious organisms, but if sufficient colostrum is obtained soon after birth, its resistance is greatly increased. The calf should be born in clean surroundings. Permitting it to be born in a stall where other cows have given birth since it was thoroughly cleaned and disinfected, frequently results in serious infections. The navel of the newborn calf should be painted with tincture of iodine to prevent the entrance of infectious organisms.

Noninfecting diarrhea (common scours or common diarrhea) is the most frequently encountered health problem in young dairy calves prior to weaning. The major causes are associated with overfeeding of milk or milk replacer, use of unsanitary utensils, abrupt changes in feed temperature or other details of milk or milk-replacer feeding. Stressful conditions, eating of bedding, excessive consumption of dry feed, or nutritional deficiencies also may cause common diarrhea.

If prompt attention is given to correcting the cause of diarrhea, mortality usually is not high. However, noninfectious scours slows the growth rate and weakens the calf making it more susceptible to other diseases. When common diarrhea is observed, the amount of milk or milk replacer should be reduced and other treatment initiated immediately. Since the calf often appears dull and has a reduced appetite even before diarrhea is evident, early treatment is feasible. Many treatments for noninfectious diarrhea have proven useful including therapeutic levels of antibiotics and/or sulfa drugs. Special attention should be directed to good sanitation and removing stressful conditions such as dampness or drafts. Since one effect of diarrhea is dehydration, the water consumption should not be reduced. Electrolytes are sometimes used to counteract some dehydration effects. In some situations, dairymen and veterinarians elect to switch calves having common diarrhea from milk replacer to whole milk. Fermented colostrum as a milk replacer does not increase the incidence of common scours.

Infectious diarrhea (white scours), also known as calf septicemia, generally is the most frequent cause of death in calves during the first several days of life. The term white scours comes from the grayish-white color of the feces which have a very thin consistency and a characteristic very foul odor. Other symptoms include loss of appetitie, general dullness, and severe dehydration as indicated by sunken eyes. White scours is caused by infectious organisms, often perhaps by viruses followed by *Escherichia coli*.

Since the death rate from infectious diarrhea usually is very high, prevention is most important. The two main defenses are good sanitation and a calf with high resistance to infection. If one calf develops white scours, the clean-up and disinfecting procedures should be extremely thorough, including the place where the calf was born and all feeding equipment. Resistance to infections includes a healthy, properly nourished dam; birth in a clean, dry, draft-free place; consumption of adequate colostrum; good nutrition; low level antibiotic feeding; satisfactory housing; good feeding conditions, and other essential management procedures.

When white scours develops, the amount of milk or milk-replacer should be reduced. Likewise, treatment should be prompt, including use of antibiotics, and/or sulfa drugs, and other measures as prescribed by a veterinarian. Electrolyte solutions are sometimes used to counteract dehydration effects. Blood transfusions may also be beneficial. Many other products including vitamin A are often given as part of the treatment. It is wise to have a carefully planned procedure and all needed products on hand so white scours can be treated promptly. These plans can be made on a cooperative basis between the herdsman and the veterinarian.

Salmonella bacterial infections can produce an explosive, uncontrollable diarrhea (Jarrett, 1977). Frequently a very high percentage of the calves die within 24 to 48 hours after being afflicted. The problem is most likely to occur when calves are frequently purchased. The recommended actions relative to feeding, colostrum, sanitation, and other details of good calf raising minimize the likelihood of such an infection occurring.

Pneumonia is a major cause of death in young calves. Often it develops in calves weakened by diarrhea, inadequate nutrition, or exposure to some form of stress such as dampness, drafty conditions, or sudden temperature changes. Many of the deaths from pneumonia occur in calves between 3 and 8 weeks of age with a much lower incidence after 8 weeks. Symptoms of pneumonia include elevated temperatures, rapid breathing, coughing, dullness, an unpleasant nasal discharge, weakened condition, and loss of appetite. Since pneumonia is caused by infectious organisms, care should be taken to avoid exposing other calves. Treatment should include removing any environmental cause, keeping the calf dry and reasonably warm, possibly with a heat lamp in cold weather, and such other measures as may be developed in cooperation with the veterinarian.

Internal parasites can be a problem if young calves have access to land areas exposed to other calves or cattle without sufficient time for the organisms to be destroyed by natural forces. The most satisfactory way to deal with internal parasites in young calves is to avoid conditions that permit exposure.

14.7. FEEDING PROCEDURES

Generally, the young calf is permitted to suckle colostrum from the dam for at least several hours to as much as 2 or 3 days after birth. It is perhaps better for the calf to stay with the dam for at least one full day after birth, because of initial problems in training the calf to drink milk, and the need of colostrum for at least 24 hours, and preferably 2 or 3 days (see Section 14.2.a). From an operational standpoint, often it is advantageous to remove the calf after about one day.

Milk or milk replacer is fed by a number of procedures. One effective way is the use of nipple pails. This method has the advantage of being easier to train the calf to take milk, or milk replacer, than does use of open pails. With the nipple-pail system, it is easier to avoid the extreme hunger which sometimes occurs before the calf begins drinking from an open pail. Following extreme hunger, excessive intake of milk or milk replacer may lead to diarrhea.

With the open-pail procedure for feeding milk or milk replacer it is easier to clean the utensils and maintain good sanitation than with nipple pails. However, more careful attention is needed in training the calf to drink as soon as possible. When large amounts of milk are fed, more diarrhea may occur with open-pail feeding. With good management, the incidence of diarrhea is about the same for open-bucket and nipple-pail feeding procedures. When large numbers of calves are fed, substantial labor may be saved with an automatic feeder. Although available in various types, they are not widely used on typical moderate sized dairy farms.

As discussed in Section 14.2.b, all healthy calves of each breed can be fed the same amount of milk or milk replacer from the time they are taught to drink until weaning (Miller and Clifton, 1962). This procedure reduces the labor needs and is especially helpful if an automatic feeder is employed. In recent years, substantial research has shown that whole milk or milk replacer can be fed only once per day with good results (Galton and Brackel, 1976). The advantage of the once-per-day frequency of feeding is the reduction in labor in feeding and cleaning utensils, found to be 39% less labor in one study (Galton and Brackel, 1976). Once-per-day feeding is not satisfactory for calves fed the large amounts of milk necessary in veal production (see Chapter 16).

Although procedures used in feeding starter, concentrate mixture, and hay are simpler and easier than with milk or milk replacer, satisfactory sanitation is important. Any uneaten starter should be removed daily to avoid consumption of partially "spoiled" feed. Until weaned, each calf should be fed and watered individually in such a way that infectious organisms from one calf cannot be transmitted to another. Beginning sometime after weaning, calves can be fed satisfactorily in small groups of comparable sized animals.

14.8. HOUSING AND MANAGEMENT OF CALVES

Individually housing calves until a few weeks after they are weaned from milk, greatly diminishes the spread of infectious organisms. Young calves housed together often suck each other. It is generally believed that if the udder of a baby calf is sucked by another, there is a much greater probability of mastitis when she becomes a cow. Many types of housing are suitable for raising young calves. The general needs include avoiding excessive drafts, dampness, sudden drastic changes in temperature, and high humidity. Except for immediately after birth and when weakened by illness, young calves tolerate fairly low temperatures quite effectively. Likewise, if humidity and dampness are not problems, comparatively high temperatures are acceptable. Housing should be economical when both the capital cost per calf per year and labor are considered.

Individual outside calf pens involving a simple roof and sides, one end with a dry platform for the calf to lie on, and an open area on the other end can be used very effectively throughout the year for raising calves in much of the United States (Fig. 14.1) (Muller and Owens, 1976; Van Horn *et al.,* 1976). The initial cost is very low, but labor costs may be comparatively high. If on the ground, the pen should be moved frequently in a special pattern to avoid parasitic infestations. Often, but not invariably, the amount of time involved in feeding is higher than in a central calf barn. The individual calf pen exposes the caretaker much more to unpleasant weather. Usually the calf is better able to tolerate cold weather than the caretaker.

A calf barn or other facility need not be extremely expensive, but it is important that good sanitation be maintained (Fig. 14.2). Frequently calves stay much

Fig. 14.1 Simple but effective individual pens being used to raise calves in Connecticut. (Courtesy of W. J. Miller, Univ. of Georgia.)

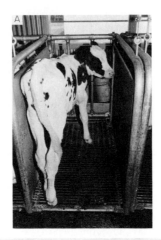

Fig. 14.2 Many different types of arrangements in calf barns give good results. These photographs show healthy calves in (A) metal pens and (B) more economical wooden pens in outside facility, southern California. (A. Courtesy of C. K. Walker, Univ. of Maine; B. Courtesy of W. J. Miller, Univ. of Georgia.)

healthier and perform better with fewer death losses in a new barn than in an older one. Subsequently health problems, poor performance, and high mortality often develop. To avoid these problems, it is desirable to have all the calves out of the barn for two or three weeks periodically, and at least once each year during which time thorough cleaning and disinfecting programs are instituted. Also, a strict sanitation program should be maintained throughout the year.

Water that is clean and fresh should be readily available to calves so that intake is not restricted, beginning at least by the time starter consumption is initiated.

Management. Several other important details of managing dairy calves are important. The young calves should be dehorned by an acceptable procedure such as the electric dehorner or the caustic potash method. Likewise, a good record-keeping system is needed. A routine vaccination program for calves for such diseases as blackleg, brucellosis (between 2 and 6 months of age), leptospirosis, malignant edema, and shipping fever is an important part of calf raising (Smith, 1977).

REFERENCES

Baumwart, A. L., L. J. Bush, M. Mungle, and L. D. Corley (1977). *J. Dairy Sci.* **60,** 759-762.

Foley, J. A., and D. E. Otterby (1978). *J. Dairy Sci.* **61,** 1033-1060.

Fosgate, O. T., and R. Harrington (1977). *Hoard's Dairyman* **122,** No. 13, 829.

Galton, D. M., and W. J. Brakel (1976). *J. Dairy Sci.* **59,** 944-948.

Herlich, H., and F. W. Douvres (1977). *J. Dairy Sci.* **60,** 283-288.

Huber, J. T. (1975). *J. Dairy Sci.* **58,** 441-447.

Jarrett, J. A. (1977). *Hoard's Dairyman* **122,** No. 6, 408-409.

Jenny, B. F., S. E. Mills, and G. D. O'Dell (1977). *J. Dairy Sci.* **60,** 942-946.

McClurkin, A. W. (1977). *J. Dairy Sci.* **60,** 278-282.

Miller, W. J., and C. M. Clifton (1962). *Ga., Agric. Exp. Stn., Circ.* [N.S.] **33.**

Miller, W. J., Y. G. Martin, and P. R. Fowler (1969). *J. Dairy Sci.* **52,** 672-676.

Muller, L. D., and M. J. Owens (1976). *Hoard's Dairyman* **121,** No. 19, 1107 and 1138.

Muller, L. D., G. L. Beardsley, and F. C. Ludens (1975). *J. Dairy Sci.* **58,** 1360-1364.

National Research Council (NRC) (1978). "Nutrient Requirements of Dairy Cattle," 5th rev. ed. Natl. Acad. Sci., Washington, D.C.

Ramsey, H. A., and T. R. Willard (1975). *J. Dairy Sci.* **58,** 436-441.

Roy, J. H. B. (1969). *In* "Nutrition of Animals of Agricultural Importance Part 2" (D. Cuthbertson, ed.), pp. 645-716. Pergamon, Oxford.

Schingoethe, D. J. (1976). *J. Dairy Sci.* **59,** 556-570.

Smith, P. C. (1977). *J. Dairy Sci.* **60,** 294-299.

Swannack, K. P. (1971). *Anim. Prod.* **13,** 381-382.

Van Horn, H. H., M. B. Olayiwole, C. J. Wilcox, B. Harris, Jr., and J. M. Wing (1976). *J. Dairy Sci.* **59,** 924-929.

15

Feeding and Management of Heifers, Dry Cows, and Bulls

15.0. INTRODUCTION

Compared with baby calves and high-producing cows, heifers, dry cows, and bulls are relatively easy to feed. Perhaps this comparative ease is one reason these dairy animals often are neglected. Although the nutritional requirements of heifers, dry cows, and bulls are less demanding, the total amount of feed required is a substantial percentage of the total on a dairy farm.

15.1. PROVIDING THE NUTRIENT REQUIREMENTS FOR DAIRY HEIFERS OF DIFFERENT AGES

From the time of weaning at about four to six weeks of age until calving around 2 years old, the dairy heifers increase in weight about 10-fold. During this growth period the types of feeding programs needed change appreciably. Within a few weeks after weaning, it is practical to place heifers in small groups of similar size (see Section 14.7). These heifers should not nurse each other as excessive nursing may contribute to mastitis later.

By the time the calf is weaned, considerable rumen development has occurred, but more remains to be achieved. The rumen microbes should be able to synthesize the various essential amino acids, B vitamins, vitamin K, and other nutritional factors produced in the mature ruminant so that these are not required in the diet. Because of rapid growth relative to body size, the young postweaned heifer continues to need a diet comparatively high in usable energy. Usually she is unable to meet all her energy need from forages.

In addition to high energy requirements, the young postweaned heifer has a higher requirement for protein and minerals than do older heifers. The rate of growth as a percentage of body weight in the young heifer soon after weaning is much higher than in yearling heifers. Thus the proportion of the nutrients being used for growth of tissues is much greater in the younger animal. As she grows older, the percentage used for maintenance increases constantly.

Initially, the postweaned calf generally must obtain a substantial share of its nutrients from concentrates. As it grows older, the proportion that can come from forages increases very rapidly (Fig. 15.1). Typically, when forages are fed free choice, dairy heifers from 2 to 6 months of age will need about 4 to 6 lb of concentrates per day. However, the amount depends on the forage quality. With higher quality forage, less concentrate is needed. When very high quality pasture is used, well-grown dairy heifers may be able to meet their energy needs from only forage as early as 4 to 6 months of age. With more typical quality forages, some concentrates may be needed until they are 8 to 10 months old. From the

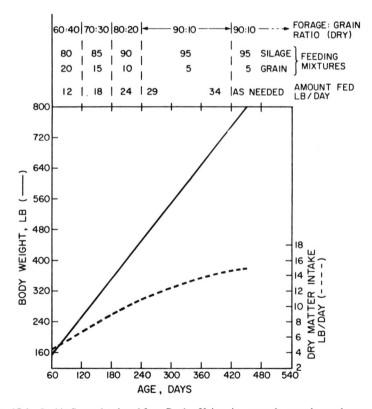

Fig. 15.1 In this figure developed from Purdue University research, note the much more rapid increase in body weight relative to dry matter intake and the decreasing proportion of grain (concentrates) relative to forage needed to obtain satisfactory growth in heifers from 2 to 15 months of age (see Noller and Moeller, 1977). (Courtesy of *Hoard's Dairyman*.)

time she is able to obtain the energy required from forage until just before calving, the dairy heifer can obtain all her digestible energy from good forages.

Beginning around 2 to 4 weeks prior to calving it is believed desirable to feed the dairy heifer concentrates. The exact time depends somewhat on the condition of the heifer and the quality of forage. If she is in poor condition, concentrate feeding should begin earlier. Even with heifers in good condition, it appears beneficial to begin feeding concentrates about 2 weeks prior to calving. The buildup should be gradual until the heifer is perhaps receiving as much as 1% of her body weight in concentrates each day.

Feeding substantial concentrates prior to calving accustoms the heifer to the type feeds needed for lactation. During this adjustment period, the rumen microorganisms adapt to utilizing the concentrates. The appetite and entire metabolic system of the heifer results in more normal metabolism during early lactation. When given substantial concentrates before calving, heifers with high potential often appear to have both higher peak milk yield and a greater lactation total. This approach when followed by liberal concentrate feeding soon after calving is sometimes called "challenge feeding" or "lead feeding."

15.2. FEEDING DAIRY HEIFERS: PRACTICAL CONCERNS

In raising dairy heifers it is desirable to meet all the nutritional requirements at the lowest practical cost. Since forages generally are lower cost sources of nutrients than concentrates, usually it is practical to give forages free choice and limit concentrates to that needed for nutrients not provided by the forage. The amount of concentrates required is determined primarily by the energy needs not met by the forages (see Chapter 2). The energy requirements for normal growth of dairy heifers of various sizes are presented in the Appendix, Table 1 (National Research Council, 1978). These recommended amounts of energy should produce heifers that are large enough to breed around 15 months of age and to calve about 24 months (Table 15.1).

Appreciable research has been conducted on the effects of both underfeeding and overfeeding heifers during the rearing period (Schultz, 1969; Swanson, 1960; Swanson and Hinton, 1964). Substantially underfeeding on energy during the growing period results in animals weighing much less with smaller differences in skeletal size (Fig. 15.2). The undersized heifers may not exhibit first estrus until they are many months older. Generally, the time of first estrus occurs at about 45% of the animal's mature weight and at about the same size regardless of the growth rate (Schultz, 1969).

If bred to calve at the same age as those fed more normally, the underfed, smaller heifers have more calving difficulties and produce less milk in the first

Fig. 15.2 Underfeeding of heifers on energy during the growing period decreases the size. Identical twin Holstein heifers with one fed normally on energy and the other fed about two-thirds normal. (A) At 18 months after breeding; (B), after calving. (Courtesy of E. W. Swanson, Univ. of Tennessee.)

lactation. When all are well fed after calving, the reduction in milk production is much less in the first lactation than the size differences. In later lactations the animals underfed as heifers produce as much as normally fed heifers with no major differences in lifetime total milk yield. Likewise, animals substantially underfeed as heifers, if well fed after calving, make up for the differences in size with much of the catchup during the first lactation.

Feeding substantially more energy than recommended (see Appendix, Table 1) usually results in excessively fat heifers. In some experiments the overfed heifers had fatty deposits resulting in serious, permanent udder damage (Schultz, 1969; Swanson and Hinton, 1962) (Figs. 15.3, and 15.4). Overfed, excessively fat heifers may perform much less satisfactorily than those fed normally, including staying in the herd a shorter length of time. Thus, overfeeding energy to heifers not only is costly, but often results in a less valuable animal. At least some of the adverse effects of overfeeding may be alleviated by breeding such

Fig. 15.3 Identical-twin Jersey heifers. One greatly overfed on energy and the other fed normally. (Courtesy of E. W. Swanson, Univ. of Tennessee.)

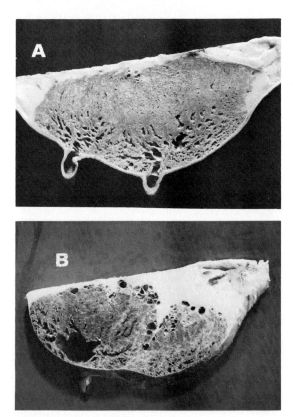

Fig. 15.4 Excessive overfeeding of heifers on energy causes permanent udder damage impairing ability to lactate. (A) Cross section of udder of animal fed normally as a heifer; (B) identical twin greatly overfattened as a growing heifer showing large amounts of fatty tissues and large underdeveloped areas in the mammary glands. Both photos taken at end of second lactation. (Courtesy of E. W. Swanson, Univ. of Tennessee.)

heifers at an earlier age. Generally heifers should be bred according to size rather than to age (Table 15.1).

Moderately underfeeding heifers on energy for relatively short periods such as a few weeks does not have major effects on the later size or subsequent performance. Through mechanisms known as compensatory growth, the heifer is able to catch up much of the reduced size by more rapid growth when adequate feed is provided. This is the reason the growth rate during the milk feeding period has little effect on size at the time of breeding or calving.

The skilled herdsman is able to judge the amount of concentrates needed to keep heifers growing at about the normal rate from the appearance of the animals. Normal weights for various ages are shown in Table 15.2. If scales are not readily available, the weights of animals can be determined by heart girth measurements (Table 15.2). However, some skill is needed to properly make these measurements.

15.2.a. Types of Forages for Dairy Heifers

Good pasture and/or hay are desirable forages for dairy heifers. Likewise, corn silage or good grass silage having a moderately high dry matter content are acceptable forages. Heifers fed high-moisture grass silages usually do not perform as well as those fed hay made from comparable forage. (Other important details related to forages of many types, and their evaluation are discussed in Chapters 9 and 12.)

When pasture is the forage for heifers, both the quality and the amount available should be adequate to supply the required nutrients. Often dairymen neglect heifers on pasture resulting in slower growth and development. If pasture is inadequate, supplemental feed such as other forage or additional concentrates are satisfactory substitutes.

TABLE 15.1
Suggested Body Weights for Breeding Dairy Heifers

Breed	Body weight (lb)
Ayrshire	600
Brown Swiss	750
Guernsey	550
Holstein	750
Jersey	500

TABLE 15.2
Normal Heart Girth Measurement and Weight of Calves and Heifers during the Growing Period[a]

Age in months	Holstein		Ayrshire		Guernsey		Jersey	
	Inches	Pounds	Inches	Pounds	Inches	Pounds	Inches	Pounds
Birth	31	96	29 1/2	72	29	66	24 1/2	56
1	33 1/2	118	32	98	31 1/2	90	29 1/2	72
2	37	161	35 1/2	132	34 1/2	122	32 1/2	102
3	40 1/4	213	38 3/4	179	38	164	35 1/4	138
4	43 1/2	272	42 3/4	236	41 1/4	217	38 1/4	181
5	47	335	45 1/2	291	44 1/4	265	41 1/2	228
6	50	396	48 1/4	340	47	304	44 1/2	277
7	52 1/2	455	51 1/4	408	49 3/4	362	47 1/4	325
8	54 3/4	508	53	447	51 3/4	410	49 3/4	369
9	57	559	55	485	53 3/4	448	51 3/4	409
10	58 3/4	609	57	526	55	486	53 1/4	446
11	60 1/2	658	58	563	56 3/4	521	55	481
12	62 1/2	714	59	583	58 1/4	549	56 1/2	520
13	63 1/4	740	60 3/4	630	59 1/4	587	57 1/2	540
14	64 1/4	774	62	666	60 1/2	615	58 1/2	565
15	65 1/4	805	63	703	61 3/4	640	59	585
16	66 1/4	841	64	731	62 1/2	674	59 3/4	611
17	67 1/4	874	65 1/4	758	63 1/2	696	60 1/2	635
18	68 1/2	912	66	781	65	727	61 1/2	660
19	69 1/4	946	66 1/2	813	65 1/2	752	62 1/2	687
20	70 1/2	985	67 1/2	841	66 1/4	780	63	712
21	71 1/2	1025	68 1/2	885	67 1/2	816	64	740

[a] From Anonymous (1961).

15.2.b. Types of Concentrates and Supplements for Dairy Heifers—Balancing the Ration

The content of protein, minerals, and vitamins needed in concentrate or supplement for heifers is determined by the amount in the forages, the heifer's needs, and the relative amounts of concentrates and forage consumed. Soon after weaning, the young heifer eats much more concentrate than forage. Thus the composition of the concentrate depends only to a small extent on the forage. The amount of concentrate needed relative to forage decreases markedly with age until all the energy needs can be obtained from forage. During this advancing time the needed composition of the concentrate depends more and more on the forage.

Generally, concentrate ingredients suitable for lactating cows are satisfactory for heifer rations. Although a somewhat different level of protein and minerals may be needed for the heifers, similar formulations are adequate. During the period when the heifer is able to obtain all her energy needs from good forage, it is important to avoid deficiencies of protein, minerals, or vitamins. If the forage has sufficient usable energy but is deficient in one or more other nutrients, these can be supplied in a supplement.

Usually salt will be needed in a supplement. A phosphorus and trace mineral supplement are also needed with many types of forages. In making the decision as to whether to provide supplemental minerals, it is pertinent to keep in mind that forages on individual farms often are very different in mineral composition from "book" or average values. Except with good pasture and harvested forages made from legumes, or well-fertilized, early cut grasses, additional protein may be needed. It is wise to have the forage analyzed in a forage testing laboratory to see if added protein is needed (see Section 12.15). If the forage does not have a good, green color, supplemental vitamin A may be needed (see Chapter 6).

15.2.c. Managing Dairy Heifers

Although heifers between 3 months of age and calving usually have far fewer health problems than baby calves, they are subject to many possible diseases and abnormalities, including some metabolic problems comparable to older cattle (see Chapter 18). The heifer should have been dehorned preferably during the latter part of the first month of life. However, if this has not been done, it should be as soon as feasible thereafter. If the heifer is to be vaccinated for brucellosis, this should be done between 2 and 6 months of age. Other routine vaccinations, if recommended by the local veterinarian, for blackleg, malignant edema, and shipping fever can be given at the same time as the brucellosis vaccine.

The importance of breeding heifers according to size is discussed in Section 15.2 (Table 15.1). Although generally it is recommended that heifers be fed and bred to calve around 24 months of age. This may need modification in some situations. Seriously underfed heifers will not be large enough to breed for 24-month calving (see Section 15.2). The needs of the milk market may dictate a somewhat older age of calving. Usually calving substantially under 24 months can only be achieved by very liberal feeding throughout the growing period. Breeding heifers to calve when too small greatly increases calving difficulties. Allowing them to become too old, large, and/or fat prior to breeding and calving increases problems such as more severe udder edema (Hayes and Albright, 1976). In addition to the rule of thumb of breeding according to the desired size for the breed, often better results are attained if heifers are not bred until after the second or third heat period.

Heifers can be housed and kept in groups (Fig. 15.5). Especially for the smaller ones, only a limited number, perhaps no more than about 12 for young

Fig. 15.5   This is a well-designed heifer barn on a Virginia dairy farm which provides shelter, a covered feed bunk, and sunlight. The exposure in such a barn is important to provide protection from adverse weather and to obtain the sanitizing effects of direct sunlight. (Courtesy of W. J. Miller, Univ. of Georgia.)

heifers, should be in each group. With older heifers, much larger groups may be practical, especially on pasture.

Although they may need some protection from rain, snow, and wind, dairy heifers can tolerate considerable cold with little or no adverse effects. While shade for protection from the sun may be needed in some areas, generally heifers are able to tolerate comparatively high temperatures. Prior to calving, it is desirable to accustom the ''springing'' heifer to the routine of the milking barn so that the initial milking procedure will go smoothly and perhaps require less labor as well as decrease the stress on the animal.

15.3. FEEDING AND MANAGEMENT OF DRY COWS

Cows having a dry period of 6–8 weeks produce substantially more milk in the following lactation than those not given a dry period (Butcher, 1976; Coppock *et al.*, 1974). A substantially longer time decreases total milk production for the two lactations. Various reasons have been proposed to explain the beneficial effects of the dry period including the need to replace reserves of body nutrients and to rebuild secretory tissues in the udder. Most evidence suggests that replenishing nutrients in tissues, while important, probably is not the major reason for the beneficial effects of the dry period. For instance, when only half the udder is given a dry period, that half produces substantially more milk in the next lactation.

The dry cow should be fed sufficiently to be in good condition, but not overly fat, at calving. Very fat cows, at calving, have more disease and other difficulties than those in good condition. The problems of excessively fat cows have been given the name, "fat cow syndrome," (Morrow, 1976) (see Section 18.12). Cows fed energy rich diets and having a very long dry period are most often afflicted. Some of the problems include a higher incidence of milk fever, ketosis, displaced abomasum, retained placenta, udder edema, mastitis, and death.

If the cow has been fed well, she may have gained appreciable weight during the latter part of the lactation. Thus a high rate of gain will not be needed in the dry period. The dry cow in reasonably good condition at the end of the lactation is able to meet her energy needs from good quality forage (Fig. 15.6). If she is in poor condition at the beginning of the dry period, or if the forage is of low quality, some concentrates may be needed.

It is important that the dry cow be given the needed protein, minerals, vitamins, and other nutrients (see Appendix, Tables 2 and 3). If any of these are deficient in the forage, they should be provided either in the concentrate or in a supplement. The nutrients, other than energy, are needed by the cow for optimum health and to rebuild body reserves. Of comparable importance, all nutrients are required to develop a healthy fetus. When the dry cow is not fed adequate amounts of any one of many of the essential nutrients, the calf may be weakened.

The percentage of protein and a few minerals needed in rations for dry cows is somewhat lower than for lactating cows. Most practical diets for dry cows should contain supplemental salt and usually it should be trace mineralized. Likewise, supplemental phosphorus and calcium often will be needed (see Section 18.1 for the effect of supplemental calcium and phosphorus on subsequent milk fever).

As discussed in Section 15.1 for dairy heifers, it is believed that dry cows should be given concentrates beginning about 2 weeks before calving. This permits the rumen microbes and the metabolic systems of the cow to adapt before the need for substantial amounts of concentrates in early lactation (Swanson and Hinton, 1962). The dry cow should give birth to her calf in clean, comfortable surroundings. This is beneficial to the health of both the calf and the cow. A satisfactory health and disease control program is important on dairy farms. It should be developed with adequate veterinary assistance and often involve key routine procedures for dry cows such as infusions to reduce subsequent mastitis.

15.4. FEEDING AND MANAGEMENT OF DAIRY BULLS

Beginning at a fairly early age the bull calf has a higher growth rate than the heifer and thus requires more energy and other essential nutrients (see Appendix, Table 1). However, the nutrient content of the ration should be similar for the

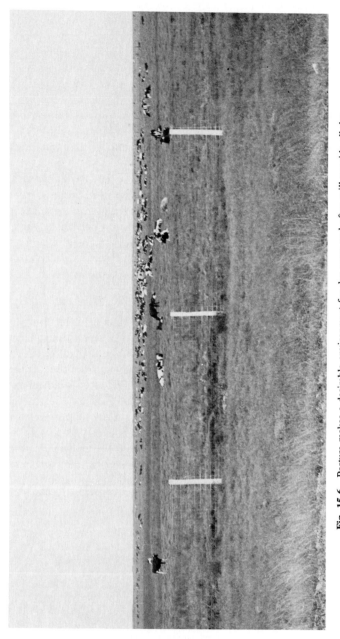

Fig. 15.6 Pasture makes a desirable environment for dry cows and often will provide all the needed nutrients, except for a few minerals. Dry cows on a large Florida dairy. (Courtesy of W. J. Miller, Univ. of Georgia.)

young bull and the young heifer. Thus comparable feeding programs are needed.

Underfeeding young bulls, as with heifers, reduces growth rate and delays sexual maturity and the time they can be used for breeding. However, even substantial underfeeding of energy does not seriously impair the breeding ability of the bull at a later age. Although overfeeding will speed sexual development to some degree, long continued overfeeding impairs the subsequent breeding ability. Generally, bulls should grow rapidly, but not so fast as to become overly fat. Feeding yearling bulls a high energy ration free choice for 140 days in beef performance testing programs apparently does not impair subsequent reproductive capability. Especially as they mature, excess fat should be carefully avoided. The well-nourished, lean bull is a more effective breeding animal.

Just as with young heifers, young bulls need concentrates during the first few months of age. With increasing age and size they can obtain more and more of the needed nutrients from forages. The information presented in Sections 15.1, 15.2, 15.2.a, and 15.2.b, relative to feeding heifers, is equally pertinent to bulls. Beyond about 5 to 10 months of age bulls can obtain the energy needed from high quality forage but all the required nutrients should be provided. Any deficiencies in the forage should be supplied by concentrates or supplements.

Because they do not have the large calcium need for milk as does the cow, the requirement of the mature bull is much lower than for the cow (see Appendix, Tables 1, 2, and 3). It is widely believed, although not firmly established, that some of the feet, leg, and back problems of bulls are caused by excessive calcium. Alfalfa hay, the most important hay crop in the United States (see Sections 9.2.a and 9.6), is very rich in calcium with considerable range among different lots. Bulls fed largely on alfalfa hay generally receive much more calcium than their minimum needs. Often mineral mixtures contain appreciable calcium, further increasing the excess. Because of the possibility of harmful effects from too much calcium, it is believed wise to include some nonlegume forage in the ration of bulls and to use only the supplemental calcium needed to meet requirements.

Before they begin mounting, which occurs as early as 4 months of age, young bulls should be separated from heifers. Their housing and health-related needs are comparable to those of heifers. Because of their disposition and great strength, special precautions should be taken to avoid injury from bulls. Numerous dairymen have been seriously injured or killed by bulls. The importance of *handling bulls in a safe way is an absolute requirement for sound dairy management.*

REFERENCES

Anonymous (1961). *U.S. Dep. Agric., Farmers' Bull.* **2176.**
Butcher, K. R. (1976). *Univ. Ga., Coop. Ext. Serv. Dairyfax,* Dec.

Coppock, C. E., R. W. Everett, R. P. Natzke, and H. R. Ainslie (1974). *J. Dairy Sci.* **57,** 712–718.
Hayes, R. L., and J. L. Albright (1976). *Hoard's Dairyman* **121,** No. 2, 75 and 103.
Morrow, D. A. (1976). *Hoard's Dairyman* **121,** No. 12, 747.
National Research Council (NRC) (1978). ''Nutrient Requirements of Dairy Cattle,'' 5th rev. ed. Natl. Acad. Sci., Washington, D.C.
Noller, C. H., and N. J. Moeller (1977). *Hoard's Dairyman* **122,** No. 13, 811 and 821.
Schultz, L. H. (1969). *J. Dairy Sci.* **52,** 1321–1329.
Swanson, E. W. (1960). *J. Dairy Sci.* **43,** 377–387.
Swanson, E. W., and S. A. Hinton (1962). *J. Dairy Sci.* **45,** 48–54.
Swanson, E. W., and S. A. Hinton (1964). *J. Dairy Sci.* **47,** 267–272.

16

Feeding and Nutrition of Veal Calves

16.0. INTRODUCTION

Production of veal calves is a specialized aspect of the dairy industry. Although veal makes up a relatively minor part of the total industry, for some individual producers it is a very important business. There are two basic types of veal with all possible variations between. Young calves sold for veal during the first few days of life are known as bob vealers. In contrast to the relatively low quality bob veal, many calves are fed for several weeks to a high degree of finish. The finished veal may be graded and such terms as choice veal, fancy veal, or finished veal used to describe the animals. Market conditions and requirements for best prices vary substantially. Thus it is important for the grower to know the characteristics of the local market.

16.1. FEEDING VEAL CALVES

Feeding veal calves has many similarities to feeding calves for herd replacements during the first few weeks of life, but there are important differences. Veal calves should be given an ''all liquid diet,'' either whole milk or a milk replacer. No dry feed, neither concentrates nor hay is used as these result in an objectionable dark or red meat color. Likewise, veal calves should not be permitted to eat bedding material.

During the first several days of life, the veal calf should be fed about the same as the young calf being raised for herd replacements (see Chapter 14). In this period, the importance of preventing disease problems and keeping mortality low is paramount (see Section 14.1). Thus the calf should receive colostrum (Section 14.2.a), and the amount of milk or milk replacer should be limited so as not to increase the incidences of common diarrhea. Because of the large amount of milk or milk replacer required for veal calves, nipple-pail feeding is preferable to open buckets as digestive disorders are less likely to occur (see Section 14.7). Likewise, veal calves should be fed at least twice per day.

After the early "critical period" in its life, the amount of milk or milk replacer given the veal calf is gradually increased to permit maximum gains. The age at which the increase begins varies with producers. For maximum caution in preventing diarrhea some producers may not increase the feeding rate until the calf is about 3 weeks old. Others start shortly after a week of age. By the end of 4 weeks the veal calf will be receiving as much milk or milk replacer as it can take. This is in contrast to the herd-replacement calf which will be weaned about this time or shortly afterward.

Maximum possible gains are needed to obtain the high degree of fatness needed at an early age. Generally a choice veal calf has an obvious "pinch" of fat around the tailhead and internally some fat covering of the kidneys and feathering between the ribs.

It takes about 10 lb of whole milk to produce a pound of gain in a veal calf. Since the total dry matter content of whole milk is only about 12.5%, it is apparent that the young calf has a very high feed efficiency. The amount of energy in a milk replacer required for a pound of gain is comparable to that of whole milk. Thus, around 1.3 to 1.5 lb of milk-replacer powder is needed for each pound of growth.

A milk replacer for veal calves should be quite similar to that used for herd replacement calves with some crucial differences including a higher fat content. Whereas, 10% fat is adequate for herd-replacement calves, about 16–25% is needed for vealers. Adequate emulsification of the fat is important. Lecithin is a good emulsifying agent. It is crucial that the size of the fat globules remain small and that the powder readily mix in water. Often animal fats have given better results than vegetable fats in milk replacers. Generally, hydrogenation of vegetable fats to make them more saturated improves calf performance. Highly unsaturated fats have given poor results in research studies.

Historically, the meat of calves fed only milk has been distinguished from that of those given dry feed, hay or concentrate, by the pale color. The characteristic pale color, due to iron deficiency anemia (see Section 5.22), is one of the trade marks of "good veal." If the iron deficiency anemia of the veal calf is too severe, growth and performance will be reduced. Studies in Scotland have shown that between 20 and 40 ppm iron in the dry matter of the milk replacer will give maximum performance while retaining the pale, anemic color (Webster *et al.,* 1975).

As with milk replacers for other calves, those for vealers should contain antibiotics, vitamins A, D, and probably E, all the essential minerals, and a high protein content (see Appendix, Table 1). The protein percentage can be reduced somewhat as the veal calf becomes larger (see Appendix, Table 1). Best results, normally, have been obtained when milk solids make up the bulk of the replacer (see Section 14.2.e).

16.2. MANAGEMENT OF VEAL CALVES

Veal finishing operations can be divided into two major types. Some dairymen finish some or all excess calves born in their herd. Other veal growers purchase calves, often from "auction barns." With either type of grower, the amount of profit or loss from each calf is dependent on several factors. One of the most important is starting with the right kind of calf. For best results, healthy, large, blocky calves should be selected. If the calves are purchased, it is desirable that the navel be dry and have no indication of infection. It is usually recommended that calves doing poorly be culled early in the feeding program.

Problems and procedures in controlling and preventing diseases of veal calves are similar to those with other calves (see Section 14.6). However, when calves are purchased from many sources, chances of introducing disease organisms are greatly enhanced. Likewise, one cannot be sure a purchased calf has received adequate colostrum. Accordingly, with purchased calves, a routine, strong, prophylactic treatment program is wise if not manditory (Warner, 1969). This may consist of a broad spectrum antibiotic given both orally and systemically soon after purchase and continued for several days. Substantial amounts of vitamins A, D, and often E are frequently administered.

Because of the high incidences of infections in purchased calves, there should be a routine program for dealing with sick calves. It is widely recommended that temperatures be taken daily during the first 2 or 3 weeks. When the temperature exceeds 103°F, treatment including reduced feed intake should begin immediately (see Section 14.6).

With purchased calves, the emphasis on strict sanitation and minimization of opportunity for spreading infections is most important. Often it is desirable to purchase enough calves to fill the barn at about the same time. Between groups the barn can be thoroughly cleaned and sterilized. A period of time between groups is beneficial.

The most satisfactory weight for selling finished veal calves varies with the market and may range from less than 200 lb to more than 300 lb. Beginning with a large calf, 200 lb should be attained in 7 to 8 week and 300 lb in 12 to 14 weeks (Warner, 1969).

Housing requirements for veal calves are similar to those for raising other calves. Because of the much longer liquid feeding period, the labor requirements usually make individual outside pens uneconomical. Too high temperatures appear to cause problems including more pneumonia (Webster *et al.*, 1975). Prolonged exposure to excessively warm temperatures induces shedding of hair. The calf may lick the hair and develop hair balls in the abomasum which can cause digestive problems. Temperatures under 70°F are desirable. A good barn should be well ventilated but free from excessive drafts. Too low relative humidity (below 50%) can dry respiratory passages increasing susceptibility to infections.

16.3. ECONOMICS OF VEAL PRODUCTION

Profits or losses from growing veal calves are quite dependent on a good feeding and management program and on keeping mortality low. Likewise, success requires a suitable economic environment. Cost relationship must permit one to pay all the costs with a reasonable margin remaining.

There are reasons for suspecting profit opportunities in veal finishing may be less frequently available in the future than has been true historically. One is the generally increasing relative prices for whole milk, dried skimmed milk powder, and other milk solids. Because these milk products are major input items for veal growth, the feed cost may continue to rise. Likewise, the value of dairy calves for producing dairy beef (see Chapter 17) has resulted in generally rising prices for suitable beginning calves. A part of this is related to the ever decreasing number of dairy cows in the United States and many other countries, and the wider realization that the dairy calf makes a very efficient beef animal. With the relative increased cost of both feed and animals, production cost of veal may continue to increase more than that of other meat. Thus veal is a luxury food.

REFERENCES

Warner, R. G. (1969). *Feed Manage.* **20** (No. 8) pp. 22-24.
Webster, A. J. F., H. Donnelly, J. M. Brockway, and J. S. Smith (1975). *Anim. Prod.* **20,** 69-75 (Rowett Research Inst. Reprint No. 745).

17

Raising and Feeding Dairy Beef for Meat Production

17.0. INTRODUCTION

Although dairy cattle are kept primarily for milk production, ultimately most are utilized for meat and are an important income source on dairy farms. The proportion of the total beef supply coming from dairy cattle varies widely in different parts of the world and among countries. In the United States dairy cattle provide about one-fifth of the beef (see Section 1.0), and in some countries, including a number in Western Europe, the majority of the beef is from dairy breeds. Often the term "dairy beef" is used to denote steers of dairy breeds grown and finished in much the same way as beef steers. In the broader sense, however, dairy beef also includes cull cows, heifers, bulls, and calves including veal calves.

The price of beef is an important factor in whether or not to cull individual dairy cows. When beef prices are high, the culling rate increases on the average dairy farm. Even so, most culling is for low production, mastitis, lack of reproduction, disease problems, undesirable disposition, and other factors related to producing milk. Although their beef value is substantially influenced by size and finish, generally no special effort is made to prepare cull cows for slaughter. The term cull, for dairy cows, refers to use for dairy purposes and not for beef. Many discarded cows and bulls are quite desirable meat animals, often being utilized as processed meat including hamburger. Large, old bulls known as "balogna bulls" frequently bring a good price because of the excellent yield of salable products.

Although fairly important, meat from cows and breeding bulls is produced somewhat incidently to the production of milk. Most female calves are needed as herd replacements. Thus, the main optional choice in meat production from dairy cattle is how to utilize the excess bull calves. Contrary to some popular opinion, young dairy animals, especially the larger breeds such as the Holstein, are highly desirable for beef production (Hartman, 1975; Ramsey *et al.*, 1963). For instance, the Holstein steer grows rapidly, has excellent feed efficiency, and yields

341

Fig. 17.1   Dairy steers make excellent beef animals. This photo shows steers of both dairy and beef breeding grazing at the Ruakura Animal Research Station, New Zealand. (Courtesy of W. J. Miller, Univ. of Georgia.)

high quality meat. As a producer of beef, he is fully equal to the best of the specialized beef breeds (Fig. 17.1).

The cost of keeping a beef cow for a year just to produce a calf to be used as a beef animal is considerable. Thus using surplus, male dairy calves is an efficient and low cost way to produce beef. This may be of crucial importance in further decreasing veal production (see Section 16.3). Even though overall efficiency suggests that surplus dairy calves contribute more to the human food needs as beef animals than as vealers, the real determination will be made by the economics of the market place.

17.1. CHARACTERISTICS OF DAIRY-BEEF ANIMALS

Although dairy steers are excellent animals for producing high quality meat with good feed efficiency, historically the United States beef market has been built around the characteristics of the specialized European beef breeds. Since these market considerations have an impact on the profitability of producing beef, feeding programs for dairy beef must recognize them even when they are not basically sound or in the best interest of consumers. The trend toward increased emphasis on lean meat, rather than excessive fat, gives the dairy steer a more favorable place. Both human health and cost of production would benefit from considerable further reduction in the amount of fat required for a "choice" grade animal (Cunha *et al.*, 1975). Such a change also would put the dairy steer in a more competitive position.

The suitability of dairy steers and heifers for beef production varies with the breed and individual animal. The larger breeds, especially the Holstein and

Brown Swiss, grow much more rapidly than smaller breed animals such as the Jersey and Guernsey. From the standpoint of flavor, tenderness, nutritional value, and overall desirability, the quality of meat from any of the major dairy breeds is equal to that of the specialized beef breeds (Ramsey *et al.,* 1963). With comparable feeding conditions, the carcass of dairy steers will contain less fat but will receive a similar grade by current standards. However, market grades, which are based primarily on confirmation, are lower for dairy steers than those of specialized European beef breeds. Dairy steers have less trimmable fat which requires considerable feed to produce (Cunha *et al.,* 1975) and is undesirable as human food.

17.2. FEEDING PROGRAMS FOR DAIRY-BEEF ANIMALS OF DIFFERENT AGES

In contrast to animals being raised for herd replacements or breeding bulls, excessive fattening of young dairy-beef animals does not cause subsequent problems. The key question is how to produce market animals at the maximum profit. The most desirable feeding program for growing and finishing dairy-beef animals is determined by relative prices of feeds and of the different type animals at slaughter. Primarily, these price relationships determine the proportion of concentrates and forages which should be fed.

17.3. FEEDING THE YOUNG DAIRY-BEEF CALF

Typically, the most practical way to feed the young calf for dairy beef is about the same as for herd replacement animals (see Chapter 14). This involves a limited amount of milk replacer or whole milk plus free-choice concentrates and hay as soon as the calf will eat these dry feeds. Some dairy beef growers purchase baby male dairy calves and suckle them on nurse cows during the first few weeks. Two to several calves may be raised on the same cow during a lactation. This method is effective in raising calves, but often is only suitable for a small producer with adequate family labor. Young dairy-beef calves raised on limited milk or milk replacer grow less rapidly than those receiving more milk from nurse cows. The feed cost, however, is substantially lower with less milk and there is relatively little effect on the animal at a normal slaughter weight.

Until about 3 months of age, the young dairy-beef calf can be fed quite similarly to the herd-replacement calf. After that the amounts of grain which should be fed depends on the relative cost of forages and concentrates. If the cost of nutrients in forage is much lower than in concentrates, usually it is wise to limit concentrates to about the amounts fed herd-replacement animals.

17.4. FEEDING THE GROWING DAIRY-BEEF ANIMAL

The total amount of nutrients required to produce dairy-beef animals is less with a maximum rate of growth until the animal is ready for slaughter. The reason is the lower maintenance requirements due to less time needed to reach the final weight. A highly digestible feed is required to achieve maximum rates of gain. Although this can include a substantial percentage of very high quality forage (Scruggs, 1975), especially pasture, usually a high proportion of concentrates is needed. When the cost of nutrients in forages is substantially lower than from concentrates, often it is most profitable to grow dairy-beef animals primarily on forages until they reach 500 to 800 lb. This approach is comparable to that of young heifers or bulls raised for breeding purposes (see Chapter 15).

17.5. FEEDING DAIRY BEEF DURING THE FINISHING PERIOD

The most profitable way to feed dairy-beef animals during the finishing period prior to slaughter depends on the cost of feeds and the market place for different type animals. Contrary to widely accepted dogma, highly nutritious and delicious beef can be obtained from dairy steers finished on high quality forage, especially pasture (Anonymous, 1976; Ramsey et al., 1963). However, the rates of gain generally are lower on all-forage diets. Likewise, the degree of fatness and the official grade are usually lower when no grain is fed.

Dairy steers of the larger breeds perform quite well in the feed lot when given a high-concentrate diet. In the United States for many years, this has been the traditional way to finish beef cattle, but is only practical when concentrates are comparatively low in cost. Whether future prices will justify feeding large amounts of grains and other concentrates to finish beef steers, either of dairy or beef breeding, is now unknown. The use of less grain would not materially lower the value of the beef for the consumer (Cunha et al., 1975).

17.6. OTHER SPECIAL CONSIDERATIONS

17.6.a. Effects of Age and Finish on Feed Efficiency

The feed efficiency of growing cattle is greatly influenced by the stage of growth. Young calves are very efficient in converting feed into body gains. Veal calves make a pound of gain for about 1.25 pounds or dry matter in whole milk (see Section 16.1). As cattle grow older and larger, the amount of feed required per pound of gain continues to increase for two reasons. As they become larger, the proportion of the feed used for maintenance increases and the proportion of the gain represented by fat increases. Fat contains far more calories than protein, bone, and water—the other main constituents of the body. Much more feed is needed to produce a pound of fat tissue than any other component. Because of the reduced feed efficiency, the degree of fatness or finish most profitable for dairy steers is the minimum needed to obtain a satisfactory grade and price per pound.

17.6.b. Dairy-Beef Crossbred Cows

One of the key factors determining rate of gain and weaning weight of beef calves is the amount of milk produced by the calf's dam. With specialized beef breeds, milk production often is quite low. Crossbred cows with some dairy breeding, especially Holstein or Brown Swiss, frequently are superior in the beef-cow herd.

17.6.c. Future of Dairy Beef

Regardless of cost relationships among feeds, the fundamental competitive advantages of using surplus bull calves make this the lowest cost way to produce desirable beef. However, the amount of beef produced from dairy steers is limited by the number of cows in dairy herds. Distribution of the advantages of the dairy-beef animal among the dairymen producing the calf, the feeder(s) growing it to market weight, the packer, other handlers, and the consumer is determined by the market place. The distribution varies with time and location.

REFERENCES

Anonymous. (1976). *Prog. Farmer* May, p. 66.

Cunha, T. J., A. Z. Palmer, R. L. West, and J. W. Carpenter (1975). *Feedstuffs* **47**, No. 13, 37, 38, and 47.

Hartman, D. A. (1975). *Hoard's Dairyman* **120**, No. 12, 745, 748, and 749.

Ramsey, C. B., J. W. Cole, B. H. Meyer, and R. S. Temple (1963). *J. Anim. Sci.* **22**, 1001–1008.

Scruggs, C. G. (1975). *Prog. Farmer* August, pp. 27–30.

18

Nutritional and Metabolic Disorders of Dairy Cattle

18.0. INTRODUCTION

How dairy cattle utilize nutrients, the nutrients required, feeds which supply the nutrients, and the feeding of various types of dairy cattle were discussed in earlier chapters. If everything always went smoothly, this information should be adequate to understand the feeding and nutrition of dairy cattle. Unfortunately, several nutritional and metabolic disorders frequently occur. Accordingly, successful dairy cattle feeders must be prepared to cope with these problems. Thus an understanding of the abnormalities is essential. While not a veterinary guide, this chapter is devoted to a brief discussion of some of the major metabolic and nutritional related disorders of dairy cattle.

18.1. MILK FEVER (PARTURIENT PARESIS)

Milk fever is very common among high-producing dairy cows and is of considerable practical importance. The term "milk fever" is a misnomer, as body temperature is normal or below normal. Generally, milk fever occurs during the first three days after calving, but occasionally is observed earlier or later. It rarely develops in "first calf heifers," has a very low frequency after the second calving, but is much more prevalent in mature and older cows.

Susceptibility to milk fever is genetically related (Miller, 1970). The incidence is much higher among Jerseys than with other breeds, with Guernseys apparently having more than Holsteins. Likewise, cows which have milk fever after one calving are much more likely to have it again. Much of this milk fever susceptibility of certain cows is believed to be caused by genetic differences.

The symptoms of milk fever are progressive including decreased or loss of appetite, a general uneasiness or apprehensiveness and possibly hyperexcitability, dull eyes, an unsteady gait, and weakness. These are followed by the cow being unable to rise and taking a characteristic position of lying on her sternum with her head tucked into the flank, coma and if untreated, usually death (Fig. 18.1). The symptoms of milk fever are caused by too little calcium in the blood serum and other extracellular fluids. Normally the blood serum contains 9–12 mg of calcium per 100 ml. If the level decreases to about 5 mg per 100 ml, milk fever symptoms usually develop. The amount of calcium entering and leaving the blood serum is quite large relative to the amount there at any one time. Because of the great importance of calcium for numerous functions, the blood serum content is regulated very closely (see Section 5.16). Normally, with major changes in the amount of calcium eaten or used, the content in the blood serum varies very little.

With the onset of lactation, the quantity of calcium going into the milk is very large relative to that in the blood serum. The calcium secreted into colostrum in one hour is equal to about half the total in the blood serum at any one time (Miller, 1970). Since this is about the decrease necessary to produce milk-fever symptoms, it is evident that calcium in the blood serum must be continually replenished to avoid milk fever.

Normally, the homeostatic control mechanisms keep the blood serum relatively constant (see Section 5.16 for more detail). After calving, in some cows, these control mechanisms fail to counter the sudden large drain of calcium taken from the blood for milk synthesis. Thus milk fever is caused by failure in the regulation of calcium metabolism (Miller, 1970). The basic reason for the failure of the regulatory system has not been established.

Without treatment a very high percentage of cows having milk fever will die. An intravenous injection (about 250 to 500 ml of solution containing 8–12 gm of Ca) of calcium gluconate (borogluconate) is the most frequently used treatment. Often the mixture also will contain magnesium and possibly some phosphorus and even potassium. Some scientists believe udder inflation, a much older treatment, has a sounder physiological basis (Mayer et al., 1967). Injections of

Fig. 18.1 Milk fever cow. (Courtesy of R. H. Whitlock, Univ. of Georgia.)

calcium often cause excessively high blood calcium for a period of time, probably inhibiting the normal adjustments in the calcium regulatory mechanism which was the original cause of the milk fever. Theoretically, the relapse ratio should be lower after udder inflation, but udder inflation may contribute to mastitis. Antibiotics and good sanitation can minimize this problem.

In contrast to the very high death rate without treatment, about 75% of cows recover from milk fever with only one treatment of calcium gluconate or udder inflation. The high recovery rate is somewhat surprising as the amount of calcium injected is only sufficient for a short period of time. Thus, with only one treatment, apparently the regulatory mechanisms controlling calcium metabolism usually begin functioning normally again.

In some instances, more than one treatment is required. Likewise, some cows may be unresponsive, often possibly due to other complications. Also, sufficient nerve or muscle damage may occur during the milk fever period to lead to the "downer cow syndrome." The far higher incidence of milk fever in mature cows compared with first calf heifers appears to be associated with the much smaller amount of readily exchangeable calcium in the bone of older cows. Mature cows are less able to mobilize calcium from the bone than young ones.

Milk fever represents an important economic loss to the dairyman. In addition to the cost of treatment and those individuals which become downer cows, or die, cows having milk fever stay in the herd for a shorter period of time (Payne, 1964). Because of the losses, it is highly desirable to prevent milk fever. Contrary to the obvious, feeding excessive amounts of calcium during the dry period

TABLE 18.1

Effects of Feeding Varying Levels of Calcium for 2 Weeks Prior to Calving on Incidence of Milk Fever in Jersey Cows[a]

Herd	Calcium treatment	Ca[b] (gm/day)	P (gm/day)	Milk fever Cases	Milk fever %
1	High	40	27	2/17	12
	Deficient	18	28	0/14	0
2	High	46	51	6/14	43
	Deficient	15	30	0/10	0
3	Very high	162	60	7/11	64
	Deficient	14	25	0/6	0
4	Very high	173	60	5/18	28
	Deficient	13	27	0/7	0

[a] Adapted from Goings (1976) and Wiggers et al. (1975).

[b] The estimated calcium requirement for these cows would have been about 30 gm/day (see Appendix, Table 2). NRC (1978).

increases the incidence of milk fever (Goings, 1976; Miller, 1970). Feeding a relatively low-calcium and high-phosphorus diet for a few weeks before calving followed by more normal amounts after calving has reduced the incidence of milk fever (Table 18.1) (Goings, 1976; Goings *et al.*, 1974; Miller, 1970; National Research Council, 1978; Wiggers *et al.*, 1975) (see Section 5.16).

Massive amounts of vitamin D (20 million units per day) fed for 3 to 7 days immediately prior to calving decreases the number of milk fever cases. If fed longer than 7 days, this vitamin D causes toxicity effects including calcium deposition in soft tissues. If the treatment is discontinued for more than about 1 day prior to calving, the beneficial effects disappear and the incidence may even increase. The vitamin D metabolite, 25-hydroxy-vitamin D_3, with its lower toxicity than vitamin D, appears to offer promise as a way to prevent milk fever (NRC, 1978).

It is generally believed that feeding a normal calcium to phosphorus ratio of between about 1.3 to 1 and 2.0 to 1 during most of the year is beneficial in preventing milk fever. Very wide or very narrow ratios of calcium to phosphorus are believed to increase the incidence of milk fever. The absolute amounts of calcium and phosphorus are important. Certainly, during the entire year, a deficiency of either calcium or phosphorus should be avoided. Likewise, a substantial excess of either may contribute to more milk fever. In Ohio research, carefully regulating the calcium and phosphorus intakes to about recommended levels has reduced the incidence of milk fever (Hibbs *et al.*, 1977).

The genetic susceptibility of cows to the disorder is also an important determinant of the amount of milk fever. If milk fever becomes a major problem on a farm, this factor may need careful consideration.

18.2. GRASS TETANY

Occasionally large numbers of lactating dairy cows die from grass tetany over a very short period of time. Grass tetany is caused by too little magnesium in the blood serum and other extracellular fluids (see Section 5.18). Most grass tetany occurs in lactating cows grazing highly succulent pasture in cool seasons. The time of year varies in different parts of the world. In the United States grass tetany is a less serious problem in dairy cattle than with beef animals because most dairy cattle are given substantial amounts of supplemental feeds along with succulent pasture. If pasture is the only or the major feed, under the high-risk conditions, grass tetany can be a serious problem with United States dairy cows just as it is in many other countries with highly developed pastures.

The practical way to avoid grass tetany is to provide an adequate amount of magnesium to dairy cows, especially when conditions are conducive to the development of this condition. Many other aspects of grass tetany are discussed in Section 5.18.

18.3. BLOAT

As discussed more fully in Chapter 1, a tremendous amount of fermentation occurs in the rumen and reticulum with some of the products formed being gases, mainly carbon dioxide and methane. Normally these gases are removed by eructation (belching). In certain situations, however, the gases are not eliminated, creating bloat (Figs. 18.2, and 18.3). Although many questions remain for research workers on the basic causes of bloat, it is generally accepted that inability to remove the gases rather than excess gas formation is the primary cause of b¹oat.

In practical livestock production, losses from bloat can be considerable, with published reports of bloat occurring in at least 23 countries (Clarke and Reid,

Fig. 18.2 (A) A bloated animal (left) and its identical twin, given poloxalene, not bloated. (B) Frothy rumen contents shooting out when cap was removed from fistula illustrating the pressure existing in rumen of the bloated animal (see Bartley, 1965). (Courtesy of E. E. Bartley, Kansas State University.)

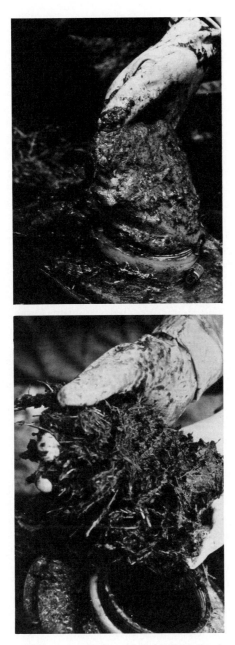

Fig. 18.3 Top photo. Rumen contents from bloated animal showing frothy nature due to trapped gases. Bottom photo. Rumen contents from animal given poloxalene did not contain frothy foam (see Bartley, 1965). (Courtesy of E. E. Bartley, Kansas State Univ.)

1974). It has been estimated that the total loss is as much as 100 million dollars per year in the United States.

Bloat may occur either in cattle grazing legumes or those fed large amounts of concentrates. Historically, the greatest practical problem has been pasture or legume bloat (Cole and Boda, 1960), but in comparatively recent years, feedlot bloat resulting from feeding a diet containing too little forage or roughage often is a serious problem. Whereas legume bloat may kill cattle within hours after the feeding begins, feedlot bloat usually develops slowly over weeks and often becomes chronic (Clarke and Reid, 1974). Both legume and feedlot bloat are the frothy or foamy type in which the gas is trapped and does not escape easily (Clarke and Reid, 1974) (Fig. 18.3). This probably is the major reason the excess gas is not removed normally. Free-gas bloat also can occur due to an esophageal obstruction.

Certain legumes, including alfalfa, white clovers, of which ladino is one, red clover and alsike clover, if grazed as the only forage, often cause severe bloat. Bloat is most likely to occur during the rapidly growing, highly succulent phase (Clarke and Reid, 1974). In some situations, a high percentage of a herd may bloat within a very short time period, but more often fewer animals are involved. Not all legumes present an important bloat problem. Lespedezas, trefoils, crimson clover, arrow leaf clovers, hop clovers, sweet clover, kudzu, peas, and soybeans usually are not bloat-producing forages. Generally, grasses are not very bloat provocative; but some, including very young, highly succulent wheat, cause bloat under certain conditions (Bartley et al., 1975).

The degree of bloat ranges from very mild, causing minor discomfort, to acute bloat resulting in death if the pressure is not relieved. Chronic mild bloat can lower the performance of animals due, at least partially, to reduced feed consumption. In addition to the problems associated with excess pressure, severe, acute, bloat can produce rumen paralysis.

When frothy bloat occurs, often it is necessary to get rid of some of the gases to avoid death. Several ways of eliminating the gases have been used with some success (Cole and Boda, 1960). If the condition is not too severe, walking the animal uphill or vigorous exercise may be helpful. Likewise, inserting a tube or hose down the esophagus to the reticulorumen can relieve the pressure. Tying a stick in the mouth to stimulate salivation also can be beneficial. If the bloat is mild, antifoaming agents such as oils, fats, detergents, and silicones are effective in eliminating the excess gas in frothy bloat. These may have to be administered into the rumen via drench or stomach tube.

The physical treatment of last resort with acute bloat is the release of gas from the rumen by making a puncture directly through the left side of the animal near the middle of the triangle formed by the backbone, hipbone and last rib. This can best be done with an instrument known as the trocar. When the trocar is inserted, the cannula should be left in place to allow the gaseous material to move out over a period of time. This type of pressure release may be followed by infection or peritonitis. Thus veterinary assistance is advisable.

Although treatment often is necessary to prevent serious losses when bloat occurs, prevention is more desirable. Frequently, it is not practical to avoid the use of bloat-causing legumes. In many areas alfalfa and the bloat-producing clovers are highly productive and very nutritious forages which also fix large quantities of nitrogen in the soil. When bloat is likely to be a problem, feeding some dry forage or the inclusion of grasses in the pasture will eliminate or greatly diminish the amount of bloat. Likewise, feeding forage by green chopping reduces the probability of bloat relative to grazing. When grazing bloat-producing pastures, often it is advisable to use a bloat-preventing substance, many of which are antifoaming agents (Clarke and Reid, 1974; Cole and Boda, 1960). These antifoaming agents reduce the stability of the frothy material permitting the gas to escape (Figs. 18.2, and 18.3). Certain antibiotics are effective bloat-preventing agents for short periods, but they rapidly lose their value with continued use (Cole and Boda, 1960). Of the antifoaming agents, the nonionic surfactant poloxalene is highly effective against legume bloat, but less useful with feedlot bloat (Bartley *et al.*, 1965 and 1975; Clarke and Reid, 1974). One of the key problems with substances such as poloxalene is obtaining sufficient consumption at needed intervals to prevent bloat. The possible methods of administration include incorporating into the salt or feeding with concentrates. Another concern in a bloat preventative is to avoid substances which might be transmitted into the milk or meat causing a residue problem. Poloxalene apparently presents no residue problem and its effectiveness does not diminish rapidly with time (Bartley, 1965).

Feedlot bloat can be prevented by inclusion of adequate coarse forage or roughage material in the ration. Some concentrates appear to be much more bloat producing than others. A feedlot ration which produces bloat in dairy cattle also may contribute to other digestive problems such as displaced abomasum, acute indigestion, founder, or adversely affect fat content of milk.

Some animals are much more susceptible to bloat than others. It is believed that a part of this tendency is due to genetic differences in cattle (Clarke and Reid, 1974).

Many drugs and other substances have been shown to produce bloat experimentally but most are of greater academic interest than of practical importance. Although much research has been conducted, the substances in bloat-producing forages responsible for the bloat are only partially defined.

Shortly after death most cattle bloat rapidly due to the continued fermentation within the reticulorumen and the failure of gases to be eliminated in dead animals. This type of bloating should not be confused with that occurring in otherwise healthy animals.

18.4. KETOSIS (ACETONEMIA)

In many respects ketosis contrasts sharply with the three nutritional and metabolic disorders just discussed. Unlike milk fever, grass tetany, and bloat, ketosis seldom causes death and none are sudden. The basic cause(s) of ketosis is not well understood and may not be clear-cut when fully understood. Further, there are no sure ways of either preventing or treating this metabolic problem. Fortunately a high percentage of cows having ketosis recover spontaneously without treatment.

Ketosis is a metabolic disorder in which the concentrations of ketone bodies are elevated in several body fluids. These ketone bodies, B-hydroxybutric acid, acetoacetic acid, and acetone, are normal metabolites in animals, but with ketosis unusually high amounts are found in the blood, urine, and milk. Likewise, acetone can be readily detected on the breath of a ketotic cow and often in the milk.

When an animal is fasted or starved, elevated concentrations of ketone bodies normally develop in the body fluids. During starvation, the animal obtains energy for maintenance largely from utilization of body stores of fat. The ketone bodies are by-products from the incomplete conversion of fat to energy and other major end products.

Ketosis in dairy cows is classified as either primary or secondary. In secondary ketosis, the animal may have an elevated temperature indicating infection. Frequent causes of secondary ketosis include retained placenta, metritis, displaced abomasum, and hardware (Schultz, 1969). With primary ketosis, temperatures are not elevated and there is no apparent indication of a secondary cause, but often distinguishing between primary and secondary ketosis is difficult. In dairy cows, primary ketosis usually occurs between 10 days and 8 weeks after calving (Schultz, 1969). The incidence varies widely under different conditions. Apparently more ketosis occurs in the Northern United States than in Southern areas, and there are more cases during the winter feeding period. Excessively fat cows are more susceptible to ketosis (see Section 18.12). Underfeeding of concentrates after calving is believed to increase the probability of ketosis developing because an underfed cow must obtain a higher proportion of her energy needs from body fat stores. High-producing cows are believed to be more prone to develop ketosis (Schultz, 1971).

Reduction or loss of appetite is the first symptom of primary ketosis (Schultz, 1969). Often the cow is gaunt, has reduced rumen activity, and is somewhat constipated. Dullness is characteristic of most ketotic cows. However, a small percentage of cows with ketosis have the nervous type in which the cow is highly excitable. Diagnosis of ketosis generally is on the basis of excessive amounts of ketone bodies in the urine, blood, and/or milk, with failure to find other problems which would indicate ketosis as a secondary effect.

Along with the decreased appetite, milk production of ketotic cows is reduced, but fat content of milk may be very high. If ketosis persists, the cow may lose considerable weight and produce much less milk for the whole lactation. Cows having ketosis in one lactation are more prone to this disorder in later lactations. Other changes occurring in ketotic cows include decreased blood sugar, increased free fatty acids in the blood, lower liver glycogen, and elevated liver fat content.

Over a long period of time, numerous treatments have been developed for ketosis. Many were enthusiastically promoted soon after development, but most have since been discarded as of little value. Perhaps the oldest definitely beneficial treatment is the intravenous injection of glucose. Treatment with hormones including adrenocorticotropic hormone (ACTH) or glucocorticoids has had some apparent success. The oral administration, as feed or drench, of propylene glycol or sodium propionate may be beneficial. With all treatments there is a comparatively high incidence of relapses or noncures. Frequent spontaneous recoveries complicate the evaluation of a treatment.

As with most other metabolic disorders, preventing ketosis is more desirable than treatment. How this can be done is not clearly defined. It is believed that avoiding excessive fattening during the dry period, feeding adequate concentrates in early lactation, providing sufficient good forage, meeting all nutrient needs, avoiding hay crop silages high in butyric acid, and eliminating any stressful conditions will reduce the incidence of ketosis (Schultz, 1971). Sometimes propylene glycol or sodium propionate are fed in problem herds.

18.5. NITRATE TOXICITY

Certain plants, including corn, cereals such as oats, and weeds, under some conditions accumulate large amounts of nitrates. Most of the nitrate is in the stems with very little in grain. The species of plants involved normally do not contain harmful amounts of nitrate. When the soil has huge quantities of nitrogen such as on feedlot areas, following use of very large amounts of manure, or from fertilizer spillage, nitrate accumulation may occur. Most problems develop where drought or other stress conditions retard growth and prevent normal maturity and seed formation.

Since moderate amounts of nitrates are readily utilized by ruminants as a source of nonprotein nitrogen, toxicity becomes a problem only with excessive quantities. Inspired by the deaths of large numbers of cattle in some individual herds over short time periods, extensive research was conducted on acute nitrate toxicity many years ago.

Following consumption of high amounts of nitrates in a short time, some of the nitrate is reduced to nitrite by the rumen microbes. Nitrite is much more toxic than nitrate. When absorbed into the blood, the nitrite converts hemoglobin to methemoglobin which is not able to carry oxygen to the tissue. If the proportion of the hemoglobin converted to methemoglobin is sufficiently high, the animal does not obtain sufficient oxygen. Symptoms such as difficult, labored and rapid breathing; staggering; and cyanosis of membranes develop. If the toxicity is sufficiently severe, death may occur in 1 to 4 hours after symptoms are evident. Excitement, vigorous exercise, or stressful conditions aggravate the problem and increase the death losses.

When acute nitrate toxicity is suspected, a quick indication can be obtained from the blood color. Methemoglobin gives the blood a chocolate-brown color in contrast to the normal red. If detected early, intravenous injections of methylene blue may prevent death. The methylene blue converts the methemoglobin back to hemoglobin. Recovery may occur spontaneously over a period of time when nitrate toxicity is not too severe. Both the diagnosis and definition of the effects of the acute nitrate toxicity are much simpler than for the mild but chronic toxicity. Major differences of opinions exist among scientists both on the extent and results of chronic nitrate toxicity. Several years ago, it was widely believed that chronic nitrate toxicity was a serious problem in numerous dairy cattle with effects ranging from abortions and other reproductive problems to low milk yield and widespread vitamin A deficiencies. Additional evidence, while not fully delineating the problem, indicates that it is less important than thought earlier. Cattle appear to adapt, at least partially, to higher than normal intakes of nitrates. Probably a part of this adaptation is associated with changes in the rumen fermentation.

Considerable attention has been given to the amount of nitrates which can be safely consumed in the feed and water and the degree of toxicity resulting from different intake levels. The effects of nitrate are influenced by several other factors. Well-fed cattle, especially those receiving substantial amounts of concentrates, can tolerate more nitrate than those that are underfed. Excessive

TABLE 18.2
Different Ways of Expressing the Nitrate Content of Feeds and Conversion of Data[a]

Nitrate	Nitrate–N content (%)	To convert to nitrate (NO_3) equivalent multiply by
Potassium nitrate (KNO_3)	13.8	0.61
Nitrate (NO_3)	22.6	1.0
Nitrate nitrogen (N)	100	4.4

[a] To convert data in ppm to %, move decimal point four places to the left.

amounts of protein, particularly highly degradable proteins, may decrease the tolerance to nitrates.

The amounts of nitrates can be expressed in different ways, often leading to confusion. For instance, in earlier studies, nitrate often was calculated as potassium nitrate (KNO_3) (Table 18.2). Nitrate also is expressed as nitrate nitrogen (N) or as the nitrate (NO_3). One gm of nitrate nitrogen (N) is equivalent to 7.2 gm of potassium nitrate (KNO_3) or 4.4 gm of nitrate (NO_3). Although the amount of nitrate required to cause a given degree of toxicity is quite variable, certain guidelines have been widely publicized and used (Table 18.3) (Crowley, 1970; Dolge, 1967; Prewitt, 1975).

18.6. DISPLACED ABOMASUM

In recent years, the diagnosed incidence of displaced abomasums has substantially increased and can affect 15–20% of the cows in a herd during a year (Buck, 1976). Usually the abomasum is displaced under and to the left of the rumen. In about 20% of cases, it moves to the right of the normal position. Displaced abomasum occurs most often in older and larger dairy cows within a few weeks after parturition, but occasionally the condition develops in heifers, bulls, or steers. The incidence of displaced abomasum is much higher in cows fed a high-concentrate, low-forage diet and in those fed only finely chopped forage. Occasionally, it occurs in animals fed normal amounts or even all forage diets. Apparently "going off feed," infections, mastitis, metritis, milk fever, and other situations causing reduced "abomasal tone" can lead to a displaced absomasum.

The severity of symptoms associated with displaced abomasum varies widely among individuals. Typical effects include reduced appetite and feed intake, lower milk production, loss of weight, and mild secondary ketosis. The signs

TABLE 18.3
Guidelines on Use of Feeds Containing Nitrates[a]

Nitrate in feed dry matter expressed as nitrate (NO_3) (%)	Use of feed
Below 0.44	Generally safe under all conditions
0.44 to 0.88	Usually safe when fed in balanced rations, but should make up no more than 50% of the total dry matter intake of pregnant animals
0.88 to 1.5	Limit to less than 50% of dry matter intake and feed with caution in well balanced rations
Over 1.5	Potentially toxic; feed with great caution

[a] Adapted from Prewitt (1975) and other sources.

closely resemble those of ketosis. A skilled veterinarian usually can differentiate the displaced abomasum from other abnormalities having similar symptoms or signs. In a minority of cases, the displaced abomasum can be corrected by rolling the cow on her back and massaging the abomasum toward the correct position, but more frequently surgery is required.

Due both to lost production and veterinary expenses, the displaced abomasum can be an important economic loss, easily reaching $150 per cow (Buck, 1976). Thus, feeding and management practices which minimize the incidence are desirable. These include avoiding excessive concentrates in the dry period, feeding adequate long forage (especially long hay, or pasture) with only limited corn silage in the dry period, and preventing excessive fattening.

18.7. DIARRHEA

Diarrhea is a major problem in the very young calf (see Chapters 14 and 16). Although relatively less important in other dairy cattle, diarrhea can develop as a result of a variety of nutritional and metabolic causes. These include excessive levels of dietary magnesium, vitamin A deficiency, other severe nutritional deficiencies or excesses, or a sudden change to much more laxative feeds. For example, abruptly switching dairy animals from dry feeds to highly succulent pastures such as young clover may produce diarrhea. Diarrhea is also caused by a number of infectious agents and internal parasites.

18.8. ACUTE INDIGESTION, FOUNDER, LACTIC ACID ACIDOSIS, LAMINITIS

When unadapted cattle consume a large quantity of concentrates over a short period, acute indigestion often develops with profound effects (Miller and O'Dell, 1969). This condition is called a number of names including founder, laminitis, and lactic acid acidosis. The sudden introduction of large amounts of readily fermentable carbohydrates, especially starches, into an unadapted rumen causes a major shift in the rumen microbes to those producing large amounts of lactic acid (Huber, 1975). Several major changes may follow including a very low rumen pH and high histamine production. The cattle go off feed, become generally weak, and appear depressed. Frequently there is severe damage to the lamina in the feet causing the typical syndrome of foot problems known as laminitis often with permanent damage (Maclean, 1971). Likewise, acidosis, dehydration of tissues, and severe diarrhea may result. Animals with severe cases may die. If they recover, the damage to the mucosal tissue of the rumen may have decreased absorptive ability causing an extended period of poor performance. Often liver abscesses are a secondary effect.

Some concentrates have a much greater tendency to cause the acute indigestion, founder, or lactic acid acidosis syndrome than others. Wheat is much worse than corn. Because of the metabolic problems associated with sudden feeding of large amounts of concentrates, cattle should be adapted to such rations over a period of time, perhaps as long as three or four weeks. Large quantities of concentrates should not be given animals which have been "off feed" or deprived of feed for a period of time. This is one of the reasons for beginning the feeding of a moderate amount of concentrates to heifers and cows prior to calving (see Sections 15.1 and 15.3).

18.9 RUMEN PARAKERATOSIS, LIVER ABSCESSES

When excessive levels of concentrates are fed, several digestive and metabolic problems often develop (see Section 10.1.b). These include reduced fat content in the milk, cattle going "off feed," rumen parakeratosis, and liver abscesses. Feeding all the forage too finely ground can produce similar effects (see Section 8.4). Rumen parakeratosis is characterized by a darkening color, an enlargement, and hardening of the papillae of the rumen wall (Siegmund *et al.*, 1973). The keratinized epithelial cells of the papillae are less effective in absorbing nutrients resulting in less efficient animal performance. In fattened cattle, a higher incidence of liver abscesses often accompanies rumen parakeratosis. More research has been conducted on rumen parakeratosis and liver abscesses in finishing beef cattle than in dairy cattle. It is believed that the effects on performance may be even more serious in lactating cows because of the longer time they would remain in the herd.

18.10. DEPRAVED APPETITE

Often dairy cattle develop an appetite for many nonfeed materials such as dirt, wood, and many other substances. In a few instances these depraved appetites have been associated with a definite nutritional deficiency such as chewing on old bones by cattle fed inadequate phosphorus. More frequently the depraved appetite does not appear to have a specific cause. Some of the chewing more likely results from boredom and then becomes a habit.

There is a widespread belief that cattle develop an appetite for materials which have something needed. This has sometimes been translated into the belief that given a free choice of materials containing all the nutrients needed, the cattle will choose a balanced diet. Specifically, this has been used for minerals. Most of the research data available indicates that dairy cattle will not choose the amount of each nutrient needed (see Sections 5.6.e and 13.2.b).

18.11. PRUSSIC ACID TOXICITY, OTHER TOXIC SUBSTANCES IN FEEDS

A number of grasses, including sudan grass *(Sorghum vulgare sudanesis)*, johnson grass *(Sorghum halepense)* and common sorghums *(Sorghum vulgare)* sometimes develop toxic levels of prussic acid (hydrogen cyanide), which is a very deadly poison (Siegmund *et al.*, 1973). The toxic levels of prussic acid are most likely to occur following stress conditions such as drought, frost, wilting, or disease. Some wild plants such as wild cherry *(Prunus serotina)* also may have toxic concentrations of prussic acid in some situations, especially in the wilted state. Prussic acid poisoning inhibits certain enzyme systems necessary for the transport of oxygen, causing a deficiency for the tissues. If there is sufficient poison, death may occur in a short period. With small amounts the animal may recover or take a somewhat longer time to die.

When conditions conducive to prussic acid poisoning are suspected, the feed should be tested or an animal of low value can be grazed and any toxicity observed. Forage containing excessive amounts of prussic acid, can be preserved as hay with little danger of poisoning. Likewise, ensiling greatly reduces the prussic acid content.

Many other toxic substances are sometimes found in feeds. Among these some of the minerals such as lead, arsenic, fluorine, molybdenum, and selenium, which are most often present in toxic amounts, are discussed in Chapter 5. Several forage plants such as sweet clover, Phalaris grass, and fescue, sometimes have toxic amounts of harmful substances (Siegmund *et al.*, 1973). Others may harbor toxins such as ergot or mycotoxins. Numerous other plants, including many wild ones, contain toxic substances (Siegmund *et al.*, 1973), but usually these are not eaten by well-fed dairy cattle. The probability of their being consumed is much greater with hungry cattle.

The list of toxic or harmful chemicals which can accidently find their way into cattle feed is exceedingly long (Siegmund *et al.*, 1973). Thus care should be taken to avoid the entry of unwanted substances into dairy cattle feed or water.

18.12. FAT COW SYNDROME

In recent years a combination of metabolic, digestive, infectious, and reproductive conditions affecting the overly fat cow near the time of calving have

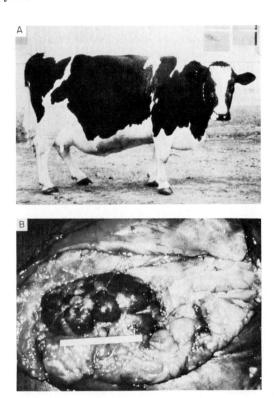

Fig. 18.4 When fed too much energy for extended periods, cows become too fat, leading to greatly increased metabolic and health problems. (A) Excessively fat cow. (B) Kidney from fat cow. (Courtesy of D. A. Morrow, Michigan State Univ.) (Morrow, 1976a.)

become known as the "fat cow syndrome" (Fig. 18.4) (Morrow, 1976a,b; NRC, 1978). Cows affected by this widely observed condition have a greatly increased susceptibility to stress and diseases, especially around calving time. Generally this syndrome develops in cows fed substantially more utilizable energy than needed for an extended period. Although observed in those on various feeding programs, it has occurred most frequently in cows fed large amounts of corn silage throughout the lactation and dry periods. Often the cows also have received substantial quantities of concentrates both in late lactation and in the dry period. A low level of milk production and a long dry period increase the probability of excessive obesity. Deficiencies of other nutrients such as too little protein and/or certain minerals may be a factor in the development of the syndrome.

In one study, cows having the fat cow syndrome had a greatly increased incidence of ketosis, retained placenta, and deaths (Table 18.4) (Morrow,

TABLE 18.4
Disease Conditions and Losses Associated with the Fat Cow Syndrome
When Conditions Were Favorable for the Syndrome and after
Preventive Procedures Were Effected[a]

| | (% of cows calving) | |
Condition	Fat cow syndrome conditons	After preventive procedures initiated
Ketosis	38	3
Retained placenta	62	13
Mastitis	6	2
Milk fever	5	2
Deaths	25	3

[a] Adapted from Morrow (1976a). 120 cows calved during each of the periods.

1976a). Likewise, both milk fever and mastitis were increased. The syndrome also has been identified with increased occurrence of uterine infections, displaced abomasum, and "downer" cows. Often the response of cows to treatment for the fat cow syndrome is poor with many dying or performing poorly, perhaps due to severe liver and kidney damage (Morrow, 1976b).

The best approach to the fat cow syndrome is to avoid the condition by not feeding excessive amounts of usable energy, especially during late lactation and the dry period. Likewise, excessively long dry periods should be avoided. Cows in good condition at the end of the lactation often will obtain sufficient energy from forage alone in the dry period. Where feasible, it may be desirable to provide pasture or hay during this time. Whatever the forage program, the dry cows should be fed an adequate amount of the nonenergy essential nutrients. If the feeding program requires that dry cows be given corn silage as the only forage, sufficient protein and essential minerals are important. The quantity of corn silage should be limited to the extent necessary to avoid excessive obesity.

18.13. OTHER NUTRITIONAL AND METABOLIC
PROBLEMS OF DAIRY CATTLE

Although the above problems are among the more important of those related to the nutrition and metabolism of dairy cattle, they do not represent a complete list. For instance, at times the feeding program causes undesirable flavor(s) in milk. Examples including feed flavors when cows consume wild onions or other plants are well known to dairymen. In some situations, feeding silage in restricted places prior to milking produces an undesirable milk flavor. Similarly, milk from cows having ketosis has a characteristic "off flavor" caused by acetone.

Many flavor problems in milk are more difficult to diagnose than those discussed above. Some, including oxidized milk, may be related to nutritional inadequacies. For instance, the incidence of oxidized milk is higher when cows are consuming stored feed than when they are given pasture. Often inclusion of pasture or alfalfa meal in the ration has reduced this problem.

REFERENCES

Bartley, E. E. (1965). *Hoard's Dairyman* **110,** No. 23, 1373 and 1395.

Bartley, E. E., H. Lippke, H. B. Pfost, R. J. Nijweide, N. L. Jacobson, and R. M. Meyer (1965). *J. Dairy Sci.* **48,** 1657–1662.

Bartley, E. E., G. W. Barr, and R. Mickelsen (1975). *J. Anim. Sci.* **41,** 752–759.

Buck, G. R. (1976). *Hoard's Dairyman* **121,** No. 21, 1252–1253.

Clarke, R. T. J., and C. S. W. Reid (1974). *J. Dairy Sci.* **57,** 753–785.

Cole, H. H., and J. M. Boda (1960). *J. Dairy Sci.* **43,** 1585–1614.

Crowley, J. W. (1970). *Hoard's Dairyman* **115,** No. 16, 879.

Dolge, K. L. (1967). *Feedstuffs* **39** (No. 26), p. 40.

Goings, R. L. (1976). *Proc. Cornell Nutr. Conf. Feed Manuf.* pp. 118–121.

Goings, R. L., N. L. Jackson, D. C. Beitz, E. T. Littledike, and K. G. Wiggers (1974). *J. Dairy Sci.* **57,** 1184–1188.

Hibbs, J. W., W. E. Julien, and H. R. Conrad (1977). *Hoard's Dairyman* **122,** No. 15, 911 and 931.

Huber, T. L. (1975). *In* "Buffers in Ruminant Physiology and Metabolism" (M. S. Weinberg and A. L. Sheffner, eds.), pp. 96–106. Church and Dwight Co., New York.

Maclean, C. W. (1971). *Vet. Rec.* **89,** 34–37.

Mayer, G. P., C. F. Ramberg, Jr., and D. S. Kronfeld (1967). *J. Am. Vet. Med. Assoc.* **151,** 1673–1680.

Miller, W. J. (1970). *Proc. Ga. Nutr. Conf. Feed Ind.* pp. 32–42.

Miller, W. J., and G. D. O'Dell (1969). *J. Dairy Sci.* **52,** 1144–1154.

Morrow, D. A. (1976a). *Hoard's Dairyman* **121,** No. 12, 747.

Morrow, D. A. (1976b). *J. Dairy Sci.* **59,** 1625–1629.

National Research Council (NRC) (1978). "Nutrient Requirements of Dairy Cattle," 5th rev. ed. Natl. Acad. Sci., Washington, D.C.

Payne, J. M. (1964). *Vet. Rec.* **76,** 1275–1282.

Prewitt, L. R. (1975). *Hoard's Dairyman* **120,** No. 19, 1119.

Schultz, L. H. (1969). *Feed Manage.* **20** (No. 2), pp. 24–27.

Schultz, L. H. (1971). *J. Dairy Sci.* **54,** 962–973.

Siegmund, O. H., C. M. Fraser, J. Archibald, D. C. Blood, J. A. Henderson, D. G. Howell, and R. L. Kitchell (Editorial Board) (1973). "The Merck Veterinary Manual," 4th ed. Merck and Co., Inc., Rahway, New Jersey.

Wiggers, K. D., D. K. Nelson, and N. L. Jacobson (1975). *J. Dairy Sci.* **58,** 430–431.

19

Integrating the Feeding and Nutrition of Dairy Cattle into Practical Feeding Programs for Individual Farms

19.0. INTRODUCTION

Most of the first eighteen chapters of this book have been devoted to the nutrient needs of dairy cattle, feeds as sources of nutrients, how different classes of dairy cattle are fed, and some of the more important nutritional problems. Other considerations also are important in practical feeding. The overall plan should furnish the required nutrients and include a practical way of putting the feeds before the animals as needed. The whole farm program should be economically sound, have an adequate amount of capital, and generate a satisfactory profit. Sufficient but not excessive labor, with adequate skills, training, and dedication is a key item for a successful operation.

Innumerable combinations of feeds and feeding programs will supply the nutrients needed by dairy cattle. Thus additional requirements and priorities can be and should be considered in a good program. A dairy farm operation can be viewed as a type of "manufacturing business" in which milk and meat are produced for sale usually at wholesale commodity prices. To be competitive, a dairy farm must obtain the needed inputs such as feed, labor, equipment, and other items at a reasonable cost. Likewise the operation should function smoothly and continuously. Among many other essentials is the necessity for having the feeds needed throughout the year.

19.1. CHOOSING FEED TO BE GROWN ON THE FARM AND THAT TO BE PURCHASED

It is convenient to think of a dairy farming operation as two independent but related businesses. The first is the care, feeding, milking, and handling of the

364

animals. The second concerns obtaining the feed supplies needed. The most practical approach in obtaining the feed supplies on different farms varies from growing almost all the feed ingredients to purchasing everything. Generally first priority is given to growing forage rather than concentrates. Reasons for this priority include the higher cost of transporting forages, and less readily available sources of high quality forages in commercial channels (see Chapters 9 and 12). A dependable supply of high quality forage is readily available commercially in only a few areas. Alfalfa hay grown under irrigation in Southwestern United States is one such example.

If feeds other than forages are grown on the dairy farm, usually the second priority goes to grains rather than protein supplements. This is partially related to the greater degree of processing needed with the protein supplements relative to grains. Choosing the proportion of feeds to be grown on the dairy farms sometimes is quite simple and obvious, but in other situations the decision is quite complex. If land use is costly and/or unavailable, it may be advisable to purchase most or all the feeds. In contrast, when land and other inputs are available, a large part of the needed feed may be most efficiently produced on the farm.

A major factor in whether to purchase or produce feeds sometimes centers around the amount of capital available. For a given size dairy herd, much more capital, labor, and managerial effort are needed if all the feed is grown on the farm. Alternately, the operator may be able to choose between a larger herd or growing a higher percentage of the feed needed.

19.2. CHOICE OF FORAGES DEPENDENT ON LOCAL CONDITIONS

In Chapter 9, considerable discussion was devoted to various types of forages for dairy cattle. Soil, climate, land availability, and other considerations, including cost of production may limit the practical choices to a very few. One forage or feed may have such overwhelming advantages as to be the only practical choice for certain purposes. For instance, corn silage as a major part of the feeding program, is such a choice in many areas. When substantial amounts of suitable land with a low utility value for other purposes are available, appreciable pasture may be profitably utilized. On other areas and farms, the lower yields, often inherent in pasture production, may severely limit its practical use.

In choosing a forage program, equipment and facility investment required in growing, harvesting, storage, and feeding are major considerations. Often it is more practical to keep the program simpler than would be indicated if investment cost were ignored. For instance, some dairymen choose to limit harvested forages to silages in order to avoid the need of owning hay-making, storage, and feeding equipment. Others use only hay. Either approach can be successful, if it fits the individual farm situation and is well done.

19.3. SUPPLEMENTING FEEDS GROWN ON THE
FARM

Perhaps the most typical dairy farm system is one in which most or all the forage is grown on the farm with a part or all of the concentrates purchased. In such a system, there should be both a long-range and a short-range plan for providing the concentrates.

19.3.a. The Long-Range Program

The best approach to providing concentrates and supplemental feeds depends on the particular farm situation. In the long-range, overall plan, several basic questions must be answered. These include how much of the concentrates will be grown and how much purchased, and whether the concentrates are to be mixed on the farm or purchased as already mixed feed or some combination. The amounts of grains that should be grown on the dairy farm are dependent on the effective costs involved and available resources, including land. Whether available land should be used this way is partially related to its value for alternate uses. Likewise, labor, equipment, and other input factors must be considered.

19.3.b. Mixing Concentrates on the Farm Versus
Purchasing

The decision concerning mixing the concentrates on the farm relative to purchasing already mixed feeds should depend on the approach which will result in the most profitable operation. Although no one answer is applicable to all dairy farms, a few general aspects are often important. Usually it is not practical to formulate dry milk replacers on individual dairy farms (see Chapter 14). These are utilized in low volumes and the necessary technology, ingredients and manufacturing procedures make mixing them on farms impractical.

When substantial amounts of grains are grown on the dairy farm, there are more advantages to mixing the concentrates on the farm or at a nearby local mixer than when all the ingredients must be purchased. The effective price of purchased grain usually is materially higher than its net value at the farm if it were to be sold. Normally, it is unwise to sell grain produced on the farm and then purchase other grain in the form of concentrates.

When grain is grown on the farm, often a high-protein, minerals, and vitamins supplement designed to combine with grains is purchased. With this approach, the individual dairyman is not as responsible for the technical details in formulating the concentrate. A portion of the responsibility is assumed by the organization preparing the supplement. If grain is not produced, many dairymen purchase the concentrate already mixed. Advantages to this approach include savings associated with not owning feed mixing equipment, reduced labor, lower capital requirements, and less technical know how. Usually, mixing and quality control are better with purchased mixed concentrates. The total cost of the feed may or may not be materially higher than for the ingredients necessary to mix the feed on the farm. Costs are not invariably higher for the mixed feed for a number of reasons. Often the feed company can purchase ingredients at a lower cost than an individual dairyman. Also, they can use more by-products or other ingredients that provide nutrients at a lower cost. Further, the feed manufacturer is able to change formulations more frequently in response to price changes.

In situations where the concentrates are to be fed mixed with the silage, dairymen may buy one or more ingredients and/or a combination of mixed concentrates. Depending on the availability of supplies, ingredients such as beet pulp, citrus pulp, high-moisture corn, and other specialty items may be purchased in substantial quantities for inclusion in the complete ration.

19.3.c. Storage of Concentrates

If protected from weather, insects, and rodents, most grains, and many by-product feeds can be stored in suitable structures for extended periods with minimum deterioration. Storage of grains and other concentrate ingredients requires considerable space and investment cost. Since grains are harvested seasonally, during much of the year a storage cost either on or off the farm is a normal expense. Concentrates are one of the largest expense items on a dairy farm. Thus if the ingredients are stored and owned for a considerable time before feeding, appreciable capital is required. When capital is in short supply, the lower amount of capital needed may be one reason for purchasing concentrates rather than mixing on the farm from stored ingredients. However, the use of the capital is an expense item, regardless, as prices on an average advance with time after harvest.

Concentrates should be fed within a relatively short period of time after being mixed. Several adverse changes occur when ground and mixed feeds are in storage including destruction of vitamins, reduction of palatability, and even spoilage. Some ingredients deteriorate much more rapidly in storage than others (see especially Chapter 10). For example, rice bran which contain a high level of fat may become rancid. Likewise, rancidity appears to be of much more concern with peanut meal than with soybean or cottonseed meal. Among the minor ingredients, vitamin A presents a serious deterioration problem. Accordingly, many commercial vitamin A products have been stabilized. Even so, potency is reduced during storage, especially in the presence of light or when temperatures and humidity are high.

19.3.d. The Short-Range Program—Least Cost
Formulation

For many years considerable attention has been devoted to least cost formulation (see Section 13.5.b). Because different feeding programs influence production and other costs, it is not appropriate to think just in terms of the "absolute" least cost. Rather, the best, overall profit picture is desired. After allowing for effects on production and other factors that influence profits, feeding programs should be designed for the lowest cost way to meet the needs of the animals. Usually least cost formulation is used less in the whole feeding program, especially the forage part (see Section 13.5.b). One reason is that exact values are seldom available for the cost of producing forages. Likewise, forage quality greatly affects the level of performance as well as the amount of concentrates required. Accordingly, value of forages vary widely (see Chapters 9 and 12). As discussed above, the forage part of the ration usually is determined as a part of the long-range feeding program. Especially where either the mixed feeds or the ingredients are purchased, the lowest cost way of supplying the concentrate needs can be determined by a least cost type of formulation procedure. In such a procedure consideration should be devoted to reliability of ingredient supplies.

19.4. FEEDING SYSTEMS AND SPECIAL CONCEPTS

On a dairy farm the whole feeding program must be organized into some system or systems. Although all possibilities cannot be discussed in this chapter, a brief mention of a few of the more widely used approaches seems desirable.

19.4.a. Separate Feeding of Forages and Concentrates

Over wide areas and long periods of time, forages and concentrates most often have been fed separately to dairy cattle (see Sections 13.2.c and 13.6.b). Before dairy farms were mechanized to the present degree, this, with some exceptions was about the only feasible way. The most important exception was feeding the concentrates on top of the silage in the feed bunk or manger. With pasture or long hay, it is necessary to feed concentrates separately. Excellent results have been achieved with separate feeding of forages and concentrates on numerous farms (see Sections 13.6 and 13.6.b).

19.4.b. Mixing Forage and Concentrates

In recent years, mixing concentrates and silage to form a complete feed has become a widely used practice on numerous dairy farms (Fig. 19.1). There is also some use of a complete feed composed of concentrates and a dry forage or

Fig. 19.1   Preparing a complete mixed ration using a front-end loader for moving concentrates to the mixer wagon. (Courtesy of W. J. Miller, Univ. of Georgia.)

roughage (see Section 13.2.c). The great increase in mechanization made practical the mixing of silage and concentrate as a complete feed. One of the main advantages is a reduction in labor (Table 13.1). The mixture provides the rumen microbes with more uniform feed material resulting in a more even fermentation, and therefore a potential for better performance. Often on small dairy farms it is not practical to mix the silage and concentrates into a single complete feed. The additional investment needed in mixing equipment and extra lots for several groups of cows can make it quite impractical with a small herd (see Table 13.1 and Sections 13.2.c, 13.6, and 13.6.b for more complete discussion).

19.5. SPECIAL MANAGEMENT TOOLS IN DAIRY CATTLE FEEDING

In attempting to find better ways of feeding dairy cattle, with changing conditions many special management approaches have been used or widely promoted. Some have been quite useful but more often they have been discarded due either to lack of acceptance or because of insufficient benefits. Among the most widely advocated special management tools is linear programming.

19.5.a. Linear Programming

The concept of linear programming, often in association with the use of computers, has been applied to numerous aspects of the dairy farm. With the feeding program, it has been used as a way to find the least cost concentrate ration (see Section 19.3.d). Obtaining an answer which would appear to maximize profit is the major advantage of linear programming. The greatest weakness is lack of sufficient accuracy for much of the needed input data.

Information critical to the outcome of many decisions is a matter of judgment and evaluation in which more precise calculations are of limited importance. One of the most serious handicaps of linear programming is failure to appreciate that uncertain information going into the computations gives answers with limited accuracy. This applies especially to aspects related to long-term feeding programs.

19.5.b. Other Approaches and Considerations

Frequent reference has been made to the greatly increased mechanization on dairy farms. Certainly more mechanization was essential to cope with the much higher labor costs compared with the price of milk and cattle. Caution is needed, however, to avoid impractical or over-mechanization.

Numerous dairymen have made serious financial mistakes in some aspects of mechanization. Obsolescence often is the key item in determining the appropriate rate of depreciation. For instance, equipment or facilities may become almost worthless due to better developments long before they cease to operate as planned. Along with obsolescence, conversion of a dairy farm to another use or the loss of the key manager can erase the economic value of otherwise fine facilities. When introduced, new equipment and approaches often have unrecognized disadvantages such as excessive repair costs, and failure to provide promised results. Ideally, the calculated depreciation rate should be sufficiently high to pay for the new investment in specialized facilities and equipment within a very small number of years. Otherwise, the investment may be unwise.

One of the cardinal principles of good salesmanship is to talk to the client in terms of solving a problem or filling a need. As a prospective purchaser, it is important to evaluate the item in terms of how effectively it will meet the need and at what cost. Serious financial problems may follow the purchase of expensive but ineffective equipment.

Perhaps it is risky to cite specifics, especially where they only fit some situations, but examples are needed to illustrate concepts. For instance, the most practical silo often is a horizontal one, either a bunker or a trench, due partially to the much lower initial cost (see Section 9.7.f). With good unloaders and mixer wagons, when properly managed, often labor and other operational costs for these silos are lower or competitive with much more expensive tower silos. Because of the smaller investment cost in the horizontal silo, the dairyman has much greater flexibility in future financial planning (Fig. 19.2).

Another example pertinent to keeping investment costs low is the use of mixer wagons and fence line feed bunks rather than the elaborate stationary mixing equipment of augers, conveyor belts, and high-cost feeding facilities. In the Southeastern United States using pine woods as a shelter and a partial answer to waste disposal greatly reduces the cost compared to elaborate buildings. Dairy

Fig. 19.2 On many farms, well-designed and managed horizontal silos substantially decrease the investment cost without causing important disadvantages, thus making the dairy farm much more profitable. (Courtesy of W. J. Miller, Univ. of Georgia.)

cattle can tolerate considerable cold and appreciably warm weather with relatively small effects on performance.

Other things being equal, an investment in a resource with many potential uses gives more flexibility and is preferred to one in highly specialized equipment or facilities. Expressed another way, a much higher depreciation rate is needed on specialized equipment and facilities. For instance, multiuse tractors and front-end

Fig. 19.3 Land does not depreciate as do buildings and equipment. In New Zealand, where this photo was taken, large amounts of nutrients for dairy cattle are obtained from pasture. (Courtesy of R. L. Kincaid, Washington State Univ.)

loaders have many uses. Thus, relative to augers and specialized feed conveying equipment, loss of value due to the farm being used for a different purpose or to obsolescence may be less with the tractor and front-end loader.

Of items with multiuse and low possibility of absolescence, land is the prime example (Fig. 19.3). The United States Internal Revenue Service recognized the basic concept that land does not depreciate, but buildings and equipment do. Admittedly, the concept is not infallible, but as a general rule, if additional land at reasonable prices will substitute for depreciable equipment or facilities, usually it is a more sound investment.

REFERENCE

National Research Council (NRC) (1978). "Nutrient Requirements of Dairy Cattle." Natl. Acad. Sci., Washington, D. C.

APPENDIX

USING APPENDIX TABLES OF NUTRIENT REQUIREMENTS
AND FEED COMPOSITION

Tables 1–5 in this Appendix were taken from the Nutrient Requirements of Dairy Cattle (National Research Council, 1978). Tables 1 and 2 can be used to calculate feeding programs for all classes of dairy cattle. Often Table 3 is more useful as values are expressed as the recommended nutrient content of the ration dry matter. In contrast to Tables 1 and 2, Table 3 gives values for all the nutrients for which the quantitative values are available. Table 1 is designed to permit calculations of the nutrient needs of both large (L) and small (S) breeds of dairy cattle growing at various rates. In this table, feed dry matter (DM) is listed for each class as a guide and is based on nearly maximum forage intakes under practical conditions. Always in formulating rations, it is essential to have a sufficient nutrient concentration to permit the animal to meet its needs without exceeding its maximum ability to consume dry matter. The maximum amounts of dry matter which lactating cows will consume are given in Table 5. Feed requirements in Table 1 were calculated from daily NE_g and NE_m requirements to determine the concentration of net energy required in feeds to equal the expected dry matter consumption (see Chapter 2). Other nutrients were calculated as described in the various chapters dealing with each nutrient. Maintenance for lactating cows and lactation requirements in Table 2 are based upon a feed concentration equivalent to 67–68% TDN. This is equal to 1.52 Mcal NE_l per kilogram DM.

Maintenance for dry, pregnant cows is based upon a feed content equal to 63% TDN that is comparable with 1.42 Mcal NE_l per kilogram. As discussed more fully in the text, especially Chapter 2, the values in Tables 1, 2, and 3 are designed for use with the feed composition values in Table 4.

Table 4 was adapted from a similar table in the Nutrient Requirements of Dairy Cattle (NRC, 1978). The NRC table was developed by the subcommittee assigned to prepare the publication. This committee obtained the information from several sources, including the exercise of best judgment for some values.

In preparing Table 4, several changes have been made from the NRC publication in an effort to make Table 4 even more useful. Some of these changes include the omission of several feeds having limited use, and the rearrangement of columns to have the values in the order of greatest ease in routine use. Several of the names designating feeds have been simplified to more closely correspond to those generally used. However, the international reference number has been retained to permit easy access to master NRC data pools.

As is discussed more fully in several chapters of the text, the nutrient content of feed ingredients is quite variable. Those presented in this table are intended to represent average values to be used as a guide. Values for individual feeds often will be substantially different.

The following are abbreviations used in the Appendix.

De	digestible energy
deg	degree
dehy	dehydrated
dist	distillers
exp	expeller
grnd	ground
IU	international units
lb	pound
ME	metabolizable energy
Mech-extd	mechanically extracted, expeller extracted
min	minimum
NE_g	net energy for gain
NE_l	net energy for lactation
NE_m	net energy for maintenance
ppm	parts per million
prot	protein
solv	solvent
TDN	total digestible energy
veg	vegetative
w	with
wo	without

TABLE 1.
Daily Nutrient Requirements of Dairy Cattle (from NRC, 1978)

Body weight (lb)	Breed size, age (wk)	Daily gain (lb)	Feed DM (lb)	Feed energy NE_m (Mcal)	NE_g (Mcal)	ME (Mcal)	DE (Mcal)	TDN (lb)	Total crude protein (lb)	Minerals (lb) Ca	P	Vitamins A (1000 IU)	D (IU)
						Growing dairy heifer and bull calves fed only milk							
55	S-1[a,b]	0.7	1.00	0.85	0.53	2.14	2.38	1.20	0.25	0.013	0.009	1.0	165
65	S-3	0.8	1.15	0.94	0.63	2.47	2.74	1.38	0.28	0.014	0.010	1.2	200
93	L-1	0.9	1.38	1.25	0.70	2.98	3.31	1.66	0.33	0.018	0.011	1.8	280
106	L-3	1.1	1.65	1.36	0.90	3.51	3.90	1.98	0.40	0.020	0.013	1.9	300
					Growing dairy heifer (F) and bull (M) calves fed mixed diets								
100		0.6	2.7	1.35	0.52	3.65	4.16	2.08	0.31	0.018	0.013	1.9	300
100(F)	S–10	0.8	2.8	1.35	0.69	3.98	4.51	2.25	0.36	0.020	0.013	1.9	300
100(M)	S–10	1.0	2.8	1.35	0.87	4.32	4.84	2.42	0.40	0.022	0.014	1.9	300
100(F)	L–3	1.2	2.8	1.35	1.05	4.58	5.10	2.55	0.44	0.024	0.015	1.9	300
100(M)	L–3	1.4	2.8	1.35	1.22	4.80	5.33	2.67	0.48	0.027	0.015	1.9	300
100		1.6	2.8	1.35	1.40	5.04	5.60	2.80	0.58	0.028	0.016	1.9	300
150		0.8	4.0	1.82	0.70	5.02	5.78	2.89	0.49	0.024	0.015	2.9	450
150(F)	S–19	1.0	4.1	1.82	0.88	5.39	6.17	3.08	0.55	0.026	0.015	2.9	450
150(M)	S–18	1.2	4.1	1.82	1.06	5.76	6.54	3.27	0.59	0.028	0.016	2.9	450
150(F)	L–9	1.4	4.1	1.82	1.24	6.07	6.85	3.42	0.64	0.031	0.017	2.9	450
150(M)	L–8	1.6	4.1	1.82	1.42	6.39	7.16	3.58	0.68	0.033	0.018	2.9	450
150		1.8	4.1	1.82	1.59	6.63	7.40	3.70	0.73	0.034	0.018	2.9	450
200		0.8	5.3	2.26	0.73	6.04	7.05	3.53	0.65	0.031	0.017	3.8	600
200		1.0	5.4	2.26	0.92	6.50	7.53	3.76	0.71	0.032	0.018	3.8	600

a Breed size: S for small breeds (e.g. Jersey), L for large breeds (e.g. Holstein).
b Age in weeks indicates probable age of S or L animals when they reach the weight indicated.

(Continued)

TABLE 1.
Daily Nutrient Requirements of Dairy Cattle (from NRC, 1978) (Continued)

Body weight (lb)	Breed size, age (wk)	Daily gain (lb)	Feed DM (lb)	NE_m (Mcal)	NE_g (Mcal)	ME (Mcal)	DE (Mcal)	TDN (lb)	Total crude protein (lb)	Ca	P	A (1000 IU)	D (IU)
200(F)	S-26	1.2	5.4	2.26	1.10	6.86	7.88	3.94	0.75	0.034	0.018	3.8	600
200(M)	S-24	1.4	5.4	2.26	1.29	7.22	8.24	4.12	0.80	0.036	0.019	3.8	600
200(F)	L-14	1.6	5.4	2.26	1.47	7.52	8.53	4.27	0.85	0.038	0.020	3.8	600
200(M)	L-12	1.8	5.4	2.26	1.66	7.89	8.90	4.45	0.90	0.040	0.021	3.8	600
200		2.0	5.4	2.26	1.84	8.27	9.28	4.64	0.94	0.041	0.022	3.8	600
					Growing dairy heifers								
300	S-38	0.8	7.9	3.07	0.83	8.12	9.64	4.82	0.88	0.034	0.021	5.8	900
300		1.0	7.9	3.07	1.04	8.60	10.11	5.06	0.92	0.036	0.022	5.8	900
300		1.2	7.9	3.07	1.25	8.98	10.51	5.25	0.96	0.037	0.023	5.8	900
300		1.4	7.9	3.07	1.46	9.47	10.97	5.48	1.00	0.039	0.024	5.8	900
300	L-23	1.6	7.9	3.07	1.66	9.89	11.40	5.70	1.04	0.041	0.025	5.8	900
300		1.8	7.9	3.07	1.87	10.26	11.76	5.88	1.09	0.043	0.026	5.8	900
400		0.8	10.5	3.81	0.96	10.16	12.18	6.09	1.15	0.040	0.026	7.7	1200
400	S-52	1.0	10.5	3.81	1.20	10.69	12.70	6.35	1.19	0.042	0.028	7.7	1200
400		1.2	10.5	3.81	1.44	11.24	13.26	6.63	1.22	0.043	0.029	7.7	1200
400		1.4	10.5	3.81	1.68	11.75	13.76	6.88	1.26	0.044	0.030	7.7	1200
400	L-32	1.6	10.5	3.81	1.92	12.17	14.18	7.09	1.30	0.045	0.030	7.7	1200
400		1.8	10.5	3.81	2.16	12.68	14.68	7.34	1.33	0.046	0.031	7.7	1200
500		0.8	12.7	4.50	1.08	11.93	14.38	7.19	1.35	0.046	0.032	9.6	1500
500	S-67	1.0	12.7	4.50	1.35	12.60	15.04	7.52	1.38	0.047	0.033	9.6	1500
500		1.2	12.7	4.50	1.62	13.21	15.65	7.82	1.41	0.048	0.034	9.6	1500
500		1.4	12.7	4.50	1.89	13.77	16.20	8.10	1.44	0.049	0.035	9.6	1500
500	L-41	1.6	12.7	4.50	2.16	14.34	16.76	8.38	1.47	0.050	0.036	9.6	1500
500		1.8	12.7	4.50	2.43	14.90	17.32	8.66	1.50	0.051	0.037	9.6	1500

	600	0.8	14.7	5.16	1.18	13.63	16.46	8.23	1.52	0.049	0.034	11.5	1800
S-81	600	1.0	14.7	5.16	1.47	14.37	17.20	8.60	1.53	0.049	0.035	11.5	1800
	600	1.2	14.7	5.16	1.76	14.97	17.79	8.89	1.56	0.050	0.036	11.5	1800
	600	1.4	14.7	5.16	2.06	15.62	18.44	9.22	1.59	0.051	0.037	11.5	1800
L-50	600	1.6	14.7	5.16	2.35	16.31	19.12	9.56	1.61	0.052	0.038	11.5	1800
	600	1.8	14.7	5.16	2.65	16.90	19.70	9.85	1.64	0.053	0.039	11.5	1800
	700	0.8	16.2	5.79	1.26	15.02	18.14	9.07	1.64	0.050	0.036	13.5	2100
S-95	700	1.0	16.4	5.79	1.57	15.86	19.02	9.51	1.67	0.051	0.037	13.5	2100
	700	1.2	16.4	5.79	1.88	16.53	19.68	9.84	1.69	0.052	0.038	13.5	2100
	700	1.4	16.4	5.79	2.20	17.22	20.36	10.18	1.72	0.053	0.039	13.5	2100
L-59	700	1.6	16.4	5.79	2.51	17.86	21.00	10.50	1.74	0.054	0.040	13.5	2100
	700	1.8	16.4	5.79	2.83	18.55	21.68	10.84	1.76	0.054	0.041	13.5	2100
	800	0.6	16.2	6.40	1.00	15.02	18.14	9.07	1.58	0.048	0.036	15.4	2400
S-109	800	0.8	17.6	6.40	1.33	16.33	19.72	9.86	1.75	0.054	0.038	15.4	2400
	800	1.0	18.0	6.40	1.66	17.34	20.80	10.40	1.80	0.055	0.039	15.4	2400
	800	1.2	18.0	6.40	1.99	18.00	21.46	10.73	1.82	0.056	0.040	15.4	2400
L-68	800	1.4	18.0	6.40	2.32	18.79	22.24	11.12	1.83	0.056	0.041	15.4	2400
	800	1.6	18.0	6.40	2.66	19.52	22.96	11.48	1.85	0.057	0.042	15.4	2400
	900	0.4	15.9	6.99	0.70	14.74	17.80	8.90	1.51	0.045	0.035	17.3	2700
S-133	900	0.6	17.4	6.99	1.04	16.14	19.49	9.74	1.68	0.050	0.039	17.3	2700
	900	1.0	19.2	6.99	1.74	18.58	22.28	11.14	1.88	0.056	0.042	17.3	2700
	900	1.2	19.2	6.99	2.09	19.35	23.04	11.52	1.90	0.057	0.043	17.3	2700
L-78	900	1.4	19.2	6.99	2.44	20.20	23.88	11.94	1.90	0.057	0.044	17.3	2700
	900	1.6	19.2	6.99	2.78	20.91	24.58	12.29	1.92	0.058	0.045	17.3	2700
	1000	0.4	17.1	7.57	0.73	15.86	19.16	9.58	1.60	0.049	0.040	19.2	3000
	1000	0.6	18.7	7.57	1.09	17.34	20.94	10.47	1.78	0.054	0.042	19.2	3000
	1000	1.0	20.2	7.57	1.82	19.75	23.64	11.82	1.95	0.060	0.045	19.2	3000
L-88	1000	1.2	20.2	7.57	2.18	20.56	24.44	12.22	1.96	0.060	0.046	19.2	3000
	1000	1.4	20.2	7.57	2.55	21.39	25.26	12.63	1.97	0.060	0.046	19.2	3000
	1000	1.6	20.2	7.57	2.91	22.16	26.02	13.01	1.98	0.060	0.046	19.2	3000
	1100	0.4	18.3	8.13	0.76	16.97	20.50	10.25	1.70	0.051	0.042	21.2	3300
	1100	0.8	20.9	8.13	1.53	19.79	23.82	11.91	1.98	0.060	0.045	21.2	3300

(Continued)

TABLE 1.
Daily Nutrient Requirements of Dairy Cattle (from NRC, 1978) (Continued)

Body weight (lb)	Breed size, age (wk)	Daily gain (lb)	Feed DM (lb)	Feed energy					Total crude protein	Minerals (lb)		Vitamins	
				NE$_m$ (Mcal)	NE$_g$ (Mcal)	ME (Mcal)	DE (Mcal)	TDN (lb)	(lb)	Ca	P	A (1000 IU)	D (IU)
1100	L-98	1.2	20.9	8.13	2.29	21.69	25.70	12.85	1.99	0.060	0.046	21.2	3300
1100		1.6	20.9	8.13	3.06	23.39	27.38	13.69	2.00	0.060	0.046	21.2	3300
1200		0.4	19.4	8.68	0.79	17.98	21.72	10.86	1.79	0.053	0.042	23.1	3600
1200	L-110	0.8	21.6	8.68	1.58	20.90	25.06	12.53	2.01	0.060	0.044	23.1	3600
1200		1.2	21.6	8.68	2.38	22.64	26.78	13.39	2.02	0.060	0.044	23.1	3600
1200		1.6	21.6	8.68	3.17	24.40	28.52	14.26	2.04	0.061	0.046	23.1	3600
1300		0.4	20.5	9.21	0.82	19.01	22.96	11.48	1.88	0.054	0.040	25.0	3900
1300		0.8	21.9	9.21	1.63	21.63	25.84	12.92	2.01	0.058	0.040	25.0	3900
1300		1.2	21.9	9.21	2.45	23.62	27.82	13.91	2.01	0.058	0.042	25.0	3900
1300		1.6	21.9	9.21	3.26	25.39	29.56	14.78	2.02	0.058	0.042	25.0	2900
						Growing dairy bulls							
300		1.0	7.9	3.07	1.01	8.53	10.04	5.02	0.93	0.036	0.022	5.8	900
300	S-34	1.4	7.9	3.07	1.41	9.36	10.86	5.43	1.02	0.039	0.024	5.8	900
300		1.8	7.9	3.07	1.82	10.12	11.62	5.81	1.11	0.043	0.026	5.8	900
300	L-20	2.0	7.9	3.07	2.02	10.51	12.00	6.00	1.15	0.045	0.027	5.8	900
300		2.2	7.9	3.07	2.22	10.89	12.38	6.19	1.20	0.047	0.028	5.8	900
400		1.0	10.5	3.81	1.10	10.58	12.60	6.30	1.20	0.041	0.027	7.7	1200
400	S-44	1.4	10.5	3.81	1.54	11.45	13.46	6.73	1.29	0.044	0.029	7.7	1200
400		1.8	10.5	3.81	1.98	12.34	14.34	7.17	1.37	0.047	0.031	7.7	1200
400		2.0	10.5	3.81	2.20	12.76	14.76	7.38	1.41	0.048	0.032	7.7	1200
400	L-28	2.2	10.5	3.81	2.42	13.19	15.18	7.59	1.46	0.050	0.033	7.7	1200
500		1.0	12.7	4.50	1.18	12.24	14.68	7.34	1.41	0.048	0.032	9.6	1500
500	S-55	1.4	12.7	4.50	1.65	13.27	15.70	7.85	1.49	0.051	0.033	9.6	1500
500		1.8	12.7	4.50	2.12	14.30	16.72	8.36	1.56	0.053	0.035	9.6	1500

500		2.0	12.7	4.50	2.36	14.72	17.14	8.57	1.60	0.054	0.036	9.6	1500
500	L–34	2.2	12.7	4.50	2.60	15.18	17.60	8.80	1.64	0.056	0.037	9.6	1500
600		1.0	14.8	5.26	1.27	13.97	16.82	8.41	1.59	0.053	0.036	11.5	1800
600	S–65	1.4	14.8	5.26	1.78	15.22	18.06	9.03	1.64	0.055	0.038	11.5	1800
600		1.8	14.8	5.26	2.29	16.27	19.10	9.55	1.71	0.057	0.040	11.5	1800
600		2.0	14.8	5.26	2.54	16.77	19.60	9.80	1.75	0.058	0.041	11.5	1800
600	L–41	2.2	14.8	5.26	2.79	17.24	20.06	10.03	1.79	0.060	0.042	11.5	1800
700		1.0	17.0	6.02	1.38	15.86	19.14	9.57	1.76	0.056	0.040	13.5	2100
700	S–75	1.4	17.0	6.02	1.93	17.13	20.40	10.20	1.82	0.057	0.042	13.5	2100
700		1.8	17.0	6.02	2.48	18.37	21.62	10.81	1.87	0.059	0.043	13.5	2100
700		2.0	17.0	6.02	2.76	18.91	22.16	11.08	1.91	0.060	0.044	13.5	2100
700	S–47	2.2	17.0	6.02	3.04	19.48	22.72	11.36	1.94	0.061	0.045	13.5	2100
800		1.0	18.7	6.78	1.48	17.42	21.02	10.51	1.90	0.058	0.043	15.4	2400
800	S–85	1.4	18.7	6.78	2.07	18.89	22.48	11.24	1.93	0.058	0.044	15.4	2400
800		1.8	18.7	6.78	2.66	20.24	23.82	11.91	1.98	0.060	0.045	15.4	2400
800		2.0	18.7	6.78	2.96	20.85	24.42	12.21	2.01	0.061	0.045	15.4	2400
800	L–54	2.2	18.7	6.78	3.26	21.41	24.98	12.49	2.04	0.062	0.046	15.4	2400
900		1.0	20.0	7.55	1.62	19.15	23.00	11.50	1.96	0.060	0.045	17.3	2700
900	S–95	1.4	20.0	7.55	2.27	20.68	24.52	12.26	1.99	0.061	0.047	17.3	2700
900		1.8	20.0	7.55	2.92	22.18	26.00	13.00	2.02	0.062	0.049	17.3	2700
900		2.0	20.0	7.55	3.24	22.99	26.80	13.40	2.04	0.063	0.050	17.3	2700
900	L–60	2.2	20.0	7.55	3.56	23.39	27.20	13.60	2.08	0.064	0.050	17.3	2700
1000		1.0	21.0	8.33	1.73	20.62	24.66	12.33	2.00	0.061	0.046	19.2	3000
1000	S–106	1.2	21.0	8.33	2.08	21.42	25.45	12.73	2.01	0.062	0.047	19.2	3000
1000	L–67	1.6	21.0	8.33	2.77	23.02	27.04	13.52	2.04	0.063	0.048	19.2	3000
1000		2.0	21.0	8.33	3.46	24.44	28.44	14.22	2.08	0.064	0.050	19.2	3000
1000		2.2	21.0	8.33	3.81	25.10	29.10	14.55	2.10	0.065	0.050	19.2	3000
1100	S–118	0.8	22.0	8.94	1.45	20.93	25.17	12.58	2.06	0.062	0.049	21.2	3300
1100		1.2	22.0	8.94	2.17	22.70	26.92	13.46	2.07	0.062	0.049	21.2	3300
1100	L–74	1.6	22.0	8.94	2.90	24.40	28.60	14.30	2.09	0.063	0.050	21.2	3300
1100		1.8	22.0	8.94	3.26	25.17	29.40	14.70	2.10	0.064	0.050	21.2	3300
1100		2.0	22.0	8.94	3.62	25.87	30.06	15.03	2.12	0.064	0.051	21.2	3300

(Continued)

379

TABLE 1.
Daily Nutrient Requirements of Dairy Cattle (from NRC, 1978) (Continued)

Body weight (lb)	Breed size, age (wk)	Daily gain (lb)	Feed DM (lb)	Feed energy					Total crude protein (lb)	Minerals (lb)		Vitamins	
				NE_m (Mcal)	NE_g (Mcal)	ME (Mcal)	DE (Mcal)	TDN (lb)	(lb)	Ca	P	A (1000 IU)	D (IU)
1200	S-129	0.6	22.6	9.55	1.13	20.96	25.32	12.66	2.08	0.063	0.049	23.1	3600
1200		1.0	23.0	9.55	1.88	22.95	27.37	13.69	2.13	0.064	0.050	23.1	3600
1200	L-82	1.4	23.0	9.55	2.63	24.81	29.22	14.61	2.13	0.064	0.050	23.1	3600
1200		1.8	23.0	9.55	3.38	26.43	30.82	15.41	2.16	0.065	0.051	23.1	3600
1300		0.6	23.7	10.14	1.16	21.97	26.54	13.27	2.16	0.064	0.049	25.0	3900
1300		1.0	23.7	10.14	1.94	24.03	28.58	14.29	2.16	0.064	0.050	25.0	3900
1300	L-92	1.4	23.7	10.14	2.72	25.99	30.52	15.26	2.16	0.065	0.050	25.0	3900
1300		1.8	23.7	10.14	3.49	27.72	32.24	16.12	2.17	0.066	0.051	25.0	3900
1400		0.2	21.5	10.71	0.40	19.94	24.08	12.04	1.89	0.064	0.049	26.9	4200
1400		0.6	24.3	10.71	1.19	22.88	27.60	13.80	2.19	0.065	0.050	26.9	4200
1400	L-102	1.0	24.3	10.71	1.99	25.08	29.74	14.87	2.18	0.065	0.051	26.9	4200
1400		1.4	24.3	10.71	2.79	26.96	31.60	15.80	2.18	0.065	0.050	26.9	4200
1500		0.2	22.5	11.28	0.41	20.86	25.20	12.60	1.97	0.064	0.050	28.8	4500
1500	L-116	0.6	24.9	11.28	1.22	23.95	28.74	14.37	2.21	0.065	0.050	28.8	4500
1500		1.0	24.9	11.28	2.03	26.01	30.78	15.39	2.20	0.066	0.051	28.8	4500
1500		1.4	24.9	11.28	2.84	27.86	32.62	16.31	2.20	0.066	0.051	28.8	4500
1600		0.2	23.6	11.84	0.41	21.89	26.44	13.22	2.05	0.064	0.050	30.8	4800
1600	L-140	0.6	25.5	11.84	1.22	24.83	29.74	14.87	2.23	0.066	0.051	30.8	4800
1600		1.0	25.5	11.84	2.04	26.99	31.88	15.94	2.22	0.066	0.051	30.8	4800
1700	L-163	0.2	24.6	12.40	0.41	22.82	27.56	13.78	2.14	0.064	0.051	32.7	5100
1700		0.6	26.1	12.40	1.23	25.62	30.64	15.32	2.27	0.066	0.051	32.7	5100
1700		1.0	26.1	12.40	2.05	27.78	32.78	16.39	2.26	0.066	0.051	32.7	5100
Growing veal calves fed only milk													
75	—	1.10	1.5	0.97	0.90	3.16	3.51	1.8	0.38	0.015	0.011	1.4	225
100	L-1.0	1.75	2.3	1.37	1.51	5.03	5.59	2.8	0.57	0.018	0.013	1.9	300

125	L-3.0	2.00	2.7	1.59	1.75	5.83	6.48	3.2	0.65	0.024	0.015	2.4	375
150	L-4.8	2.20	3.0	1.82	1.96	6.57	7.30	3.7	0.72	0.029	0.018	2.9	450
175	L-6.4	2.30	3.3	2.05	2.09	7.12	7.91	4.0	0.74	0.033	0.020	3.4	525
200	L-8.0	2.40	3.6	2.26	2.23	7.68	8.53	4.3	0.77	0.035	0.021	3.8	600
225	L-9.5	2.50	3.8	2.47	2.38	8.25	9.17	4.6	0.81	0.037	0.022	4.3	675
250	L-10.9	2.65	4.1	2.67	2.58	8.94	9.93	5.0	0.86	0.039	0.023	4.8	750
275	L-12.3	2.75	4.4	2.87	2.74	9.53	10.59	5.3	0.89	0.040	0.024	5.3	825
300	L-13.6	2.80	4.6	3.07	2.86	10.02	11.14	5.6	0.92	0.041	0.025	5.8	900
325	L-14.9	2.85	4.9	3.26	2.97	10.48	11.65	5.8	0.94	0.042	0.026	6.3	975

Maintenance of mature breeding bulls

1200	—	—	18.3	9.98	—	17.01	20.54	10.3	1.58	0.042	0.036	23.1	—
1400	—	—	20.6	11.20	—	19.07	23.04	11.5	1.76	0.049	0.040	26.9	—
1600	—	—	22.7	12.38	—	21.08	25.46	12.7	1.93	0.057	0.045	30.8	—
1800	—	—	24.9	13.53	—	23.05	27.84	13.9	2.09	0.064	0.049	34.6	—
2000	—	—	26.9	14.64	—	24.94	30.12	15.1	2.25	0.071	0.053	38.5	—
2200	—	—	28.9	15.72	—	26.77	32.34	16.2	2.41	0.079	0.057	42.3	—
2400	—	—	30.8	16.78	—	28.58	34.52	17.3	2.56	0.086	0.061	46.2	—
2600	—	—	32.7	17.82	—	30.35	36.66	18.3	2.71	0.093	0.064	50.0	—
2800	—	—	34.6	18.84	—	32.09	38.76	19.4	2.85	0.099	0.068	53.9	—
3000	—	—	36.5	19.84	—	33.79	40.82	20.4	3.00	0.107	0.072	57.7	—

TABLE 2.
Daily Nutrient Requirements of Lactating and Pregnant Cows (from NRC, 1978).

Body weight (lb)	Feed energy				Total crude protein (lb)	Calcium (lb)	Phosphorus (lb)	Vitamin A (1000 IU)
	NE_l (Mcal)	ME (Mcal)	DE (Mcal)	TDN (lb)				
Maintenance of mature lactating cows[a]								
700	6.02	10.00	11.66	5.84	0.71	0.028	0.023	24
800	6.65	11.06	12.89	6.45	0.77	0.032	0.026	28
900	7.27	12.08	14.08	7.05	0.83	0.035	0.028	31
1000	7.86	13.07	15.23	7.63	0.89	0.038	0.030	35
1100	8.45	14.04	16.36	8.19	0.95	0.040	0.032	38
1200	9.02	14.99	17.47	8.75	1.01	0.043	0.034	41
1300	9.57	15.91	18.55	9.29	1.06	0.046	0.037	45
1400	10.12	16.82	19.61	9.82	1.12	0.048	0.039	48
1500	10.66	17.72	20.65	10.34	1.17	0.051	0.041	52
1600	11.19	18.60	21.67	10.85	1.22	0.053	0.043	55
1700	11.71	19.46	22.68	11.36	1.27	0.056	0.045	59
1800	12.22	20.31	23.67	11.86	1.32	0.059	0.047	62
Maintenance plus last two months gestation of mature dry cows								
700	7.82	13.01	15.12	7.60	1.32	0.047	0.033	24
800	8.65	14.38	16.71	8.40	1.45	0.053	0.038	28
900	9.45	15.71	18.25	9.17	1.57	0.059	0.042	31
1000	10.22	17.00	19.76	9.93	1.69	0.064	0.045	35
1100	10.98	18.26	21.22	10.66	1.80	0.070	0.050	38
1200	11.72	19.50	22.65	11.38	1.92	0.075	0.053	41
1300	12.44	20.70	24.05	12.08	2.03	0.080	0.057	45
1400	13.16	21.88	25.43	12.78	2.13	0.085	0.060	48
1500	13.85	23.05	26.78	13.45	2.24	0.090	0.064	52
1600	14.54	24.19	28.10	14.12	2.34	0.095	0.067	55
1700	15.22	25.32	29.41	14.78	2.44	0.100	0.071	59
1800	15.88	26.42	30.70	15.42	2.54	0.105	0.075	62
Milk production—nutrients per pound of milk of different fat percentages								
Fat (%)								
2.5	0.27	0.45	0.52	0.260	0.072	0.0024	0.0017	—
3.0	0.29	0.49	0.56	0.282	0.077	0.0025	0.0017	—
3.5	0.31	0.53	0.61	0.304	0.082	0.0026	0.0018	—
4.0	0.34	0.56	0.65	0.326	0.087	0.0027	0.0018	—
4.5	0.36	0.60	0.69	0.344	0.092	0.0028	0.0019	—
5.0	0.38	0.63	0.73	0.365	0.098	0.0029	0.0019	—
5.5	0.40	0.67	0.78	0.387	0.103	0.0030	0.0020	—
6.0	0.42	0.71	0.82	0.410	0.108	0.0031	0.0021	—
Body weight change during lactation—nutrients per lb. weight change								
Weight loss	−2.23	−3.74	−4.33	−2.17	−0.32			
Weight gain	2.32	3.88	4.52	2.26	0.50			

[a] To allow for growth of lactating cows, increase the maintenance allowance for all nutrients except Vitamin A by 20% during the first lactation and 10% during the second lactation.

382

TABLE 3.
Recommended Nutrient Content of Rations for Dairy Cattle (from NRC, 1978).

Nutrients (concentration in the feed dry matter)	Lactating cow rations				Nonlactating cattle rations					Maximum concentrations (all classes)
Cow wt. (lb)	≤ 900	1100	1300	≥1550	Dry pregnant cows	Mature bulls	Growing heifers and bulls	Calf starter concentrate mix	Calf milk replacer	
Daily milk yields (lb)	<18 / <24 / <31 / <40	18–29 / 24–37 / 31–46 / 40–57	29–40 / 37–51 / 46–64 / 57–78	>40 / >51 / >64 / >78						
Ration No.:	I	II	III	IV	V	VI	VII	VIII	IX	Max.
Crude protein (%)	13.0	14.0	15.0	16.0	11.0	8.5	12.0	16.0	22.0	—
Energy										
NE_l (Mcal/lb)	0.64	0.69	0.73	0.78	0.61	—	—	—	—	—
NE_m (Mcal/lb)	—	—	—	—	—	0.54	0.57	0.86	1.09	—
NE_g (Mcal/lb)	—	—	—	—	—	—	0.27	0.54	0.70	—
ME (Mcal/lb)	1.07	1.15	1.23	1.31	1.01	0.93	1.01	1.42	1.71	—
DE (Mcal/lb)	1.26	1.34	1.42	1.50	1.20	1.12	1.20	1.60	1.90	—
TDN (%)	63	67	71	75	60	56	60	80	95	—
Crude fiber (%)	17	17	17	17[a]	17	15	15	—	—	—
Acid detergent fiber (%)	21	21	21	21	21	19	19	—	—	—
Ether extract (%)	2	2	2	2	2	2	2	2	10	—
Minerals[b]										
Calcium (%)	0.43	0.48	0.54	0.60	0.37	0.24	0.40	0.60	0.70	—
Phosphorus (%)	0.31	0.34	0.38	0.40	0.26	0.18	0.26	0.42	0.50	—
Magnesium (%)[c]	0.20	0.20	0.20	0.20	0.16	0.16	0.16	0.07	0.07	—
Potassium (%)	0.80	0.80	0.80	0.80	0.80	0.80	0.80	0.80	0.80	—
Sodium (%)	0.18	0.18	0.18	0.18	0.10	0.10	0.10	0.10	0.10	—
Sodium chloride (%)[d]	0.46	0.46	0.46	0.46	0.25	0.25	0.25	0.25	0.25	5

(Continued)

TABLE 3.
Recommended Nutrient Content of Rations for Dairy Cattle (from NRC, 1978) (Continued)

	Lactating cow rations				Nonlactating cattle rations					Maximum concentrations (all classes)
Cow wt. (lb)		**Daily milk yields (lb)**			Dry pregnant cows	Mature bulls	Growing heifers and bulls	Calf starter concentrate mix	Calf milk replacer	
≤900	<18	18–29	29–40	>40						
1100	<24	24–37	37–51	>51						
1300	<31	31–46	46–64	>64						
≥1550	<40	40–57	57–78	>78						
Nutrients (concentration in the feed dry matter)										Max.
Ration No.:	I	II	III	IV	V	VI	VII	VIII	IX	Max.
Sulfur (%)[d]	0.20	0.20	0.20	0.20	0.17	0.11	0.16	0.21	0.29	0.35
Iron (ppm)[d,e]	50	50	50	50	50	50	50	100	100	1000
Cobalt (ppm)	0.10	0.10	0.10	0.10	0.10	0.10	0.10	0.10	0.10	10
Copper (ppm)[d,f]	10	10	10	10	10	10	10	10	10	80
Manganese (ppm)[d]	40	40	40	40	40	40	40	40	40	1000
Zinc (ppm)[d,g]	40	40	40	40	40	40	40	40	40	500
Iodine (ppm)[h]	0.50	0.50	0.50	0.50	0.50	0.25	0.25	0.25	0.25	50
Molybdenum (ppm)[i,j]	—	—	—	—	—	—	—	—	—	6
Selenium (ppm)	0.10	0.10	0.10	0.10	0.10	0.10	0.10	0.10	0.10	5
Fluorine (ppm)[j]	—	—	—	—	—	—	—	—	—	30
Vitamins[k]										
Vit. A (IU/lb)	1450	1450	1450	1450	1450	1450	1000	1000	1720	—
Vit. D (IU/lb)	140	140	140	140	140	140	140	140	270	—
Vit. E (ppm)	—	—	—	—	—	—	—	—	300	—

[a] It is difficult to formulate high-energy rations with a minimum of 17% crude fiber. However, fat test depression may occur when rations with less than 17% crude fiber or 21% ADF are fed to lactating cows.

[b] The mineral values presented in this table are intended as guidelines for use of professionals in ration formulation. Because of many factors affecting such values, they are not intended and should not be used as a legal or regulatory base.

[c] Under conditions conducive to grass tetany (see Chapter 5), should be increased to 0.25 or higher.

384

[a] The maximum safe levels for many of the mineral elements are not well defined; estimates given here, especially for sodium chloride, sulfur, iron, copper, manganese, and zinc are based on very limited data; safe levels may be substantially affected by feeding conditions.

[e] The maximum safe level of supplemental iron in some forms is materially lower than 1000 ppm. As little as 400 ppm added irons as ferrous sulfate has reduced weight gains (Standish et al., 1969) (see Chapter 5).

[f] High copper may increase the susceptibility of milk to oxidized flavor (see Chapter 5).

[g] Maximum safe level of zinc for mature dairy cattle is not less than 1000 ppm.

[h] If diet contains as much as 25% strongly goitrogenic feed on dry basis, iodine provided should be increased two times or more.

[i] If diet contains sufficient copper, dairy cattle tolerate substantially more than 6 ppm molybdenum (see Chapter 5).

[j] Maximum safe level of fluorine for growing heifers and bulls is lower than for other dairy cattle. Somewhat higher levels are tolerated when the fluorine is from less available sources, such as phosphates (see Chapter 5). Minimum requirement for molybdenum and fluorine not yet established.

[k] The following minimum quantities of B complex vitamins are suggested in milk replacer: niacin, 2.6 ppm; pantothenic acid, 13 ppm; riboflavin, 6.5 ppm; pyridoxine, 6.5 ppm; thiamin, 6.5 ppm; folic acid, 0.5 ppm; biotin, 0.1 ppm; vitamin B_{12}, 0.07 ppm; choline, 0.26%. It appears that adequate amounts of these vitamins are furnished when calves have functional rumens (usually at 6 weeks of age) by a combination of rumen synthesis and natural feedstuffs (see Chapter 6).

TABLE 4.
Average Composition of Commonly Used Dairy Cattle Feeds (dry matter basis)

	Reference number	Dry matter (%)	DE (Mcal/lb)	ME (Mcal/lb)	Growing cattle NE_m (Mcal/lb)	Growing cattle NE_g (Mcal/lb)	Lactating cows NE_l (Mcal/lb)	TDN (%)	Crude protein (%)	Cellulose (%)	Crude fiber (%)	Lignin (%)	Acid detergent fiber (%)	Cell walls (%)
Alfalfa, fresh	2-00-196	27	1.22	1.03	0.59	0.28	0.62	61	19.0	26	28	9.0	35	45
Alfalfa hay, suncured														
Early veg. 1st cut	1-00-048	89	1.36	1.17	0.68	0.42	0.70	68	23.4	22	21	6.4	28	38
Early veg.	1-00-050	89	1.30	1.11	0.64	0.37	0.67	65	21.7	23	24	7.6	31	41
Late veg.	1-00-054	89	1.24	1.05	0.60	0.33	0.63	62	19.9	24	27	8.6	34	44
Early bloom	1-00-059	90	1.16	0.97	0.56	0.27	0.59	58	17.2	28	31	10.1	38	48
Full bloom	1-00-068	88	1.08	0.88	0.52	0.20	0.54	54	15.0	30	35	11.6	42	52
Mature	1-00-071	91	1.04	0.85	0.50	0.16	0.52	52	13.5	32	37	12.4	44	55
Alfalfa meal														
15% protein	1-00-022	93	1.22	1.03	0.59	0.31	0.62	61	16.3	29	33	12.0	41	51
17% protein	1-00-023	93	1.24	1.05	0.60	0.33	0.63	62	19.7	24	27	10.6	35	45
Alfalfa silage, wilted. 25–40% dry matter (energy value on dry matter basis is same as alfalfa hay of same maturity).														
Apple pomace, dried	4-00-423	89	1.38	1.19	0.69	0.43	0.71	69	4.9		17			
Bahia grass, grazed	2-00-464	30	1.04	0.85	0.50	0.16	0.52	52	7.9		32			
Bakery waste	4-00-466	92	1.78	1.59	0.99	0.65	0.93	89	11.9		1			
Barley grain	4-00-549	89	1.66	1.47	0.89	0.59	0.87	83	13.9		6		7	19
Pacific coast	4-07-939	89	1.64	1.45	0.88	0.58	0.86	82	10.7		7		9	21
Barley hay, suncured	1-00-495	87	1.14	0.95	0.56	0.25	0.58	57	8.9		26			
Barley straw	1-00-498	88	0.98	0.78	0.48	0.11	0.49	49	4.1	37	42	11.0	59	80
Beet, mangels, roots	4-00-637	11	1.56	1.37	0.82	0.54	0.81	78	11.4		8			
Beet pulp														
Dried	4-00-669	91	1.56	1.37	0.81	0.54	0.81	78	8.0		22	5.0	34	59
Wet	4-00-671	10	1.56	1.37	0.81	0.54	0.81	78	9.0		20			
W molasses, dried	4-00-672	92	1.56	1.37	0.81	0.54	0.81	78	9.9		17			
Beet tops w crowns, silage	3-00-660	21	1.08	0.88	0.52	0.20	0.54	54	12.7		13			
Bermuda grass, coastal. hay														
Late veg.	1-20-900	91	1.06	0.87	0.51	0.18	0.53	53	9.5		31	9.0	33	75
Mature	1-00-716	91	0.96	0.76	0.47	0.09	0.48	48	6.0		34	12.0	35	80
Bluegrass														
Canada														
Grazed	2-00-764	31	1.40	1.21	0.70	0.44	0.72	70	17.0		26			
Hay	1-00-762	93	1.30	1.11	0.63	0.37	0.67	65	11.6		29			
Early veg.	1-20-889	97	1.41	1.23	0.71	0.45	0.73	71	17.3		26			

	Calcium (%)	Phosphorus (%)	Magnesium (%)	Sodium (%)	Chlorine (%)	Potassium (%)	Sulfur (%)	Cobalt (ppm)	Copper (ppm)	Iron (ppm)	Manganese (ppm)	Zinc (ppm)	Vitamin A (1000 IU/lb)	Vitamin E (ppm)
Alfalfa, fresh	1.72	0.31	0.27	0.20	0.47	2.03	0.39	0.090	9.9	300	50	18	44	152
Alfalfa hay, suncured														
Early veg. 1st cut													34	
Early veg.	2.12	0.30	0.26	0.22	0.34	2.26	0.63	0.090	13.4	200	39	17	34	
Late veg.	2.45	0.30	0.25	0.22	0.34	2.75	0.29			250	34		33	26
Early bloom	1.25	0.23	0.30	0.15	0.38	2.08	0.30	0.090	21.7	200	32	17	15	
Full bloom	1.28	0.20	0.31	0.14		1.86	0.26	0.560	11.7	170	34	24	5	
Mature	1.17	0.17	0.35	0.15		1.97	0.20	0.090	13.4	200	33	17	1	
Alfalfa meal														
15% protein	1.32	0.24	0.31	0.08	0.48	2.50		0.190	11.2	330	31	22	12	
17% protein	1.43	0.26	0.39	0.10	0.52	2.68		0.390	10.6	490	31	17	22	
Alfalfa silage, wilted, 25–40% dry matter (energy value on dry basis is same as alfalfa hay of same maturity).														
Apple pomace, dried	0.13	0.12	0.07	0.14		0.49	0.02						0	
Bahia grass, grazed	0.45	0.19	0.25			1.45				300	8		33	
Bakery waste													1	
Barley grain	0.05	0.37	0.15	0.03	0.20	0.45	0.18	0.110	9.1	90	19	17	0	
Pacific coast	0.05	0.36	0.14	0.02	0.17	0.60		0.100	9.1	60	18	16	0	
Barley hay, suncured	0.21	0.30	0.19	0.14		1.49	0.17	0.060	4.1	300	39		10	
Barley straw	0.24	0.09	0.19	0.14	0.68	2.28	0.17	0.070	10.1	300	17			
Beet, mangles, roots	0.19	0.19	0.19	0.66	1.23	1.98	0.19			190			0	
Beet pulp														
Dried	0.75	0.11	0.30	0.19	0.04	0.23	0.22	0.100	13.7	330	38	10	0	
Wet	0.90	0.10	0.14			0.20						10	0	
W molasses, dried	0.61	0.11	0.14	0.40		1.78	0.42	0.230	16.0	210	26	10	0	
Beet tops w crowns, silage	2.32	0.20	1.07	0.54		5.79	0.57							
Bermuda grass, coastal. hay														
Late veg.	0.46	0.18	0.17	0.44		1.57							23	
Mature												20	11	
Bluegrass														
Canada														
Grazed	0.39	0.38	0.16			2.04							69	
Hay	0.30	0.29	0.33	0.11		1.59	0.13				79		45	
Early veg.	0.30	0.29	0.33	0.11		1.59	0.13				93		61	

TABLE 4.
Average Composition of Commonly Used Dairy Cattle Feeds (dry matter basis) (Continued)

| | | | | | Growing cattle | | Lactating cows | | | | | | | Acid detergent | Cell |
	Reference number	Dry matter (%)	DE (Mcal/lb)	ME (Mcal/lb)	NE_m (Mcal/lb)	NE_x (Mcal/lb)	NE_l (Mcal/lb)	TDN (%)	Crude protein (%)	Cellulose (%)	Crude fiber (%)	Lignin (%)	fiber (%)	walls (%)
Kentucky, grazed														
Early veg.	2-00-777	30	1.44	1.25	0.72	0.47	0.74	72	17.3	20	25	3.8		
Early bloom	2-00-779	36	1.38	1.19	0.69	0.43	0.71	69	14.8	28	28	4.6		
Bonemeal, steamed	6-00-400	95	0.32	0.12	0.26	0.12		16	12.7		2			
Brewers grains														
Dried	5-02-141	92	1.31	1.13	0.65	0.39	0.68	66	26.0		16	5.0	23	42
Wet	5-02-142	24	1.34	1.15	0.66	0.40	0.69	67	26.0		16	5.0	23	42
Brome grass														
Grazed														
Early veg.	2-00-892	32	1.36	1.17	0.67	0.42	0.70	68	14.6	27	24	3.0	31	60
Mature	2-00-898	56	1.30	1.11	0.64	0.37	0.67	65	9.0	34	33	4.5	38	67
Hay														
Late veg.	1-00-887	88	1.24	1.05	0.60	0.33	0.63	62	10.5		33	4.7	40	68
Late bloom	1-00-888	90	1.08	0.88	0.52	0.20	0.54	54	7.4		40	7.5	44	72
Carrot roots, fresh	4-01-145	12	1.64	1.45	0.88	0.58	0.86	82	10.1		9			
Citrus pulp														
Dried	3-01-234	20	1.65	1.47	0.89	0.59	0.87	83	7.3		16		23	23
Silage	4-01-237	90	1.54	1.35	0.80	0.53	0.80	77	6.9		14			
Clover														
Alsike														
Grazed, early veg.	2-01-314	19	1.30	1.11	0.64	0.37	0.67	65	24.1		17			
Hay, suncured	1-01-313	88	1.20	1.00	0.58	0.30	0.61	60	14.7		29			
Crimson														
Grazed, early veg.	2-20-890	18	1.26	1.07	0.62	0.34	0.64	63	17		28			
Hay, suncured	1-01-328	87	1.20	1.00	0.58	0.30	0.61	60	16.9		32			
Ladino														
Grazed, early veg.	2-01-380	21	1.40	1.21	0.70	0.44	0.72	70	24.7		14			
Hay, suncured	1-01-378	91	1.22	1.03	0.59	0.31	0.62	61	23.0		19	6.6	32	36
Red														
Fresh														
Early bloom	2-01-428	20	1.36	1.17	0.67	0.42	0.70	68	21.1		19			
Full bloom	2-01-429	28	1.28	1.09	0.63	0.36	0.66	64	14.9		30			
Hay, suncured	1-01-415	88	1.18	0.99	0.57	0.28	0.60	59	14.9	26	30	10.0	41	56

	Calcium (%)	Phosphorus (%)	Magnesium (%)	Sodium (%)	Chlorine (%)	Potassium (%)	Sulfur (%)	Cobalt (ppm)	Copper (ppm)	Iron (ppm)	Manganese (ppm)	Zinc (ppm)	Vitamin A (1000 IU/lb)	Vitamin E (ppm)
Kentucky, grazed														
Early veg.	0.56	0.47				2.27							81	156
Early bloom	0.46	0.39	0.11			2.01							51	
Bonemeal, steamed	30.51	14.31	0.67	0.48	0.01	0.19	0.13	0.100	17.2	880	32	400	0	
Brewers grains														
Dried	0.29	0.54	0.15	0.28	0.13	0.09	0.34	0.100	22.2	270	41	106	0	25
Wet	0.29	0.54	0.15	0.28	0.13	0.09	0.34	0.100	22.2	270	41	106	0	25
Brome grass														
Grazed														
Early veg.	0.59		0.18										83	
Mature	0.30		0.18										15	
Hay														
Late veg.	0.30	0.35	0.09	0.02		2.32	0.20						12	
Late bloom	0.30	0.35	0.09	0.02		2.32	0.20						7	
Carrot roots, fresh	0.37	0.34	0.17	1.00	0.50	2.50	0.17		11.1	110	31		161	
Citrus pulp														
Dried	2.04	0.15	0.16			0.62	0.07			160		16		
Silage	2.07	0.13	0.16	0.10		0.77		0.16	6.3	170	7	16		
Clover														
Alsike														
Grazed, early veg.	1.31	0.25	0.32	0.46	0.78	2.54	0.17						70	
Hay, suncured									6.0	460	117		7	
Crimson														
Grazed, early veg.	1.42	0.18	0.27	0.39	0.63	1.54	0.28						43	
Hay, suncured										700	171.3		4	
Ladino														
Grazed, early veg.	1.38	0.24	0.29	0.39	0.63	2.82	0.28						64	
Hay, suncured								0.150	8.8	600	208.7	17	11	
Red														
Fresh														
Early bloom	2.26	0.38	0.51	0.20		2.49	0.17			300			45	
Full bloom	1.01	0.27	0.51	0.20		1.96	0.17			300			38	
Hay, suncured	1.49	0.25	0.45	0.18	0.32	1.66	0.17	0.150	11.2	210	73.3	17	6	1912

(Continued)

TABLE 4.
Average Composition of Commonly Used Dairy Cattle Feeds (dry matter basis) (Continued)

	Reference number	Dry matter (%)	DE (Mcal/lb)	ME (Mcal/lb)	Growing cattle		Lactating cows		Crude protein (%)	Cellulose (%)	Crude fiber (%)	Lignin (%)	Acid detergent fiber (%)	Cell walls (%)
					NE_m (Mcal/lb)	NE_g (Mcal/lb)	NE_l (Mcal/lb)	TDN (%)						
Coconut (copra meal)														
Meal mech-extd.	5-01-572	93	1.62	1.43	0.86	0.57	0.84	81	21.9		13			
Meal solv.	5-01-573	92	1.48	1.29	0.75	0.49	0.77	74	23.1		16			
Corn (maize)														
Aerial part (fodder)	1-02-775	82	1.30	1.11	0.64	0.37	0.67	65	8.9	28	26	3.0	33	55
Wo ears wo husks, dried (corn stover)	1-02-776	87	1.18	0.99	0.57	0.28	0.60	59	5.9	25	34	11.0	39	67
Wo ears wo husks, as silage	3-02-836	27	1.16	0.97	0.56	0.27	0.59	58	7.2	25	32	12.0	40	68
Cobs, ground	1-02-782	90	0.93	0.74	0.46	0.07	0.47	47	2.8	28	35	7.0	35	89
Dist. grains														
Dried	5-02-842	92	1.68	1.49	0.90	0.60	0.88	84	29.5		13			43
W solubles, dried	5-02-843	92	1.76	1.57	0.98	0.64	0.92	88	29.8		10	6.5		
Dist. solubles, dried	5-02-844	93	1.76	1.57	0.97	0.64	0.92	88	28.9		4	2.2		
Ears w husks, ground, dried	4-02-849	87	1.60	1.41	0.84	0.56	0.83	80	9.3		9			
Ears w husks, silage	3-02-839	43	1.44	1.25	0.72	0.47	0.74	72	8.8		12			
Gluten, meal, 60% protein	5-02-318	91	1.68	1.49	0.90	0.60	0.88	84	65.9		3			
Gluten feed	5-02-903	90	1.64	1.45	0.87	0.58	0.86	82	25.0		9			
Grain, yellow	4-21-018	89	1.76	1.57	0.97	0.64	0.92	88	10.0		2		3	9
Hominy feed	4-02-887	91	1.84	1.65	1.05	0.68	0.97	92	11.8		6		12	
Silage														
Well-eared	3-02-823	35	1.40	1.21	0.70	0.44	0.72	70	8.0		24		31	51
Not well-eared	3-08-600	35	1.30	1.11	0.64	0.37	0.67	65	8.4		32			
Sweet cannery residue, ensiled	3-07-955	29	1.44	1.25	0.72	0.47	0.74	72	8.8		27			
Cotton seed														
Hulls	1-01-599	90	0.76	0.56	0.39	0.00	0.37	38	4.3	60	50	23.0	71	90
Meal exp. 41% prot.	5-01-617	93	1.54	1.35	0.80	0.53	0.80	77	44.0		13	6.0	20	28
Meal solv. 41% prot.	5-01-621	93	1.50	1.31	0.77	0.50	0.78	75	44.8		13			
Whole	5-13-749	93	1.95	1.77	1.17	0.73	1.03	98	24.9		18		29	39

	Calcium (%)	Phosphorus (%)	Magnesium (%)	Sodium (%)	Chlorine (%)	Potassium (%)	Sulfur (%)	Cobalt (ppm)	Copper (ppm)	Iron (ppm)	Manganese (ppm)	Zinc (ppm)	Vitamin A (1000 IU/lb)	Vitamin E (ppm)
Coconut (copra meal)														
Meal mech-extd.	0.23	0.66	0.23	0.04		1.65	0.37	0.140	15.2	1420	70.6		0	
Meal solv.	0.18	0.66	0.39	0.04	0.03	1.32	0.37	0.140	10.4	750	71.8		0	
Corn (maize)														
Aerial part (fodder)	0.43	0.23	0.29	0.03	0.19	0.95	0.14		7.7	100	68.1		1	
Wo ears wo husks, dried (corn stover)	0.60	0.09	0.45	0.07		0.92	0.17		5.1	200			1	
Wo ears wo husks, as silage	0.38	0.42	0.31	0.03		1.65							3	
Cobs, ground	0.12	0.04	0.07			0.84	0.47	0.130	7.3	230	6.2		0	
Dist. grains														
Dried	0.10	0.40	0.07	0.10	0.08	0.20	0.46	0.080	48.6	200	20	35	0.5	
W solubles dried	0.16	0.79	0.07	0.98	0.18	0.50	0.32	0.330	48.6	200	20	86	1	40
Dist. solubles, dried	0.38	1.47	0.69	0.59	0.28	1.87	0.40		88.9	590	79	92	0	55
Ears w husks, ground, dried	0.05	0.26	0.17	0.05		0.56	0.22	0.300	8.8	80	28	18	1	
Ears w husks, silage	0.06	0.27											1	
Gluten, meal, 60% protein	0.18	0.51	0.05	1.00	0.12	0.03	0.44	0.050	31.0	1480	8	29	3	20
Gluten feed	0.33	0.86	0.32	1.06	0.24	0.67	0.24	0.230	53.0	600	26	52	2	15
Grain, yellow	0.03	0.31	0.13	0.01	0.05	0.35	0.14	0.040	3.6	30	6	21	0.5	25
Hominy feed	0.06	0.58	0.26	0.09	0.06	0.59	0.03	0.066	16.1	70	16		1	
Silage														
Well-eared	0.27	0.20	0.28	0.01		1.05	0.08	0.0.0	13.2	640	34	21	8	
Not well-eared	0.34												2	
Sweet, cannery residue, ensiled													2	
Cotton seed														
Hulls	0.14	0.07	0.14	0.31	0.02	0.96	0.26	0.020	54.6	150	10	22	0	
Meal exp. 41% prot.	0.17	1.28	0.61	0.06	0.04	1.49	0.43	0.164	21.3	240	24		0	35
Meal solv. 41% prot.	0.17	1.31	0.61	0.05	0.03	1.53	0.23	0.164	21.3	240	24	91	0	15
Whole	0.15	0.73	0.35	0.31		1.20	0.26		54.6	150	10		0	

(Continued)

TABLE 4.
Average Composition of Commonly Used Dairy Cattle Feeds (dry matter basis) (Continued)

	Reference number	Dry matter (%)	DE (Mcal/lb)	ME (Mcal/lb)	Growing cattle NEm (Mcal/lb)	Growing cattle NEg (Mcal/lb)	Lactating cows NEl (Mcal/lb)	TDN (%)	Crude protein (%)	Cellulose (%)	Crude fiber (%)	Lignin (%)	Acid detergent fiber (%)	Cell walls (%)
Cowpea, hay	1-01-645	90	1.26	1.07	0.61	0.34	0.64	63	18.4		27			
Defluorinated phosphate (Table 5.16)														
Dicalcium phosphate	6-01-080	96												
Distillers grains, see Corn, Sorghum, grain variety														
Fats and oils, animal fat (not exceeding 3% of ration)	4-00-409	99	3.63	3.40	2.38	1.19	2.38	182						
Fescue, meadow (tall)														
Grazed	2-01-920	28	1.24	1.05	0.60	0.33	0.63	62	12.4		29			
Hay, suncured	1-01-912	88	1.10	0.90	0.53	0.22	0.56	55	10.5		33	6.9	43	65
Flax														
Linseed meal, solv.	5-02-048	91	1.51	1.33	0.78	0.52	0.79	76	38.6		10			
Linseed meal, exp.	5-02-045	91	1.61	1.43	0.86	0.58	0.84	81	38.8	38	10		17	25
Groundnut, see Peanut														
Hominy feed, see Corn, hominy														
Johnson grass, see Sorghum, Johnson grass														
Lespedeza														
Kobe or Korean														
Grazed														
Late veg.	2-21-023	32	1.18	0.99	0.57	0.28	0.60	59	16.4		32			
Early bloom	2-20-885	25	1.10	0.91	0.53	0.22	0.56	55	16.4		32			
Hay														
Late veg.	1-21-019	92	1.18	0.99	0.57	0.28	0.60	59	17.8		24			
Early bloom	1-21-020	93	1.10	0.91	0.53	0.22	0.56	55	15.5		28			
Full bloom	1-21-022	93	0.94	0.74	0.46	0.07	0.47	47	13.4		32			
Sericea, hay, late veg.	1-09-172	93	0.90	0.71	0.44	0.03	0.44	45	18.6		22	16.0		
Limestone, grnd, min 33% calcium	6-02-632	100												
Linseed meal, see Flax														
Mangel, see Beet, mangles														
Meat meal	5-00-385	93	1.52	1.33	0.78	0.52	0.79	76	57.1		2			
Meat and bonemeal	5-00-388	94	1.44	1.25	0.73	0.47	0.74	72	53.8		2			
Milk														
Buttermilk, dried	5-01-160	93	1.71	1.54	0.94	0.62	0.90	86	34.2					
Fresh, whole	5-01-168	12	2.60	2.34	1.64	0.91	1.64	130	25.8					
Skimmed, dried	5-01-175	94	1.71	1.54	0.94	0.62	0.90	86	36.0					

	Calcium (%)	Phosphorus (%)	Magnesium (%)	Sodium (%)	Chlorine (%)	Potassium (%)	Sulfur (%)	Cobalt (ppm)	Copper (ppm)	Iron (ppm)	Manganese (ppm)	Zinc (ppm)	Vitamin A (1000 IU/lb)	Vitamin E (ppm)
Cowpea, hay	1.34	0.32	0.43	0.27	0.17	1.99	0.35	0.070		300	485		6	
Defluorinated phosphate (Table 5.16)														
Dicalcium phosphate	23.70	18.84		2.71		0.04			6.2	1320	153	28	0	
Distillers grains, see Corn, Sorghum, grain variety														
Fats and oils, animal fat (not exceeding 3% of ration)														8
Fescue, meadow (tall)														
Grazed	0.61	0.42	0.37			2.34		0.130					61	146
Hay, suncured	0.57	0.36	0.59			1.74		0.135			24		13	136
Flax														
Linseed meal, solv.	0.43	0.91	0.66	0.15	0.04	1.52	0.44	0.140	28.2	360	41		0	18
Linseed meal, exp.	0.43	0.93	0.64	0.12	0.04	1.36	0.04	0.470	29.0	190	43	36	0	9
Groundnut, see Peanut														
Hominy feed, see Corn, hominy														
Johson grass, see Sorghum, Johnson grass														
Lespedeza														
Kobe or Korean														
Grazed														
Late veg.	1.12	0.28				1.28							47	
Early bloom	1.35	0.21	0.27			1.12				250			29	
Hay														
Late veg.	1.14	0.26	0.25			1.20				340	178		26	
Early bloom	1.23	0.25	0.26			1.00				400	256		25	
Full bloom	1.04	0.23	0.24			1.03		0.040	0.28	300	152		2	
Sericea, hay, late veg.	1.54	0.26				0.69							7	
Limestone, grnd, min 33% calcium	36.07	0.02	2.05	0.05	0.03	0.12	0.04			3500	280		0	
Linseed meal, see Flax														
Mangel, see Beet, mangles														
Meat meal	8.49	4.31	0.29	1.41	1.40	0.59	0.53	0.137	10.4	470	10	112	0	1
Meat and bonemeal	10.29	5.39	1.20	0.78	0.80	1.38	0.28	0.195	1.6	530	14	102	0	1
Milk														
Buttermilk, dried	1.07	0.73	0.10	1.04	0.38	1.35	0.09		1.1	<10		44	1	6
Fresh, whole	0.89	0.72	0.08	0.34	0.92	1.16		0.005	0.8	<10		23	7	
Skimmed, dried	1.25	1.03	0.11	0.50	0.54	1.62	0.33	0.117	1	<10		68	0	9

(Continued)

TABLE 4.
Average Composition of Commonly Used Dairy Cattle Feeds (dry matter basis) (Continued)

	Reference number	Dry matter (%)	DE (Mcal/lb)	ME (Mcal/lb)	Growing cattle NE_m (Mcal/lb)	NE_g (Mcal/lb)	Lactating cows NE_l (Mcal/lb)	TDN (%)	Crude protein (%)	Cellulose (%)	Crude fiber (%)	Lignin (%)	Acid detergent fiber (%)	Cell walls (%)
Millet, foxtail														
Grazed	2-03-101	28	1.28	1.09	0.63	0.36	0.66	64	9.5		32			
Hay, suncured	1-03-099	86	1.14	0.95	0.55	0.25	0.58	57	8.6		30			
Molasses														
Beet, min. 79 deg. brix	4-00-668	77	1.50	1.31	0.77	0.50	0.78	75	8.7					
Citrus	4-01-241	65	1.54	1.35	0.80	0.53	0.80	77	10.9					
Sugarcane, molasses, dehy	4-04-695	96	1.36	1.17	0.68	0.42	0.70	68	10.7		5			
Sugarcane, molasses, min. 79.5 deg. brix	4-04-696	75	1.44	1.25	0.73	0.47	0.74	72	4.3					
Napier grass, grazed														
Late veg.	2-03-158	15	1.26	1.07	0.62	0.34	0.64	63	11.0	33	31	10.0	45	70
Late bloom	2-03-162	23	1.04	0.85	0.50	0.16	0.52	52	7.8	35	39	14.0	47	75
Oats														
Cereal by-product (oatmeal)	4-03-303	91	1.84	1.66	1.05	0.68	0.97	92	16.2		5			
Grain, Pacific coast	4-03-309	89	1.52	1.33	0.78	0.52	0.79	76	13.6	18	12	3.4	17	31
Grain	4-07-999	91	1.54	1.35	0.80	0.53	0.80	77	10.1		12			
Groats (hulled oats)	4-03-331	91	1.86	1.67	1.07	0.69	0.98	93	17.5		3			
Hay, suncured	1-03-280	88	1.22	1.03	0.59	0.32	0.62	61	9.2		31	6.0	36	66
Straw	1-03-283	90	0.96	0.77	0.47	0.09	0.48	48	4.4	40	41	14.0	47	70
Silage														
Late veg.	3-20-898	30	1.24	1.05	0.60	0.33	0.64	62	12.8		30			
Dough stage	3-03-296	32	1.18	0.99	0.58	0.28	0.60	59	9.7		34			
Orchard grass														
Grazed, early veg.	2-03-439	24	1.34	1.15	0.66	0.40	0.69	67	18.4		27	3.0	31	55
Hay														
Early bloom	1-03-425	87	1.24	1.05	0.60	0.33	0.64	62	10.2		34			
Late bloom	1-03-428	88	1.00	0.81	0.49	0.13	0.50	50	8.4		37	6.0	39	69
Oyster shells, fine grnd, min. 33% Ca	6-03-481	100							1.0					
Pea														
Seeds	5-03-600	90	1.66	1.47	0.89	0.59	0.87	83	26.5		6			
Vine silage	3-03-596	24	1.12	0.93	0.54	0.23	0.57	56	13.1	34	30	9	49	59

	Calcium (%)	Phosphorus (%)	Magnesium (%)	Sodium (%)	Chlorine (%)	Potassium (%)	Sulfur (%)	Cobalt (ppm)	Copper (ppm)	Iron (ppm)	Manganese (ppm)	Zinc (ppm)	Vitamin A (1000 IU/lb)	Vitamin E (ppm)
Millet, foxtail														
Grazed	0.32	0.19				1.94							33	
Hay, suncured	0.33	0.19		0.10	0.13	1.94	0.16						11	
Molasses														
Beet, min. 79 deg. brix	0.21	0.04	0.30	1.52	1.92	6.20	0.61	0.500	22.9	100	6	18	0	5
Citrus	2.01	0.14	0.22	0.40	0.10	0.14		0.160	112.0	500	40	137	0	
Sugarcane, molasses, dehy	0.87	0.29	0.43	0.19		3.68	0.46	1.210	72.8	240	52	33	0	6
Sugarcane, molasses, min. 79.5 deg. brix	1.19	0.11	0.47			3.17			7.9	250	56		0	
Napier grass, grazed														
Late veg.	0.60	0.41	0.26	0.01		1.31	0.10							
Late bloom	0.35	0.30	0.26	0.01		1.31	0.10							
Oats														
Cereal by-product (oatmeal)	0.08	0.48	0.18	0.01	0.05	0.59	0.29	0.050	4.8	330	48	154	0	26
Grain, Pacific coast	0.07	0.39	0.19	0.18	0.12	0.42	0.38	0.070	6.6	80	43	33	0	37
	0.10	0.36			0.13	0.41	0.23				42		0	22
Groats (hulled oats)	0.08	0.47	0.10	0.06	0.10	0.40	0.22		7	90	52		0	16
Hay, suncured	0.26	0.24	0.75	0.17	0.52	1.23	0.30		4.4	500	120		10	
Straw	0.26	0.07	0.18	0.42	0.78	2.37	0.23	0.070	10.1	200	39		2	
Silage														
Late veg.		0.10		0.37		2.44	0.24							
Dough stage	0.47	0.33												
Orchard grass														
Grazed, early veg.	0.58	0.55	0.10		0.08	3.88			7.0		31		61	
Hay														
Early bloom													7	
Late bloom													4	
Oyster shells, fine grnd, min. 33% Ca	38.22	0.07	0.30	0.21	0.01	0.10				2900	133		0	
Pea														
Seeds	0.13	0.47		0.05		1.14				60				
Vine silage	1.31	0.24	0.39	0.01		1.40	0.25						34	

(Continued)

TABLE 4.
Average Composition of Commonly Used Dairy Cattle Feeds (dry matter basis) (Continued)

	Reference number	Dry matter (%)	DE (Mcal/lb)	ME (Mcal/lb)	Growing cattle NE$_m$ (Mcal/lb)	Growing cattle NE$_g$ (Mcal/lb)	Lactating cows NE$_l$ (Mcal/lb)	Lactating cows TDN (%)	Crude protein (%)	Cellulose (%)	Crude fiber (%)	Lignin (%)	Acid detergent fiber (%)	Cell walls (%)
Peanut hay	1-03-619	91	1.16	0.97	0.56	0.27	0.59	58	10.9		33			
Peanut meal, exp.	5-03-649	92	1.66	1.47	0.89	0.59	0.87	83	49.8		9			
Solv. 45% prot.	5-03-650	92	1.54	1.35	0.80	0.53	0.80	77	54.2		11			
Pearlmillet														
Grazed	2-03-115	21	1.24	1.05	0.60	0.33	0.64	62	9.0		31			
Silage	3-20-903	30	1.18	0.99	0.57	0.28	0.60	59	6.9		32			
Pineapple, pulp, dried	4-03-722	87	1.46	1.27	0.74	0.48	0.76	73	4.6					
Potatoes, fresh	4-03-787	25	1.58	1.39	0.83	0.55	0.82	79	9.6		2			
Rape														
Grazed, early veg.	2-03-865	18	1.40	1.21	0.70	0.44	0.72	70	16.4		13			
Seed, meal														
Exp.	5-03-870	94	1.48	1.29	0.76	0.49	0.77	74	39.6		13			
Solv.	5-03-871	90	1.38	1.19	0.68	0.43	0.71	69	43.6		13			
Redtop, fresh, full bloom	2-03-891	26	1.24	1.05	0.60	0.33	0.64	62	10		27	8.0	40	64
Rice, bran w germ	4-03-928	91	1.32	1.13	0.65	0.39	0.68	66	14		12	4.3	16	24
Rice hulls	1-08-075	92	0.22	0.15	0.24	0.00	0.07	11	3.1	33	44	16.0	72	82
Rye														
Grain	4-04-047	88	1.60	1.41	0.84	0.56	0.83	80	13.8		3			
Grazed, early veg.	2-04-013	16	1.38	1.19	0.69	0.43	0.71	69	28					
Silage	3-04-020	28	1.06	0.87	0.51	0.18	0.54	53	12.6		34			
Ryegrass														
Italian														
Grazed, early veg.	2-04-079	24	1.32	1.13	0.65	0.39	0.68	66	24.2		19			
Hay, late veg.	1-04-065	89	1.24	1.05	0.60	0.33	0.64	62	10.3		24			
Perennial, hay, early bloom	1-04-075	84	1.24	1.05	0.60	0.33	0.64	62	6		35			
Safflower														
Seeds, meal	4-07-958	93	1.78	1.60	1.00	0.65	0.93	89	19.5		31			
Exp.	5-04-109	91	1.14	0.95	0.55	0.25	0.58	57	22.8		36		41	59
Solv.	5-04-110	92	1.10	0.91	0.53	0.22	0.56	55	23.9		34			
Meal wo hulls, solv.	5-07-959	90	1.52	1.33	0.78	0.52	0.79	76	46.5		17			
Sesame meal, mech-extd.	5-04-220	93	1.50	1.31	0.77	0.50	0.78	75	51.5		5			
Sodium tripolyphosphate (Table 5.16)														

Feed	Calcium (%)	Phosphorus (%)	Magnesium (%)	Sodium (%)	Chlorine (%)	Potassium (%)	Sulfur (%)	Cobalt (ppm)	Copper (ppm)	Iron (ppm)	Manganese (ppm)	Zinc (ppm)	Vitamin A (1000 IU/lb)	Vitamin E (ppm)
Peanut hay	1.23	0.17	0.49			1.38	0.23	0.080					9	
Peanut meal, exp.	0.18	0.62			0.03	1.25	0.32						0	
Solv. 45% prot.	0.22	0.71	0.04	0.45	0.03	1.29		0.12	16.6	290	32	36	0	3
Pearlmillet														
Grazed													33	
Silage													5	
Pineapple, pulp, dried	0.24	0.12								560			0	
Potatoes, fresh	0.04	0.22	0.13	0.09	0.28	2.18	0.09		28.4	90	42		0	
Rape														
Grazed, early veg.													28	
Seed, meal														
Exp.	0.69	1.04	0.54	0.5		0.9			7.4	190	65	47	0	20
Solv.	0.67	1											0	
Redtop, fresh, full bloom	0.07	0.37	0.25	0.05		2.35	0.16			200	459		28	
Rice, bran w germ	0.09	1.62	1.04	0.03	.08	1.91	.20		14.3	210	333	3	0	66
Rice hulls		0.08				0.34							0	
Rye														
Grain	0.07	0.36	0.14	0.03	0.03	0.52	0.17		7.7	70	62	36	0	17
Grazed, early veg.	0.39	0.32											102	
Silage													10	
Ryegrass														
Italian														
Grazed, early veg.	0.62	0.34				1.56				320			73	
Hay, late veg.	0.62	0.34				1.56				320			53	
Perennial hay, early bloom	0.65	0.37				1.92							22	
Safflower														
Seeds, meal	0.25	0.67	0.36	0.06		0.79	0.06		10.7	500	20	43	0	1
Exp.	0.28	0.78	0.36	0.05		0.79	0.06		10.7	530	20	44	0	1
Solv.	0.37	0.80	0.37	0.06		0.79			10.8	560	20	44	0	1
Meal wo hulls, solv.	0.44	1.41	1.33	0.04	0.18	1.33	0.06		97.4	1100	44		0	1
Sesame meal, mech-extd.	2.18	1.39	0.86	0.17	0.06	1.29	0.46	2.22			52	107	0	
Sodium tripolyphosphate (Table 5.16)														

(Continued)

TABLE 4.
Average Composition of Commonly Used Dairy Cattle Feeds (dry matter basis) (Continued)

	Reference number	Dry matter (%)	DE (Mcal/lb)	ME (Mcal/lb)	Growing cattle NE$_m$ (Mcal/lb)	Growing cattle NE$_g$ (Mcal/lb)	Lactating cows NE$_l$ (Mcal/lb)	TDN (%)	Crude protein (%)	Cellulose (%)	Crude fiber (%)	Lignin (%)	Acid detergent fiber (%)	Cell walls (%)
Sorghum, grain variety														
Distillers grains, dried	5-04-374	94	1.64	1.45	0.88	0.59	0.86	82	33.2		13			18
Grain														
6%–9% prot.	4-08-138	88	1.61	1.43	0.86	0.57	0.84	81	7.9		2	1.3	9	
9%–12% prot.	4-08-139	88	1.60	1.41	0.84	0.56	0.83	80	11.7		2			
12%–15% prot.	4-08-140	88	1.58	1.39	0.83	0.55	0.82	79	13		2			
Fodder, suncured	1-07-960	90	1.16	0.97	0.56	0.26	0.59	58	7.4		28			
Wo heads, stover	1-04-302	85	0.96	0.76	0.47	0.09	0.48	48	4.9		33			
W heads, silage	3-04-323	29	1.10	0.91	0.53	0.22	0.56	56	8.3		26			
Johnson grass, hay, suncured	1-04-407	91	1.12	0.91	0.54	0.23	0.57	56	9.6		33			
Sorgo silage	3-04-468	26	1.16	0.97	0.57	0.27	0.59	58	6.2		29			
Sudan grass														
Grazed														
Early veg.	2-04-484	18	1.40	1.21	0.70	0.44	0.72	70	16.8		31			
Mid-bloom	2-04-485	23	1.26	1.07	0.62	0.35	0.64	63	8.7		36			
Hay	1-04-480	89	1.18	0.99	0.57	0.28	0.60	59	11.0	32	29	5.0	42	72
Silage	3-04-499	23	1.18	0.99	0.57	0.29	0.60	59	11.1		34			
Soybean														
Hay														
Mid-bloom	1-04-538	94	1.12	0.93	0.54	0.23	0.57	56	17.8		30			
Dough stage	1-04-542	88	1.20	1.01	0.58	0.30	0.61	60	16.8		28			
Hulls (Soybran flakes)	1-04-560	91	1.56	1.37	0.81	0.54	0.81	78	12.0	44	39	2.0	46	67
Silage	3-04-581	28	1.08	0.88	0.53	0.20	0.54	54	17.7		28			
Seeds (whole soybeans)	5-04-610	90	1.88	1.70	1.09	0.69	0.99	94	41.7		6			
Straw	1-04-567	88	0.88	0.68	0.44	0.004	0.44	44	5.2	38	44	13.0	54	70
Soybean meal														
Solv. 44% prot.	5-04-604	89	1.61	1.43	0.86	0.57	0.84	81	49.6	8	7		10	14
Soybean meal														
Solv. 46% prot.	5-21-119	89	1.61	1.43	0.86	0.57	0.84	81	51.8		5			
Solv. dehulled, 48% prot.	5-04-612	89	1.61	1.43	0.86	0.57	0.84	81	54.0		3			
Sudan grass, see *Sorghum, Sudan grass*														

	Calcium (%)	Phosphorus (%)	Magnesium (%)	Sodium (%)	Chlorine (%)	Potassium (%)	Sulfur (%)	Cobalt (ppm)	Copper (ppm)	Iron (ppm)	Manganese (ppm)	Zinc (ppm)	Vitamin A (1000 IU/lb)	Vitamin E (ppm)
Sorghum, grain variety														
Distillers grains, dried	0.16	0.76												
Grain														
6%–9% prot.	0.04	0.33	0.19	0.03	0.10	0.38	0.18	0.29	10.8	500	17	16	0	12
9%–12% prot.	0.03	0.33	0.19	0.03	0.10	0.38	0.18	0.29	10.8	500	17	16	0	12
12%–15% prot.	0.03	0.33	0.19	0.03	0.10	0.38	0.18	0.29	10.8	500	17	16	0	12
Fodder, suncured	0.47	0.19		0.02		1.39				2000			10	
Wo heads, stover	0.48	0.11		0.02		1.20							1	
W heads, silage	0.32	0.18	0.30	0.02	0.13	1.54	0.10	0.30	34.9	270	48		2	
Johnson grass, hay, suncured	0.71	0.31	0.35	0.01		1.35	0.10			600			8	
Sorgo silage	0.35	0.20	0.27	0.15	0.06	1.22	0.10		31.1	200	61		5	
Sudan grass														
Grazed														
Early veg.	0.43	0.41	0.35	0.01		2.14	0.11			200			36	
Mid-bloom	0.43	0.41	0.35	0.01		2.14	0.11			200			33	
Hay	0.56	0.31	0.40	0.02		1.54	0.06	0.13	36.8	200	93		11	
Silage	0.48	0.19	0.49	0.02		2.56	0.06	0.27	36.6	140	99		5	
Soybean														
Hay														
Mid-bloom	1.29	0.33	0.79	0.12		0.97	0.26			300			6	
Dough stage	1.29	0.33	0.79	0.12		0.97	0.26			300			6	
Hulls (Soybean flakes)	0.45	0.17	0.38	0.05		1.03		0.12	17.8	320	14	24	0	
Silage	1.25	0.49	0.31	0.09	0.03	0.93	0.30		9.3	400	114		14	
Seeds (whole soybeans)	0.28	0.66	0.31	0.13		1.79	0.24		17.6	90	33	18	0	37
Straw	1.59	0.06	0.92			0.53					51		0	
Soybean meal														
Solv. 44% prot.	0.36	0.75	0.30	0.31	0.03	2.21	0.49	0.100	40.8	130	31	48	0	2
Soybean meal														
Solv. 46% prot.	0.36	0.75	0.30	0.31	0.03	2.21	0.49	0.100	40.8	130	31	48	0	2
Solv. dehulled, 48% prot.	0.36	0.75	0.30	0.31	0.03	2.21	0.49	0.100	40.8	130	31	48	0	2
Sudan grass, *see Sorghum*, Sudan grass														

(Continued)

TABLE 4.
Average Composition of Commonly Used Dairy Cattle Feeds (dry matter basis) (Continued)

	Reference number	Dry matter (%)	DE (Mcal/lb)	ME (Mcal/lb)	Growing cattle NE_m (Mcal/lb)	Growing cattle NE_g (Mcal/lb)	Lactating cows NE_l (Mcal/lb)	TDN (%)	Crude protein (%)	Cellulose (%)	Crude fiber (%)	Lignin (%)	Acid detergent fiber (%)	Cell walls (%)
Sugarcane, bagasse, dried	1-04-686	92	0.56	0.36	0.32	0.00	0.26	28	1.8		48			
Sunflower meal														
Wo hulls, exp.	5-04-738	93	1.40	1.21	0.70	0.44	0.72	70	44.1	21	13	12.0	33	40
Solv. wo hulls	5-04-739	93	1.30	1.11	0.64	0.38	0.67	65	50.3		12			
Sweetclover, hay	1-04-754	87	1.14	0.95	0.55	0.25	0.58	57	14.0		36			
Timothy														
Grazed														
Late veg.	2-04-903	28	1.36	1.17	0.68	0.42	0.70	68	9.6		31			
Mid-bloom	2-04-905	28	1.24	1.05	0.60	0.33	0.63	62	9.1		34	4.0	37	64
Hay, suncured														
Late veg.	1-04-881	88	1.36	1.17	0.68	0.42	0.70	68	11.4		31	3.1	33	63
Mid-bloom	1-04-883	88	1.16	0.97	0.56	0.27	0.59	58	9.5		32	5.5	40	66
Seed stage	1-04-886	88	1.02	0.83	0.49	0.15	0.51	51	6.0	31	35	11.0	45	70
Silage, 25-40% dry matter (energy value on dry matter basis is same as for hay of same maturity).														
Tomato, pomace, dry	5-05-041	92	1.16	0.97	0.56	0.27	0.59	58	23.9	24	26	11.4	50	55
Trefoil, birdsfoot														
Hay	1-05-044	91	1.22	1.03	0.59	0.31	0.62	61	15.6		30	8.8	36	47
Grazed	2-20-786	20	1.49	1.31	0.76	0.50	0.78	75	18.2		25			
Turnip roots, fresh	4-05-067	9	1.68	1.49	0.91	0.60	0.88	84	11.3		11	10.0	34	44
Vetch hay	1-05-106	88	1.24	1.05	0.60	0.33	0.63	62	19.0		31	11.0	43	58
Wheat														
Bran	4-05-190	89	1.40	1.21	0.69	0.44	0.72	70	18.0	8	11	3.0	12	45
Grain, soft white winter	4-05-337	86	1.76	1.57	0.98	0.64	0.92	88	11.5		3		4	14
Grain screenings	4-05-216	89	1.54	1.35	0.80	0.53	0.80	77	16.0	6	8	7.9		
Grazed, early veg.	2-05-176	21	1.46	1.27	0.74	0.49	0.76	73	28.6		17	3.9	30	52
Hay, suncured	1-05-172	86	1.16	0.97	0.56	0.27	0.59	58	8.7		28	7.3	41	68
Middlings	4-05-205	90	1.60	1.41	0.84	0.56	0.83	80	18.7		8			
Shorts	4-05-201	90	1.66	1.47	0.89	0.59	0.87	83	18.6		7			
Silage, early veg.	3-05-184	26	1.24	1.05	0.60	0.33	0.64	62	11.9		27			
Straw	1-05-175	90	0.92	0.73	0.45	0.05	0.46	46	4.2	39	42	13.7	54	85
Whey, dehy	4-01-182	93	1.56	1.37	0.81	0.54	0.81	78	14.0					
Yeast														
Brewers, dried	7-05-527	93	1.56	1.37	0.81	0.54	0.81	78	48.3		3			
Torula, dried	7-05-534	93	1.60	1.41	0.84	0.56	0.83	80	51.5		3			

	Calcium (%)	Phosphorus (%)	Magnesium (%)	Sodium (%)	Chlorine (%)	Potassium (%)	Sulfur (%)	Cobalt (ppm)	Copper (ppm)	Iron (ppm)	Manganese (ppm)	Zinc (ppm)	Vitamin A (1000 IU/lb)	Vitamin E (ppm)
Sugarcane, bagasse, dried	0.90	0.29	0.10	0.20		0.50	0.10			100			0	
Sunflower meal														
Wo hulls, exp.	0.46	1.12	0.79	1.31	0.20	1.16			3.8	40	25		0	
Solv. wo hulls	0.40	1.10	0.81	1.30	0.11	1.07			3.8	40	25		0	12
Sweetclover, hay,	1.27	0.26	0.49	0.10	0.37	1.34	0.49		10.1	150	103		18	
Timothy														
Grazed														
Late veg.	0.28	0.28	0.15	0.19	0.63	2.40	0.13		11.2	200			43	
Mid-bloom	0.25	0.25	0.15	0.19		1.71	0.13			200			35	
Hay, suncured														
Late veg.	0.66	0.34	0.14	0.18		1.59	0.13			200			22	
Mid-bloom	0.41	0.19	0.16	0.01		1				140	46		10	
Seed stage	0.28	0.18	0.12										5	
Silage, 25–40% dry matter (energy value on dry matter basis is same as for hay of same maturity).														
Tomato, pomace, dry														
Trefoil, birdsfoot														
Hay	1.75	0.22	0.51	0.88		1.80		0.11	9.3	230	15		26	
Grazed	2.20	0.25				1.83		0.21						
Turnip roots, fresh	0.56	0.22	0.22	1.05		2.99	0.43		21.3	110	43		0	
Vetch hay	1.18	0.34	0.27	0.52		2.12	0.15	0.35	9.9	380	73		70	
Wheat														
Bran	0.12	1.32	0.58	0.07	0.07	1.39	0.25	.011	13.8	190	130	124	0.5	3
Grain, soft white winter	0.06	0.41	0.11	0.02	0.09	0.46	0.13	0.150	7.8	40	40	30	0	34
Grain screenings	0.17	0.40	0.18	0.10		0.58	0.22			60	16		0	
Grazed, early veg.	0.42	0.40	0.21	0.07		3.50	0.19						94	
Hay, suncured	0.14	0.18	0.12	0.28		1	0.24			200			20	
Middlings	0.12	1.01	0.41	0.19	0.03	1.08	0.18	0.100	19.6	100	132	146	0	40
Shorts	0.12	0.84	0.29	0.02	0.08	0.94	0.26	0.100	10.3	110	116		0	32
Silage, early veg.	0.27	0.27	0.62	0.07	0.07	1.39	0.24	0.044	13.8	190	130		47	3
Straw	0.21	0.08	0.12	0.14	0.30	1.11	0.19	0.040	3.3	200	40		0.5	
Whey, dehy	0.98	0.81	0.14	1.10	0.08	0.92	1.12	0.116	51.4	140	5	3	0.5	0
Yeast														
Brewers, dried	0.14	1.54	0.25	0.08	0.13	1.85	0.41	0.200	35.5	100	6	42	0	2
Torula, dried	0.63	1.81	0.14	0.01	0.02	2.02	0.37		14.4	100	14	107	0	

TABLE 5
Guidelines for Estimating Maximum Intake of Dry Matter by Lactating Cows[a]

Body wt (lb):	900	1100	1300	1500	1700
FCM production[b]		(% of body wt)			
20	2.5	2.3	2.2	2.1	2.0
30	2.7	2.4	2.3	2.2	2.2
40	3.0	2.7	2.6	2.5	2.4
50	3.3	3.0	2.9	2.7	2.6
60	3.5	3.2	3.0	2.9	2.7
70	3.8	3.5	3.2	3.0	2.9
80	4.0	3.7	3.5	3.3	3.1
90	—	3.8	3.6	3.5	3.3
100	—	4.0	3.8	3.6	3.5

[a] Adapted from National Research Council (1978), which was synthesized from Chandler and Brown (1975); Smith (1971); Swanson *et al.* (1967); Trimberger *et al.* (1963, 1972).

[b] Milk production expressed as 4% fat corrected milk (see Section 2.2.d).

Index